MALDI MS

Edited by
Franz Hillenkamp and
Jasna Peter-Katalinić

1807–2007 Knowledge for Generations

Each generation has its unique needs and aspirations. When Charles Wiley first opened his small printing shop in lower Manhattan in 1807, it was a generation of boundless potential searching for an identity. And we were there, helping to define a new American literary tradition. Over half a century later, in the midst of the Second Industrial Revolution, it was a generation focused on building the future. Once again, we were there, supplying the critical scientific, technical, and engineering knowledge that helped frame the world. Throughout the 20th Century, and into the new millennium, nations began to reach out beyond their own borders and a new international community was born. Wiley was there, expanding its operations around the world to enable a global exchange of ideas, opinions, and know-how.

For 200 years, Wiley has been an integral part of each generation's journey, enabling the flow of information and understanding necessary to meet their needs and fulfill their aspirations. Today, bold new technologies are changing the way we live and learn. Wiley will be there, providing you the must-have knowledge you need to imagine new worlds, new possibilities, and new opportunities.

Generations come and go, but you can always count on Wiley to provide you the knowledge you need, when and where you need it!

William J. Pesce
President and Chief Executive Officer

Peter Booth Wiley
Chairman of the Board

MALDI MS

A Practical Guide to Instrumentation,
Methods and Applications

Edited by
Franz Hillenkamp and Jasna Peter-Katalinić

WILEY-VCH Verlag GmbH & Co. KGaA

The Editors

Prof. Dr. Franz Hillenkamp
Prof. Dr. Jasna Peter-Katalinić
Institute for Medical Physics and Biophysics
University of Münster
Robert-Koch-Str. 31
48149 Münster
Germany

Library of Congress Card No.:
applied for

British Library Cataloguing-in-Publication Data
A catalogue record for this book is available from the British Library.

Bibliographic information published by the Deutsche Nationalbibliothek
The Deutsche Nationalbibliothek lists this publication in the Deutsche Nationalbibliografie; detailed bibliographic data is available in the Internet at <http://dnb.d-nb.de>.

Composition SNP Best-set Typesetter Ltd., Hong Kong

Printing betz-druck GmbH, Darmstadt

Bookbinding Litges & Dopf GmbH, Heppenheim

Printed in the Federal Republic of Germany
Printed on acid-free paper

ISBN 978-3-527-31440-9

Contents

MALDI MS. A Practical Guide to Instrumentation, Methods and Applications.
Edited by Franz Hillenkamp and Jasna Peter-Katalinić
Copyright © 2007 Wiley-VCH Verlag GmbH & Co. KGaA, Weinheim
ISBN: 978-3-527-31440-9

Preface

The first report on what is now referred to as Matrix-Assisted Laser Desorption/Ionization Mass Spectrometry (MALDI-MS) was published in 1985, describing the principle of the method and calling the procedure "matrix-assisted desorption". Yet, today – some 21 years later – the principle still stands as described in 1985, albeit with only minor additions or corrections. What has changed dramatically, however, is the practical implementation, including a vastly improved and diversified instrumentation, combination with complementary methods such as chromatography or electrophoresis for sample (pre)fractionation, as well as the practical guidelines utilized, especially in the area of sample preparation. During the early life of MALDI-MS, it took the (bio)scientific community almost five years to realize the great potential of the technique over a wide range of analytical problems, from the discovery or identification of (mostly) organic (macro)molecules through to structure analysis and function. Since then, the practical use of MALDI-MS has grown almost exponentially and, together with electrospray ionization (ESI), MALDI is today an indispensable laboratory tool, particularly in the life sciences.

As the subtitle suggests, this book is meant as a practical guide, and consequently much emphasis has been placed on providing practical, usable information rather than a basic understanding of the underlying mechanisms. The content of the book is arranged accordingly, concentrating essentially on the main fields of application.

Following a brief presentation of the technique's historical background, Chapter 1 contains a condensed presentation of the physico-chemical mechanisms that are active in the MALDI process. Emphasis is placed especially on the impact that these mechanisms have on practical applications, for example in the case of ion fragmentation or the analysis of non-covalently bound complexes.

Chapter 2 is devoted to all aspects of MALDI-MS instrumentation, including the ion sources, lasers, and the very different types of mass analyzer which are currently in use for MALDI-MS. The aim of this chapter is to help users select the best instrument for their given application, and consequently it contains a wealth of detailed information, because the correct choice of MALDI-MS instrument is

MALDI MS. A Practical Guide to Instrumentation, Methods and Applications.
Edited by Franz Hillenkamp and Jasna Peter-Katalinić
Copyright © 2007 Wiley-VCH Verlag GmbH & Co. KGaA, Weinheim
ISBN: 978-3-527-31440-9

key to any successful application of the method. Clearly, newcomers to the field are strongly encouraged to act with due diligence before deciding on the best instrument for their application.

All of the following chapters are devoted to the main fields of application for MALDI-MS. Protein/peptide analysis, which is discussed in Chapter 3, remains the most prevalent application, and it is in this field that the majority of technical developments of MALDI have taken place during recent years. As a consequence, the use of MALDI has led to truly revolutionary advancements in the analysis of proteins and peptides. Microprobing and biomarker detection, both of which are discussed in Chapter 4, are more recent fields of application, but they are – unfortunately – also the field where most (as yet) unwarranted claims have been made. However, when future studies are conducted with sufficient understanding and caution, very useful results can be expected in these areas. In this regard, the authors hope that this chapter will highlight not only the promises of MALDI-MS but also its limitations.

MALDI-MS analysis of nucleic acids, which is described in Chapter 5, has long been a rather neglected field of application, mainly because of the inherent structural instabilities of these analytes upon laser desorption, and problems involving transfer into the vacuum of the mass analyzer. Indeed, it was only after a series of dedicated molecular biology assays had been developed that routine and largely automated analyses became possible, though in future a more widespread use of MALDI-MS is to be expected in this field. Glycans and lipids – the subjects of Chapters 6 and 7, respectively – are two major classes of biological molecule where the use of MALDI-MS as an analytical tool has trailed that for proteins. This is not because glycans and lipids are considered less important, but rather because their analysis is more complex. This is partly due to ion formation being problematic in many cases, and partly because of their heterogeneity in biological systems. Nonetheless, such applications are expected to undergo considerable expansion in the near future.

Synthetic polymers form the predominant class of non-biological molecules to undergo routine analysis by MALDI-MS, though this is complicated by these compounds often having very broad mass distributions; this not only reduces their detection sensitivity but can also lead to strongly skewed spectra. Chapter 8 addresses the problems encountered in the analysis of very different classes of synthetic polymers and mass distributions, and concentrates on practically useful procedures and recipes. The analysis of small molecules, which is described in Chapter 9, is not strictly a MALDI-based treatment but has been included because of the importance of this field, notably in the development and screening of new pharmaceutical agents. Alternative methods used in this area, such as Desorption/Ionization On Silicon (DIOS), are closely related to MALDI and use very similar mass analyzers.

The aim of this book is to suggest possible analytical methods for a variety of molecules which are at the core of life and possess a complexity that we are only just beginning to realize. It is probably fair to say that we can only understand and make practical use of what we can measure, and in this sense the authors trust that this book will serve the large community of present and future MALDI users, hopefully to answer their questions and to advance their knowledge in their chosen field.

We, the editors, thank all of those authors who took time to make this book as complete a treatment of MALDI-MS as we believe it is. We also thank the publisher, Wiley-VCH, and in particular Dr. Andrea Pillmann and Dr. Andreas Sendtko, who not only used all of their persuasive powers to convince us that such a book was a worthwhile effort, but also had the patience to usher us through the maze of book-making.

Münster, October 2006

Jasna Peter-Katalinić
Franz Hillenkamp

List of Contributors

Dr. Stefan Berkenkamp
SEQUENOM GmbH
Mendelssohn-Str. 15D
22761 Hamburg
Germany

Dr. Lucinda Cohen
PDM, Bioanalytical Research
Pfizer Global R & D
2800 Plymouth Rd
Ann Arbor, MI 48105
USA

Dr. Eden P. Go
The Scripps Research Institute
Center for Mass Spectrometry
0550 North Torrey Pines Road BCC007
La Jolla, CA 92037
USA

Prof. Dr. Franz Hillenkamp
Institute for Medical Physics and
Biophysics
University of Münster
Robert-Koch-Str. 31
48149 Münster
Germany

Dr. Karin Hjernø
Protein Research Group
Department of Biochemistry and
Molecular Biology
University of Southern Denmark
5230 Odense M
Denmark

Prof. Dr. Ole N. Jensen
Protein Research Group
Department of Biochemistry and
Molecular Biology
University of Southern Denmark
5230 Odense M
Denmark

Michael Karas
Institute of Pharmaceutical Chemistry
University of Frankfurt
Max-von-Laue-Str. 9
60438 Frankfurt

Prof. Dr. Liang Li
Department of Chemistry
University of Alberta
Chemistry Centre W3-39
Edmonton, Alberta T6G 2G2
Canada

MALDI MS. A Practical Guide to Instrumentation, Methods and Applications.
Edited by Franz Hillenkamp and Jasna Peter-Katalinić
Copyright © 2007 Wiley-VCH Verlag GmbH & Co. KGaA, Weinheim
ISBN: 978-3-527-31440-9

Prof. Dr. Peter B. O'Connor
Mass Spectrometry Resource
Boston University School of Medicine
715 Albany St.
R806 Boston, MA 02118
USA

Prof. Dr. Jasna Peter-Katalinić
Institute for Medical Physics and
Biophysics
University of Münster
Robert-Koch-Str. 31
48149 Münster
Germany

Dr. Dijana Šagi
Sanofi-Aventis
Deutschland GmbH
Process Development Biotechnology
Industriepark Hoechst, H780
65926 Frankfurt
Germany

Dr. Jürgen Schiller
Institute of Medical Physics and
Biophysics
Medical Faculty
University of Leipzig
Härtelstr. 16–18
04107 Leipzig
Germany

Dr. Gary Siuzdak
The Scripps Research Institute
Center for Mass Spectrometry
10550 North Torrey Pines Road
BCC007
La Jolla, CA 92037
USA

Prof. Dr. Bernhard Spengler
Institute of Inorganic and Analytical
Chemistry
Mass Spectroscopy
Justus-Liebig University Giessen
Schubertstr. 60
35392 Giessen
Germany

Dr. Dirk van den Boom
SEQUENOM, Inc.
3595 John Hopkins Court
San Diego, CA 92121
USA

1

The MALDI Process and Method

Franz Hillenkamp and Michael Karas

1.1
Introduction

Matrix-assisted laser desorption/ionization (MALDI) is one of the two "soft" ionization techniques besides electrospray ionization (ESI) which allow for the sensitive detection of large, non-volatile and labile molecules by mass spectrometry. Over the past 15 years, MALDI has developed into an indispensable tool in analytical chemistry, and in analytical biochemistry in particular. This chapter will introduce the reader to the technology as it stands now, and will discuss some of the underlying physical and chemical mechanisms as far as they have been investigated and clarified to date. It will concentrate on the central issues of MALDI, necessary for the user to understand for an efficient application of the technique. An in-depth discussion of these topics would be beyond the scope of this chapter, and hence the reader is referred to recent reviews [1–3]. The details about the current state of instrumentation including the lasers and their coupling to the mass spectrometers will be presented in Chapter 2.

As with most new technologies, MALDI came as a surprise even to the experts in the field on the one hand, but also evolved from a diversity of prior art and knowledge on the other hand. The original notion had been that (bio)molecules with masses in excess of about 500–1000 Da could not be isolated out of their natural (e.g., aqueous) environment, and even less be charged for an analysis in the vacuum of a mass spectrometer without excessive and unspecific fragmentation. During the late 1960s, Beckey introduced *field desorption* (FD), the first technique which opened a small road into the territory of *mass spectrometry* (MS) of bioorganic molecules [4]. Next came *secondary ion mass spectrometry* (SIMS), and in particular static SIMS, introduced by A. Benninghoven in 1975 [5]. This development was taken a step further by M. Barber in 1981 with the introduction of SIMS of organic compounds dissolved in glycerol, which he coined *fast atom bombardment* (FAB). It was in this context, and in conjunction with first attempts to desorb organic molecules with laser irradiation, that the concept of a "matrix" as a means of facilitating desorption and enhancing ion yield was born [6]. The principle of desorption by a bombardment of organic samples with the fission

MALDI MS. A Practical Guide to Instrumentation, Methods and Applications.
Edited by Franz Hillenkamp and Jasna Peter-Katalinić
Copyright © 2007 Wiley-VCH Verlag GmbH & Co. KGaA, Weinheim
ISBN: 978-3-527-31440-9

products of the ^{252}Cf nuclear decay, later called *plasma desorption* (PD) was published by R. Macfarlane in 1974 [7]. Subsequently, the groups of Sundqvist and Roepstoff greatly improved the analytical potential of this technique by the addition of nitrocellulose, which not only cleaned up the sample but was also suspected of functioning as a signal-enhancing matrix [8].

The first attempts to use laser radiation to generate ions for a mass spectrometric analysis were published only a few years after the invention of the laser [9–11]. Vastola and Pirone had already demonstrated the spectra of organic compounds, recorded with a time-of-flight (TOF) mass spectrometer. Several groups continued to pursue this line of research, mainly R. Cotter at Johns Hopkins University in the USA and P. Kistemaker at the FOM Institute in Amsterdam, the Netherlands. For a number of years the Amsterdam group held the high mass record for a bioorganic analyte with a spectrum of underivatized digitonin at mass 1251 Da ([M + Na]$^+$), desorbed with a CO_2-laser at a wavelength of 10.6 μm in the far infrared [12].

Independently of and parallel to these groups, Hillenkamp and Kaufmann developed the *laser microprobe mass analyzer* (LAMMA), the commercial version of which was marketed by Leybold Heraeus in Cologne, Germany [13]. The instrument originally comprised a frequency-doubled ruby laser at a wavelength of 347 nm in the near ultraviolet (UV), and later a frequency-quadrupled ND:YAG-laser at a wavelength of 266 nm in the far UV. The laser beam was focused to a spot of <1 μm in diameter to probe thin tissue sections for inorganic and trace atomic ions such as Na, K, and Fe. The mass analyzer of the LAMMA instruments was also a TOF mass spectrometer, and was the first commercial instrument with an ion reflector, which had been invented a few years earlier by B.A. Mamyrin in Leningrad. The sensitivity-limiting "noise" of the LAMMA spectra were signals which were soon identified as coming from the organic polymer used to embed the tissue sections, as well as other organic tissue constituents. It was this background noise which triggered the search for a systematic analysis of organic samples and which eventually led to the discovery of the MALDI principle in 1984. The principle and its acronym were published in 1985 [14] and the first spectrum of the non-volatile bee venom mellitin, an oligopeptide at mass 2845 Da in 1986 [15]. Spectra of proteins with masses exceeding 10 kDa and 100 kDa were published in 1988 [16] and presented at the International Mass Spectrometry Conference in Bordeaux in 1988, respectively.

ESI and MALDI were developed independently but concurrently, and when their potential for the desorption of non-volatile, fragile (bio)molecules was discovered people were mostly impressed by their ability to access the high mass range, particularly of proteins. However, FAB- and PD-MS had at that time already generated spectra of trypsin at mass 23 kDa and other high-mass proteins. What really made the difference in particular for the biologists was the stunning sensitivity which, for the first time, made MS compatible with sample preparation techniques used in these fields. For MALDI, the minimum amount of protein needed for a spectrum of high quality was reduced from 1 pmol in 1988 to a few femtomoles only about a year later. Today, in favorable cases, the level is now down in the low attomole range. Many other developments – both instrumental (see Chapter 2) as well as specific sample preparation recipes and assays (see other chapters of the book) – took place during the following decade, and the joint impact of all of these together has today made MALDI-MS an indispensable tool not only in the life sciences but also in polymer analysis.

The use of a chemical matrix in the form of small, laser-absorbing organic molecules in large excess over the analyte is at the core of the MALDI principle. Several developments for laser desorption schemes took place in parallel to and following publication of the MALDI principle. These all attempt to replace the chemical matrix by a more easy-to-handle physical matrix, or a more simple combination of the two. The best known of these is the system of Tanaka and coworkers, which was first presented at a Sino-Japanese conference in 1987; details were subsequently published in 1988 [17]. The matrix comprises Co-nanoparticles suspended in glycerol as the basic system into which the analyte is dissolved, similar to the sample preparation of FAB. Several other nano- and micro-particles were tested later, with results comparable to those of Tanaka [18]. For his technique, *surface-assisted laser desorption/ionization* (SALDI), Sunner and co-workers used dry carbon and graphite substrates [19]. Another technique which has found much interest for the analysis of smaller molecules (and which is described in more detail in Chapter 9) was reported by Siuzdak [20]. This method, termed *desorption/ionization on silicon* (DIOS), uses preparations of neat organic samples on porous silicon. Several other methods and acronyms use similar systems such as nanowires or sol-gel systems. All of these methods use the substrate on which the analyte is prepared for the absorption of the laser energy, and are characterized by a sensitivity lower than that of MALDI by several orders of magnitude, as well as a strongly increased ion fragmentation which limits the accessible mass range to somewhere between 2000 and 30 000 Da, depending on the method. There is reason to believe that all of these methods are based on a thermal desorption at the substrate/analyte interface with the high internal excitation of the ions and low ion yield typical for thermal desorption processes. The very high heating and cooling rates, together with high peak temperatures of the substrates as well as the suspension of the absorbers in glycerol, apparently somewhat soften the desorption, the latter most probably through adiabatic cooling in the expanding plume; derivatization of the surfaces can up-concentrate the analyte of interest at the surfaces to increase the sensitivity. Indeed, a yoctomole (10^{-21} mole) sensitivity has been achieved in this way with a perfluorophenyl-derivatized DIOS system for a small hydrophobic peptide [21].

1.2
Analyte Incorporation

What, then is so special about the chemical matrix in MALDI? Some of its important features such as the absorption of the laser energy are easily understood, but surprisingly the overall process of the desorption and ionization has not yet been fully described, almost 20 years after the invention. As a result, the search for better matrices in general or for specific application still remain mostly empirical.

One important feature is the way in which the matrix and analyte interact in the MALDI sample. In a typical UV-MALDI sample preparation small volumes of an about 10^{-6} M solution of the analyte and a near-saturated (ca. 0.1 M) solution of the matrix are mixed; the solvent is then evaporated before the sample can be introduced into the vacuum of the mass spectrometer. Upon solvent evaporation, the matrix crystallizes to form a bed of small crystals that range in size from a few to a few hundred micrometers, depending on the matrix and the details of the preparation. The typical molar analyte to matrix ratio ranges from about 10^{-2} for small molecules to ca. 10^{-4} for large proteins. The sample preparation is discussed in more detail in Section 1.8. One of the early surprises in the MALDI development was that all of the well-functioning matrices incorporate the analyte in the crystals quantitatively (up to a maximum concentration) and in a homogeneous (on the light microscopic resolution level of 0.5 µm) distribution. This was shown for the UV-MALDI matrices 2,5-dihydroxybenzoic acid (2,5-DHB) [22], sinapinic acid [23] and 4-hydroxy-α-cyanocinnamic acid (HCCA) [24], 3-hydroxy-picolinic acid [24] and the IR-MALDI matrix succinic acid [25]. This homogeneous incorporation, in conjunction with the also homogeneous energy deposition and material ablation (for a discussion, see Section 1.3) result in the co-desorption of intact non-volatile and labile molecules with the matrix and, in addition, to their cooling of internal energy in the expanding plume of material. The mechanisms and driving force for this incorporation are still largely unknown. Horneffer et al. have shown in a systematic study of different position isomers of dihydroxybenzoic acids that only 2,5-DHB incorporates homogeneously and quantitatively, whereas other isomers such as 2,6-DHB do not incorporate at all, while some others incorporate only randomly [26,27]. Confocal laser scan images of the protein avidin, labeled with the fluorochrome Texas red for single crystals of 2,5-DHB and 2,6-DHB are shown in Figure 1.1. No obvious correlation between the incorporation and the crystal structure of these isomers was found. The state of the incorporated analyte molecules in the matrix crystals is another interesting question. Based on results obtained for the incorporation of pH-indicator dye molecules, Krueger et al. have concluded that molecules retain their solution charge state in the crystal, which implies that they also retain their solvation shell [28]. Horneffer et al. have found a high density of cavities of 10–2000 nm size in crystals of both 2,5-DHB and 2,6-DHB by electron microscopy [29]. At first sight, these cavities could be assumed to contain analyte molecules with residual solvent. However, if this is the case it is difficult to understand why 2,5-DHB – but not 2,6-DHB – incorporates analytes into these cavities; attempts to localize gold-labeled protein in the cavities of 2,5-DHB were also inconclusive.

A solventless method for sample preparation was developed originally for the MALDI-MS of synthetic polymers which are often not soluble in standard solvents [30]. In this method, matrix and analyte powders are mixed and ground in a mortar or ball-mill and then applied to a MALDI target support. In a recent report it

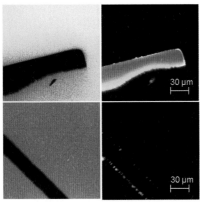

30 μm

30 μm

Fig. 1.1 Confocal laser scan fluorescence images of single crystals of 2,5-dihydroxybenzoic acid (top panels) and 2,6-dihydroxybenzoic acid (bottom panels). Both matrices were doped with the protein avidin, labeled with a Texas red (TR) fluorescent dye. The images were recorded at an x,y-plane 12 μm into the crystals. The left panels show dark shadowgraphs of the shape of the crystals against the bright green Bodipy 493/503 nm fluorescence of the immersion liquid. The right panels show the red TR fluorescence of the labeled protein.

was shown, that analyte spectra can be obtained from such preparations, even though the analyte is only chemisorbed at the matrix crystal surfaces [31]. However, the desorption is much less "soft" than MALDI-MS from samples with incorporated analytes, leading to a strongly increased metastable fragmentation of the ions and an upper mass limit for proteins of 30–55 kDa.

1.3
Absorption of the Laser Radiation

The role of the optical absorption of the matrix in the transfer of energy from the laser beam to the sample is governed by Beer's law, as described previously in 1985 [14].

$$H = H_0 * e^{-\alpha z} \tag{1.1}$$

where H is the laser fluence at depth z into the sample, H_0 is the laser fluence at the sample surface, and α is the absorption coefficient (see Chapter 2, Section 2.1 for a definition of the fluence). The absorption coefficient α equals the product of the molar absorption coefficient α_n and the concentration c_n of the absorbing molecules in the sample. The wavelength-dependent molar absorption coefficient α_n is a property of the matrix compound and, for UV-MALDI, has a maximum value of typically between 5×10^3 and $5 \times 10^4 \, l \, mol^{-1} cm^{-1}$ at the peak absorption wavelength. Molar absorption coefficients of this order of magnitude are only provided by molecules with aromatic rings; the exact wavelength of maximum absorption and its magnitude are determined by the ligands of the core ring and are tabulated in a variety of reference sources. Some care should be exercised in using the tabulated values for α_n, because they have all been measured for dilute solutions of the compounds. In the solid state of a typical MALDI sample the absorption bands are typically broadened and slightly red-shifted [32]. The concentration c_n of absorbers (chromophores) is unusually high in MALDI samples, about $10\,mol\cdot l^{-1}$, because all the solvent is evaporated before the sample is introduced into the vacuum. As a result, the absorption coefficient α ranges from about 5×10^4 to $5 \times 10^5\,cm^{-1}$. The inverse of α is called the penetration depth δ, and has values of only 20 to 200 nm. It is the depth into the sample, at which the fluence has decreased to about 30% of the value at the surface. It is also an order of magnitude estimate of the depth of material ablated (desorbed) per single laser pulse in MALDI. Because of this very shallow ablation depth, a given location of the sample can usually be irradiated many times before the material is exhausted. For the MALDI process the "density" of energy, absorbed per unit volume E_a/V of the sample is the process-determining quantity. This can be derived from Eq. (1.1) by simple differentiation to:

$$E_a/V = \alpha * H \qquad\qquad (1.2)$$

Equation 1.2 is at the core of the MALDI process. If a matrix is chosen with a sufficiently high absorption coefficient α, a relatively low fluence H_0 can be applied. Values for H_0 of 20–200 $J\,m^{-2}$ are representative for most UV-MALDI applications. As discussed in Section 2.1, pulsed lasers with a pulse width of a few nanoseconds are employed in UV-MALDI. At a fluence of 100 $J\,m^{-2}$ and a pulse width of 2 ns, the "intensity" (irradiance) of the laser beam at the sample surface is only $10^{11}\,W\,m^{-2}$ or $10^7\,W\,cm^{-2}$, not enough to induce any non-linear absorption such as non-resonant two-photon absorption. For the linear absorption the absorbed energy per unit volume can be controlled meticulously with a suitable variable attenuator in the laser beam, a feature which has turned out to be crucial for the successful MALDI of large molecules, because the desorption of non-volatile, labile molecules can only be achieved in a narrow range of energy "density". The other essential feature of this laser absorption is that the energy is transferred more or less uniformly to a macroscopic sample volume (except for the attenuation of the fluence into the sample and the fluence profile, as discussed in Section 2.1). This is very different from the situation in SIMS or PD, where incident particles create minute tracks of atomic dimensions of very high energy density in the sample, with a strong radial decline of energy density. This strongly heterogeneous energy distribution is the main reason for the limitation of these methods for the desorption of larger molecules. The fluence can also be converted into a value for the photon flux – that is, the number of photons impinging on the sample per single laser pulse. A fluence of 100 $J\,m^{-2}$ corresponds to a photon flux of 1.7×10^{16} photons per cm^2; each carrying an energy of 3.7 eV at the wavelength of

337 nm of the N_2 laser. A molar absorption coefficient of $10^4 l\,mol^{-1}\,cm^{-1}$ represents a physical absorption cross-section of the chromophore of $1.6 \times 10^{-17}\,cm^2$, resulting in an average of 0.7 photons absorbed per matrix molecule for any given laser exposure. This is, on the one hand, a very high density of excitation energy, close to the solid-state energy stored in all of the intermolecular bonds. It is, therefore not surprising that it leads to an explosive ablation of the excited sample volume. On the other hand, it renders even resonant two-photon absorption by the matrix rather unlikely. The high density of excited molecules does, however, result in a rather high rate of energy pooling in the sample, in which two neighboring excited molecules pool their energy, with one of them acquiring twice the photon energy and the other falling back to the ground state [33]. This energy pooling is an important feature in some models for the ionization of the molecules which requires at least the energy of two photons for an initial photoionization of the matrix molecules [34]. The situation is similar, but not equal, for IR-MALDI. Optical absorption in the infrared region of the spectrum represents a transition between vibrational or rotational molecular states. These transitions are typically weaker than the electronic transitions in the UV by one to two orders of magnitude. The strongest such transitions are those of the O–H and N–H stretch vibrations near a 3 μm wavelength. The absorption coefficient not only of water or vacuum-stable ice, but also of the common IR-MALDI matrix glycerol, in this wavelength region reaches peak values of $10^4\,cm^{-1}$, corresponding to a penetration depth of about 1 μm – more than 20-fold that of typical penetration depths in the UV. As a result, the ablated mass per laser exposure in IR-MALDI exceeds that of UV-MALDI by at least a factor of 10, and the sample consumption rate is accordingly higher. Typical laser fluences for IR-MALDI range from 10^3 to $5 \times 10^3\,J\,m^{-2}$. Non-linear absorption processes are even less likely for such fluences in the infrared as compared to UV-MALDI, and for the photon energy of only 0.3 eV or less even the absorption of several photons by a given chromophore or energy pooling cannot possibly excite single molecules to anywhere near their ionization energy.

1.4
The Ablation/Desorption Process

As discussed above, every laser exposure of a sample leads to the removal of a bulk volume – that is, many monolayers of matrix molecules of the sample. The term "desorption" is, therefore, somewhat ill-chosen for this process, and was so even for the field desorption for which it was originally coined. Ablation is the more correct term, and this is used interchangeably with desorption throughout this chapter. The processes of material ablation and the ionization of a minor fraction of the matrix and analyte molecules are, no doubt, intimately intertwined, and both take place on a micrometer geometrical and a nanosecond time scale. It is experimentally very difficult – if not impossible – to sort out the complex contributions of the physical processes induced by the laser irradiation in all detail. Despite this complexity, it is of considerable merit to treat the two mechanisms separately, and some basic understanding can be derived from such a discussion, this particularly, because the vast majority of material comes off as neutrals.

As was pointed out above, even at the threshold fluence for the detection of MALDI ions each laser pulse transfers an amount of energy to the sample, close to the sum of all bond energies in the solid (equivalent to the sum of the heat of fusion and evaporation). Even though this energy will in all cases lead to ablation of the excited volume, different energy dissipation processes need to be taken into account. Energy dissipation by heat conduction during the laser pulse can be neglected in all cases of UV- as well as IR-MALDI. For a penetration depth of laser radiation of 100 nm, the time constant for heat conduction of typical UV-MALDI matrices is about 10 ns – still a factor of three longer than the typical laser pulse width. In the infrared, the heat conduction time constant for 1 μm penetration depth is about 1 μs, a factor of about 10 longer than the longest pulse width of lasers (Er:YAG) used in that case. The very rapid heating of the sample by the laser radiation will also generate a thermoelastic pressure pulse in the absorbing sample volume which travels out of the excited volume with the speed of sound, carrying away part of the deposited energy. With a speed of sound in typical crystalline matrices of 2000–3000 ms^{-1} and depth of 100 nm in the UV, the acoustic time constant is less than 100 ps, much shorter than the laser pulse width of a few nanoseconds. Even though energy is constantly carried away by the pressure wave, this amounts only to a very small fraction of the total deposited energy, and the pressure in the excited volume never reaches values high enough to substantially influence the ablation process. For IR-MALDI, the situation can be very different because of the larger penetration depth. For the desorption with an Er:YAG laser the pulse width of 100 ns is long compared to the acoustic time constant, with only a negligible pressure build-up in the excited volume. The pulse width of the optical parametric oscillator (OPO) laser of only 6 ns, however, is rather close to the acoustic time constant, and the system stays close to what is called the "acoustic confinement". In this case a very high pressure of several tens of MegaPascal can build up in the excited volume. Rohlfing et al. [35] have investigated the ablation processes by measuring the recoil pressure of the ablated material with a fast acoustic transducer onto which the sample was prepared, while Leisner et al. [36] have studied the expanding plume of ablated material with high-speed time-lapse photography, both at a wavelength of 2.94 μm. Both measurements were much easier for IR-MALDI and glycerol as a matrix, because of the higher amount of material ablated. For the short pulse width and near-acoustic confinement, these authors saw pressure pulses of very high amplitude as expected, and time durations comparable to the laser pulse. For the 100-ns pulses of the Er:YAG laser, the pressure amplitude was low, but lasted for several microseconds. The plume photographs revealed that material is removed from the sample for times of up to over 100 μs in both cases. This is certainly somewhat of a surprise, because the TOF analysis had revealed that the ions are only generated during an initial phase of not longer than about 300 ns [37]. Similar experiments were conducted by Rohlfing under UV-MALDI conditions [38], using the liquid matrix nitrobenzylalcohol for better sample homogeneity and a desorption wavelength of 266 nm.

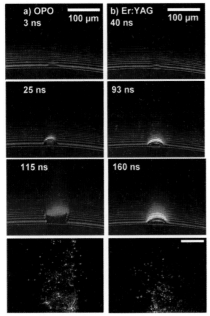

Fig. 1.2 High-speed time-lapse photographs of IR-MALDI plumes generated with an optical parametric oscillator (OPO) laser with 6-ns pulse width and an Er:Yag-laser with 100-ns pulse width. Both lasers were operated at 2.94 µm wavelength. Matrix: glycerol; time resolution 8 ns; spatial resolution 4 µm. The top three panels represent gradients of gaseous material density creating gradients of the index of refraction in the plume, recorded in a dark-field illumination mode. The lowest panel represents particle emission in the plume recorded with light scattered at 90° to the illumination beam. The thin lowest line indicates the top surface of the glycerol drop; the other striations in the dark-field images are artifacts of optical interference.

Expectedly, the recoil pressure was very low – lower even than that of the long-pulse IR-laser – because of the smaller amount of removed material. The recoil pressure pulse lasted for only less than 25 ns, the time resolution of the detection. The plume photographs revealed a material ejection for up to at least several microseconds, again much longer than the ion generation time of at most a few nanoseconds. Some typical plume photographs are shown in Figures 1.2 and 1.3. The results of these experiment can tentatively be explained by the following models. In IR-MALDI with 100 ns-long ER:YAG-laser pulses, the absorbing volume is superheated to a temperature that is substantially above the boiling temperature, followed by a volume ejection of material through boiling by heterogeneous nucleation. The situation is similar for UV-MALDI. The longer time course of material ejection in the infrared as compared to the UV is caused by a deeper penetration of the radiation into the sample, and a correspondingly higher inertia and residual heat of the excited volume. For the 6-ns pulses in IR-MALDI of the OPO-laser, the ablation process is substantially different. The strong thermoelastic wave is reflected at the sample vacuum interface, thereby reversing its phase.

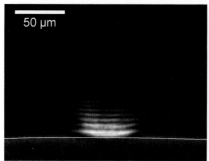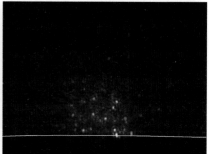

50 µm

Fig. 1.3 High-speed photographs of UV-MALDI plumes generated with a frequency-quadrupled Nd:YAG laser of 266 nm wavelength and 8 ns pulse width. Matrix: nitrobenzylalcohol; time resolution 8 ns; spatial resolution 4 µm. Left panel: dark-field image, 45 ns after laser exposure. Right panel: 90° scattered light image, 311 ns after exposure. The thin lowest line indicates the top surface of the glycerol drop; the other striations in the dark-field image are artifacts of optical interference.

It then travels back into the sample as a tensile wave, transferring the material beyond the liquid spinodal, as described by Vogel and Venugopalan for soft-tissue ablation [39]. Upon this transition, the material goes through a phase explosion by homogeneous nucleation. Even though all of these experiments were conducted on liquid samples to keep reproducibility high, they reflect, most probably, also the situation for crystalline solid samples. A contribution by the gaseous components such as CO_2 through thermal decomposition of matrix molecules is also discussed as a source of pressure build-up in the excited volume.

Theoretical models for the ionization as well as molecular modeling suggest that the ablation process generates clusters and material particles besides gaseous components. Particles are indeed seen in the plume photographs, though only at times late during the ablation, when the generation of ions is long over. Thus, this particle emission does not seem to be relevant for the MALDI ion generation. During the early expansion phase of the plume, dark-field images reveal homogeneous gradients of the index of refraction. However, the spatial resolution of the plume photographs is only a few micrometers; a distribution of clusters of different size, expected during the early phase of the plume expansion, is, therefore, not excluded by the experimental observations.

Garrison, Zhigilei [40,41] and co-workers as well as Knochenmuss [34] have modeled the ablation process using molecular dynamics simulation. Qualitatively, these simulations correctly predict many of the features observed experimentally. Particularly interesting is the consistent prediction that, while clusters are formed early during the ablation process, their internal energy does not seem to suffice for a decay by matrix evaporation, one of the assumptions of the "lucky survivor" model for ionization (see Section 1.5). It must also be observed that these simulations contain significant simplifications and, most probably more restrictive, must be scaled to very small volumes and short time regimes because of limited computation capacity. These models have become significantly refined over the past few years and will, no doubt, continue to do so. In this respect they will clearly contribute to our understanding of MALDI processes in the future.

1.5
Ionization

The mechanisms which lead to the formation of charged matrix and analyte molecules in the MALDI process are even more poorly understood than the physics of the material ablation/desorption. For a better understanding, it is important to distinguish between the ionization of matrix molecules and that of the analytes. Although no precise numbers have been determined experimentally, it is, most probably, safe to assume that the ion yield for the matrix (i.e., the ratio of ions to neutrals) is somewhere in the range of 10^{-5} to 10^{-3}. The ion yield of the analytes can be much higher, in the order of 0.1–1% for typical cases, and above 10% in exceptional cases. The intensity of the ion signals, as determined from the spectra, are not independent of each other, because charge transfer processes between the two species are, in all likelihood, taking place in the expanding plume and possibly already in the solid state upon laser irradiation. In favorable cases, the spectra even show intense analyte ion signals with negligible matrix ions, despite a typical 10^4 excess of the matrix [42].

Two models for ion formation have been proposed. The older model assumes neutral analyte molecules in the matrix crystals and a photoionization of the matrix molecules as the initial step [43,44], followed by charge transfer to the analyte molecules in the plume. The more recent "lucky survivor" model [45] assumes that proteins are incorporated into the matrix as charged species, most of which become re-neutralized within desorbed clusters of matrix and analyte.

For a laser wavelength of 337 or 355 nm (i.e., photon energies of 3.6 and 3.3 eV, respectively), at least two photons are needed for a photoionization of matrix molecules. The typical MALDI laser photon fluxes are too low for any significant resonant two-photon absorption to take place, but very efficient energy pooling between excited neighboring molecules in the crystals has been demonstrated [33]. In the gas phase, even the energy of two photons does not suffice for ionization, but in the solid phase of the crystals the ionization potential may be somewhat lower, and/or thermal energy may make up for the difference. A reaction of the photoelectron with neutral matrix molecules will give rise to negative ions beside the positive ones [44]. The frequent observation of radical matrix ions such as $M^{+\bullet}$ and/or $[M+2H]^{+\bullet}$ and a $[M-2H]^{-\bullet}$ ion besides the expected even-electron ions, among them $[M+H]^+$- or $[M-H]^-$- as well as a prominent $[2M+H-2H_2O]^+$ for 2,5-DHB and $[2MH]^+$- for CHCA is at least in agreement with the photoionization model, if not a strong indication (Fig. 1.4). As for the model prediction of the relative yield of positive versus negative analyte ions of peptides/proteins, it must be considered that the protonated matrix molecules with proton affinities between 180 and 215 kcal mol^{-1} [46] are strong gas-phase acids in comparison to basic amino acids with a much higher proton affinity (up to 245 kcal mol^{-1} for arginine) [47], result-

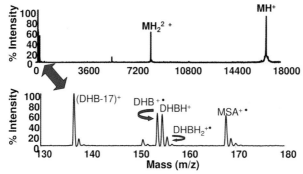

Fig. 1.4 UV-MALDI mass spectrum of myoglobin. Matrix: DHBs (2,5-dihydroxybenzoic acid (DHB) plus 2-hydroxy-5-methoxysalicylic acid (MSA); 9:1). Wavelength, 337 nm; Mass analyzer, Reflectron TOF.

ing in an efficient charge transfer and formation of positively charged analyte ions. In contrast, no such difference in basicity between negatively charged ions of matrix and peptides exists, because both are typically carboxylate anions with very close proton affinities; this should limit the yield of analyte anions.

Recently, Knochenmuss has developed a quantitative model based on the matrix photoionization model [48,49]. One argument against this model is the observation that MALDI spectra, obtained with infrared lasers at a 1.94 μm wavelength, closely resemble the UV-MALDI spectra, including radical matrix ions observed (e.g., for the succinic acid matrix). The photon energy of only 0.4 eV at this wavelength certainly excludes photoionization. Similarity of the spectra alone does, however, not suffice as proof of identical ionization processes at the two wavelengths, and indeed the ion yield for analytes in the infrared is about an order of magnitude lower than in the UV.

The more recent "lucky survivor" model of Karas et al. proposes that proteins retain their solution charge state upon incorporation into the matrix. This assumption is based on the observation that pH-indicator molecules retain their color and charge state upon crystal incorporation for acidic, neutral, or basic matrices [28]. For most common acidic matrices, almost all peptides will carry a (multiple) positive excess charge; counterions will then typically be either trifluoroacetate or matrix anions. In order to maintain the positive charges of the peptides and to keep the negative counterions separate, the analytes must be incorporated at least in a partially solvated form. In a second step, the model assumes a break-up of the crystal lattice into small clusters upon desorption, some of them with only a single analyte ion. Statistically, some of these clusters are assumed to carry a positive or negative charge by deficit, or an excess of a single counterion. In the expanding plume the clusters are assumed to lose neutral matrix and solvent molecules as well as counterions as free acids or bases after their proton-transfer neutralization with analyte (de)protonation sites. This results in a neutralization of the peptide charges except for the only remaining excess charge:

(1a) $\{(M + nH)^{n+} + (n - 1)A^-\}^{1+} \rightarrow [M + H]^+ + (n - 1)HA$

(1b) $\{(M - nH)^{n-} + (n - 1)BH^+\}^{1-} \rightarrow [M - H]^- + (n - 1)B$

These singly charged analyte ions are the lucky survivors of the neutralization process. The model elegantly explains the observation of mostly singly charged ions in MALDI spectra, and would work equally well for UV- as well as IR-MALDI. The formation of matrix/analyte clusters in the desorption process is also predicted by molecular modeling [40,41]. However, these model calculations also predict that the clusters have not enough internal energy for the proposed evaporation of the neutrals.

One of the strengths of the lucky-survivor model is that it can be equally well applied to account for the formation of negative ions from acidic conditions (2a), or the formation of positively charged ions of polyanionic (in solution) species such as nucleic acids (2b) with matrices such as 3-hydroxypicolinic acid or the trihydroxyacetophenones.

(2a) $\{[M + nH]^{n+} + (n + 1)A^-\}^- \rightarrow [M + A]^- + nHA \rightarrow [M - H]^- + (n + 1)HA$

(2b) $\{(M - nH)^{n-} + (n + 1)BH^+\}^+ \rightarrow [M + BH]^+ + nB \rightarrow [MH]^+ + (n + 1)B$

with {} as a symbol for the intermediate cluster, A^- and HA for the anion and respective conjugate acid (such as $TFA^-/HTFA$) and B and BH^+ for a base and its conjugate acid (such as NH_3/NH_4^+).

While these ionic cluster ions {.....} never show up in the MALDI spectra with the typical solutes used (e.g., trifluoroacetic acid (TFA) or ammonium salts), they have been detected for extremely strong acids [50]; their conjugate anions are extremely weak bases and thus stabilize the formed ion pairs with protonated analyte sites. Even though the relative abundance of positive to negative ions of a given analyte under typical MALDI conditions seems never to have been carefully determined, the postulate of an intermediate anion adduct $[M + A]^-$ in (2a) rationalizes the general observation that the formation of negative peptide ions out of typically acidic matrix solution is substantially less effective, because the proton-transfer step necessary to form the [M-H]⁻-ion competes with the simple dissociation of the adduct ion. Similar arguments hold for the polyanionic nucleic acids. Whilst the addition of ammonium ions result in a quantitative detection of the free acids, a substitution of the ammonium by alkyl-quaternary ammonium ions leads to the observation of adduct ions, increasing in intensity with increasing length of the alkyl chains.

Besides the processes suggested for the lucky survivor model, matrix photoionization is also probable as a parallel process, as documented by the observed matrix ions. In UV-MALDI, both processes may therefore contribute to the observed analyte ions to an as-yet unknown degree.

For analytes of very low proton affinity such as neutral carbohydrates and many synthetic polymers, cationization by Na^+, K^+ or other metal cations is usually observed in MALDI (see Chapters 6, 7 and 8). The cationization in all likelihood takes place in the expanding plume. It requires a co-desorption of the analyte, and the cations and best results are therefore obtained from sample locations where both species exist in close neighborhood, such as in the center of DHB-dried droplet preparations. Specific protocols have been developed for the MALDI of such analytes [51].

1.6
Fragmentation of MALDI Ions

Fragmentation of MALDI ions is a mixed blessing, as in all of MS. It can, on the one hand, lead to a substantial loss of spectra quality such as loss of mass resolution or even complete loss of the signal of the intact parent ion, as has been shown for the loss of sialic acids in the analysis of glycoconjugates in reflectron-TOF analyzers [52]. On the other hand, intrinsic or induced fragmentation is an indispensable tool for the acquisition of structural information in MS^n experiments. The nomenclature for the fragmentation – particularly the differentiation in post-source-decay (PSD) and in-source-decay ions – is closely related to time-of-flight analyzers, because they still constitute the majority of spectrometers used for the analysis of MALDI ions. The details of how to analyze fragment ions with different types of instrument are discussed in Chapter 2 (see Section 2.4.3 in particular).

Even though some limited ion stability – and thus fragmentation – was obvious in the early days of MALDI by peak tailing on the low mass side, and was attributed to small neutral losses of peptide and protein ions, the fact that MALDI can generate substantial prompt and metastable fragmentation of analyte ions was obscured at an early stage. This was due to two facts: the laser microprobe instrument (LAMMA 1000) used for the initial experiments indeed minimized fragmentation because of a very week acceleration field strength and low total (3 keV) ion energy; the next-generation MALDI-TOF instruments were linear instruments in which metastable fragment ions cannot be observed directly.

It was only when Kaufmann and Spengler started careful investigations of ion stability using a deceleration stage in a linear TOF instrument that the potentially high degree of metastable fragmentation was detected [53]. This was the starting point for the development of the so-called PSD analysis of metastable ions which led to today's MALDI-TOF/TOF instruments [54–56]. The character of the PSD fragmentation – that is, the classes of fragment ions observed – is in full agreement with a collisional activation process. Despite this general agreement, PSD mass spectra are often more complex than CID mass spectra, showing internal fragments and products of consecutive fragmentations, and pointing to more complex excitation mechanisms of the intramolecular degrees of freedom. Besides collisional excitation through collisions of the ions with neutrals in the plume, a direct excitation in the matrix crystal after absorption of the laser energy and excess energy of chemical reactions in the plume such as proton transfer must be considered. All of these processes depend in complex fashion on instrumental parameters such as laser fluence and focus, ion extraction field strength and delay, as well as on the choice of matrix. Information about the different contributions to ion excitation have been listed in a report which compares vacuum MALDI to atmospheric-pressure MALDI [57]; this report provides a comprehensive summary of the current knowledge of MALDI fragmentation. The effect of collisional excitation upon acceleration was demonstrated for matrix ions. These AP-MALDI investigations were later extended to determine initial MALDI plume excitation processes [58], including the application to more representative test samples than thermometer molecules, such as a model peptide and a nucleoside. Here, matrix softness yielded the best agreement with matrix proton affinity. In general, matrix proton affinities have been used with only limited success to rank matrices with respect to their effectiveness for fragmentation. In a practical approach, matrices are classified and ranked from "hard" to "soft". α-cyano-4-hydroxycinnamic acid, the matrix preferentially applied for "peptide-mass fingerprint" analyses in proteomics because of its rather homogeneous sample morphology, is one of the hardest matrices in that ranking. Because of its degree of fragmentation induction, it is also the matrix of choice for PSD- or TOF/TOF-experiments. "Super DHB" (DHBs; 2,5-DHB with a 5–10% addition of 2-hydroxy-5-methoxybenzoic acid) is on the soft end of the list and, therefore, preferentially applied for the analysis of larger proteins. By using such a soft matrix and optimizing all instrumental parameters (low extraction field, long delay times), small-neutral loss can be minimized and a good mass resolution (close to the theoretical limit) is obtainable for a linear TOF configuration, even for medium size proteins up to bovine serum albumin [59]. Common for PSD and the low-energy CID mechanisms is the randomization of the internal energy among all internal degrees of freedom before the (metastable) fragmentation. Interestingly, it was found that the order of matrices from hard to soft agrees with the initial ion velocities determined in a linear TOF. Hard matrices correlate with low initial velocities, and vice versa [60]. This points to a role for expansion cooling in the MALDI desorption process.

A very different fragmentation mechanism was first reported by R.S. Brown et al. [61,62], whereby fragment ions are formed "promptly" upon ion generation/excitation with a time delay of less than at most 100 ns – substantially less than the typical delay times in delayed-extraction TOF instruments. Therefore, these are referred to as in-source-decay (ISD) ions. ISD spectra contain signals of c- and sometimes z-type fragment ions, in addition to some a-, b- and y-ions (for fragment ion nomenclature, see Chapter 3). This type of fragmentation is observed for both positive and negative ions, indicating that ISD is independent of proton transfer, and 2,5-DHB is the matrix with the highest ISD yields [63–65]. The analogy to electron capture dissociation prompted a discussion on the possible role of electrons in ISD, although it appears clear today that ISD is mediated by hydrogen radicals [66]. ISD yields a substantially more complete fragment-ion series as compared to PSD, although hopes to use it for practical applications (e.g., in proteomics [67,68]) have not materialized, mostly because of the low intensity of the signals. An additional problem is that true MS/MS experiments cannot be carried out as a precursor selection for peptides from a mixture is not possible.

1.7
MALDI of Non-Covalent Complexes

After about 20 years' use of MALDI-MS, it is clear that the successful analysis of non-covalent interactions and complexes is rather the exception than the rule. Considering most typical MALDI protocols, this cannot be a surprise. Most "classical" matrices are organic acids and typically used in water/organic solvent mixtures, often acidified by TFA, i.e., in conditions which should result in the dissociation of most (if not all) non-covalent complexes. However, signals of non-covalent complexes have indeed been obtained from such preparations [69,70]. At least for some such systems, the dissociation seems to be sufficiently slow that a fast evaporation of the solvent conserves at least a certain fraction of the complexes. Unfortunately, adjustment of the pH and the omission of organic solvent is often not a viable alternative, as acidic matrices will be deprotonated and totally change their crystallization and incorporation properties as salts. Many salts and buffers used to adjust ionic strength/pH are, therefore, not MALDI-compatible in the desired concentration, or at least compromise the performance with respect to sensitivity and mass resolution.

The incorporation of analytes into the matrix crystals, which has been shown generally to make the desorption process softer or altogether possible [31], is another step which might lead to dissociation of the complexes. This question has been addressed in order to understand the so-called "first-shot phenomenon", which had been observed much earlier but only recently had been resolved [71,72]. For a number of selected matrices, the signals of protein–protein complexes are observed only in the spectra of first exposures of a given sample spot. Subsequent exposures yield only signals of the monomer units. By a combination of MALDI-MS and confocal laser scanning microscopy (CLSM) of complexes with fluorochromes, which exhibit a fluorescence resonance energy transfer (FRET), it was shown that in these systems the complexes dissociate upon incorporation, whereas intact complexes are precipitated at the surface of matrix crystals. The next crucial step is the intact desorption and ionization of the complex and their survival in the gas phase of the expanding plume. Such dissociation upon desorption is even more likely if the complexes are localized at the crystal surface rather than being incorporated.

The type of interaction within the complex is also a decisive parameter. From energetic considerations it is obvious that gas-phase stability of non-covalent complexes is highest for ionic interactions, followed by hydrogen bonding. Interestingly, the formation of strong ionic complexes has even been used to facilitate MALDI measurement of highly acidic and thus negatively charged biocompounds, such as oligonucleotides and heparin-derived oligosaccharides, by admixing highly basic peptides, followed by a mass determination of the intact stable complex [73]. Hydrophobic interaction should be particularly labile, because it is based on the hydrophilic environment of the solvent water, lost in the vacuum. The simple detection of non-covalent complexes by ESI shows, that the transfer of the molecule into the vacuum does not necessarily result in a dissociation, if the internal energy does not suffice for the transition to the very different conformational state. It would also appear that complexes with large contact areas between the constituents of the complex and corresponding contribution of salt- and hydrogen bridges, as well as hydrophobic interaction typical of many protein–protein complexes help to stabilize the complex. Complexes between small and large molecules, as are typical for ligand–receptor or antigen–antibody systems, are much more difficult to analyze by MALDI-MS. Their affinity depend on the exact conservation of the conformation in small epitopes of the protein, which is more easily lost in the MALDI process than the complete quaternary structure. Interestingly, spectra of the intact biotin–streptavidin complex, one of the strongest complexes known to date, have never been obtained by MALDI.

Another issue which complicates the use of MALDI for the analysis of non-covalent complexes is the formation of non-specific multimeric and adduct ions. This effect is even more pronounced for the analysis of non-covalent protein complexes, as high concentrations ($10 \, pmol \, \mu l^{-1}$ or higher) of analyte are typically used to overcome the reduced sensitivity of TOF instruments in the high mass range. Furthermore an elevated laser fluence, as well as deviation from optimal preparation protocols, are aggravating effects. It is, therefore, most important clearly to differentiate specific from non-specific interactions, and this is typically achieved by using a known non-binding/non-interacting control compound. Because of these limitations, analyses of non-covalent interactions by MALDI are typically qualitative rather than quantitative.

Within these boundary conditions, a number of reports have described the successful detection of several types of non-covalent complex [74–81]. During the early days of MALDI, high-intensity signals of non-covalent protein complexes using a nicotinic acid matrix and a laser wavelength of 266 nm were reported, for example of the tetrameric glucose isomerase [82] and a trimeric porin [83]. Comprehensive reviews have been provided by Hillenkamp [84] and Farmer and Caprioli [85] on this subject. In addition, a recent report by Zehl and Allmaier on the influence of instrumental parameters for the detection of quaternary protein structures starts from a careful review of the state of the art [86], and provides a critical discussion on the above-described problems. These authors used 2,6-dihydroxyacetophenone as (non-acidic) matrix with the addition of ammonium acetate or diammonium citrate (DAHC) to adjust solution conditions for the stabilization of protein complexes. Another only slightly acidic matrix which tolerates even high additive amounts such as DAHC is 6-azathiothymine (ATT); this was also used to investigate nucleic acids and their non-covalent complexes and adducts [87]. Superior results for double-stranded DNA were, however, obtained when using glycerol as a matrix for IR-MALDI [88].

In summary, the use of MALDI to investigate non-covalent interaction is far less straightforward than typical applications for peptides and proteins under denaturing conditions. Success is not predictable, and careful control experiments must be implemented to differentiate specific from non-specific interactions. It appears, however, that the potential of MALDI in this area is far from being fully explored. Recently, a new approach was presented investigating the formation of non-covalent complexes, based on the detection of the "intensity fading" of one complex partner [89,90], rather than of the intact complex. This approach avoids the problems related to detection of the intact complex in the high mass range, and can be carried out at analytically relevant micromolar and sub-micromolar concentration levels.

1.8
The Correct Choice of Matrix: Sample Preparation

Unfortunately or expectedly, there is no single MALDI matrix or sample preparation protocol which is suited to all analytical problems and analytes in MALDI-MS. A few of the more general considerations are discussed in the following section, though more specific information is provided in the applications chapters of this book. A representative list of commonly used matrices and their main properties is provided in Table 1.1. There are different matrices of first choice for different classes of analytes and analytical problems. For example, CHCA is used in the majority of proteomics applications for the analysis of peptide-mass-fingerprints generated by protein enzymatic digests (as discussed later). On the other hand, 2,5-DHB – and especially DHBs (i.e., DHB with an admixture of 5–10% 5-methoxy-2-hydroxybenzoic acid) – with its pronounced crystallization into large crystals of ca. 100 μm size is particularly suited to protein analysis. The reasons for this are, first, because its softness prevents strong small-neutral losses and peak tailing; and second, because the crystals incorporate the proteins, but exclude the majority of common contaminants.

A practical overview of the various matrices and preparation techniques can be found, for example, on the Internet at: http://www.chemistry.wustl.edu/~msf/damon/sample_prep_toc.html, as well as on the Internet pages of commercial suppliers of chemicals.

The "dried droplet" standard MALDI sample preparation is very simple. Here, the sample and matrix are dissolved in a common solvent or solvent system, and mixed either before deposition onto or directly on the MALDI sample support. The matrix-analyte droplet of typically 1 μL volume is then slowly dried in air, or under a forced flow of cold air. This results in a deposit of crystals which, de-

Table 1.1 A selection of commonly used MALDI matrices.

Matrix	Structure	Wavelength	Major applications
Nicotinic acid		UV 266 nm	Proteins, peptides, adduct formation
2,5-Dihydroxybenzoic acid (plus 10% 2-hydroxy-5-methoxybenzoic acid)		UV 337 nm, 353 nm	Proteins, peptides, carbohydrates, synthetic polymers
Sinapinic acid		UV 337 nm, 353 nm	Proteins, peptides
α-Cyano-4-hydroxycinnamic acid		UV 337 nm, 353 nm	Peptides, fragmentation
3-Hydroxy-picolinic acid		UV 337 nm, 353 nm	Best for nucleic acids
6-Aza-2-thiothymine		UV 337 nm, 353 nm	Proteins, peptides, non-covalent complexes; near-neutral pH
k,m,n-Di(tri)hydroxy-acetophenone		UV 337 nm, 353 nm	Protein, peptides, non-covalent complexes; near-neutral pH
Succinic acid	HOOC—CH₂·CH₂·COOH	IR 2.94 μm, 2.79 μm	Proteins, peptides
Glycerol		IR 2.94 μm, 2.79 μm	Proteins, peptides, liquid matrix

IR = infrared; UV = ultraviolet.

pending on the matrix, vary between submicrometer and several hundred micrometers in size. In many cases, surface tension leads to a non-homogeneous distribution of the individual crystals near the rim of the preparation. The best MALDI performance is usually achieved only at certain locations of the crystals, which often requires manual interference and active control by the experimenter; this is why most MALDI instruments are equipped with a microscopic observation system. The cause of these "sweet spots" has been the subject of much speculation, the commonly held notion being that of an inhomogeneous distribution of analyte within the crystals. However, this has been disproved by Horneffer et al., who found a homogeneous distribution of fluorescently labeled analyte in the crystals of a representative number of different matrix crystals by CLSM studies (see Section 1.2). A different (ionization) state of the analyte molecules in different locations, or heterogeneous orientation of the matrix crystal surfaces relative to the spectrometer axis and perpendicular to which the ions are ejected in conjunction with the limited angular acceptance of the mass spectrometer, might also cause the observed sweet spots.

As a rule of thumb, the addition of the analyte solution should not noticeably change the crystallization behavior of the neat matrix; this already indicates that the solution excess of the matrix with respect to the analyte is maintained in the crystals, and that the contaminant level is low enough. Any solute component which dramatically changes the appearance of the matrix crystals or prevents crystallization altogether – for example, low-volatility solvents such as glycerol or dimethyl sulfoxide – will prohibit a successful MALDI analysis.

Over the years, a large number of modifications of, or alternatives to, the dried-droplet technique have been developed. These many variations are often the personal preferences of MALDI users for sample preparation, and the subject may appear to be an art or a even a "black art". However, dried-droplet protocols are still the most widely used with high success. It also appears that instrumental developments using lasers with higher frequencies, together with the automation of the entire MALDI measurements, have eased the problems of heterogeneity to some extent. Among the many modifications and variations of the simple dried droplet preparation, two alternatives stand out as particularly useful and widespread, namely "surface preparation" and "anchor sample plates".

1.8.1
Surface Preparation

Surface preparation or predeposited matrix crystal layers were also often called thin layer preparation introduced to enhance sensitivity, homogeneity of the preparation, automation, and liquid chromatography (LC)-spotting.

Surface preparation was a true innovation [91]. It was shown that for the CHCA matrix, rapid evaporation of the organic solvent generates a relatively homogeneous, microcrystalline seed layer. Upon addition of the aqueous peptide analyte solution, only the very top layers of the nearly water-insoluble matrix redissolve and incorporate the analyte. The concentration of the whole analyte into only a limited depth layer at the radiation-accessible top of the preparation results in an improved sensitivity, and the structurally relatively homogeneous crystal layer improves the mass resolution, particularly in linear TOF mass spectrometers. Unfortunately, this approach is restricted to matrices which do not fully dissolve in the usually aqueous analyte solvent. However, it is generally believed that, despite its limited solubility, the matrix is partially dissolved by the analyte solution and that true incorporation of the analyte into the matrix takes place. It was, moreover, shown that contaminants such as salts could be rinsed from the surface with a splash of ice water, without any major loss in sensitivity. Hence, surface preparation became the starting point for the development of disposable MALDI targets with predeposited matrix spots. The generation of more homogeneous microcrystalline sample layers by rapid evaporation of the solvent (e.g., *in vacuo*) has also been used for a variety of other matrices.

1.8.2
Anchor Sample Plates

Anchor plates for the preparation of multiple samples have small hydrophilic islands, typically of 100–500 μm diameter, placed on a hydrophobic surface [92]. The hydrophobic surface prevents spreading of the sample solution over a large area, as otherwise observed for dried-droplet preparations. Instead, before and/or during crystallization of the matrix, the solution contracts onto these islands, thereby concentrating the matrix and analyte into a defined volume. This up-concentration is particularly useful for low-concentration analyte and matrix solutions, and also facilitates automated analyses of the fixed location samples.

A few other modifications of the sample preparation have also proven useful in specific cases:

- Mixtures of several different matrices have been reported for an improved performance, but so far only DHBs (a mixture of 90–95% 2,5-DHB and 5–10% 2-hydroxy-5-methoxybenzoic acid) has found relatively widespread application for proteins [93]. It softens the desorption and limits the small neutral loss and thereby improves mass resolution. A mixture of different trihydroxyacetophenones is sometimes used for the analysis of nucleic acids [94].
- Additives to matrix preparations are mostly used for sample clean-up. These additives do not absorb the laser radiation, but may influence the crystallization behavior of the matrix to some extent. The most frequently used method is to add DAHC as a cation scavenger to preparations of highly anionic samples such as nucleic acids [95]. The addition of ammonium phosphate to improve peptide-mass-fingerprint mass spectra by suppression of matrix cation clusters, and the use of phosphoric acid to improve DHB analysis of phosphopeptides, are two recent successful examples of this approach [96–98].

Modified surfaces, for example of sample plates, can be used for the affinity capture of analytes from crude mixtures. These can significantly enhance detection sensitivity and can be used for a simple sample clean-up. Titania (TiO_2) -coated surfaces or sol-gel systems, for example, have been shown to very selectively concentrate phosphopeptides from peptide fingerprint samples [99,100].

SELDI ("surface-enhanced laser desorption ionization") uses so-called protein chips for the detection of peptides and proteins from complex biological fluids such as blood. These protein chips contain various chromatographic media immobilized on a MALDI sample plate for the selective enrichment of constituents of the complex mixture applied. Unfortunately, a large number of unsubstantiated claims for the detection of disease-related biomarkers, mostly as a result of poor mass spectrometric performance, has discredited this approach.

Liquid matrices could avoid the undesirable heterogeneity of crystalline MALDI samples. Liquid matrices (e.g., nitrobenzylalcohol and nitrophenyloctlyether) were introduced in the early UV-MALDI reports, but never found widespread application because their performance proved to be significantly inferior to that of the solid matrices, particularly because of extensive adduct formation. A more recent development is the synthesis of ionic liquid matrices, which are synthesized by preparing a 1:1 solution of a classical organic acid matrix and an organic base [101]. A comprehensive review on the first five years of ionic-liquid matrices was recently published [102]. Unfortunately, the products with the best performance for MALDI are either solid or very highly viscous liquids. To date, these matrices seem to be mostly restricted to the MALDI-MS of small, stable analytes.

A solvent-free preparation is of particular interest for the analysis of synthetic polymers for which a common solvent with a suitable matrix is not available, and for which sizeable amounts of material are usually available [103]. In this protocol, the analyte and matrix are ground thoroughly in a mortar or ball-mill and loaded onto a MALDI target as powders, or after having been pressed into pellets. Good mass spectra can be obtained for analytes up to a mass of ca. 30 kDa, even though the analytes are not incorporated into the matrix crystals [31].

Abbreviations

CLSM	Confocal Laser Scan Microscopy
DAHC	Diammonium Citrate
DHB/2,5-DHB	(2,5-) dihydroxybenzioc acid
2,6-DHB	2,6-dihydroxybenzioc acid
DHBs	"super" DHB (mixture of 95% DHB and 5% 2-hydroxy-5-methoxy-benzoic acid
DIOS	Desorption/Ionzation On Silicon
ESI	Electrospray Ionization
FAB	Fast Atom Bombardment
FD	Field Desorption
FRET	Fluorescence Resonant Energy Transfer
HCCA	4-hydroxy-α-cyanocinnamic acid
ISD	In Source Decay
LAMMA	Laser Microprobe Mass Analyzer
MALDI	Matrix Assisted Laser Desorption/Ionization
UV-MALDI	MALDI with ultraviolet laser wavelengths
IR-MALDI	MALDI with infrared laser wavelengths
AP-MALDI	MALDI at Atmospheric Pressure
MS	Mass Spectrometry
PD	Plasma Desorption
PSD	Post Source Decay
SALDI	Surface Assisted Laser Desorption/Ionization

SELDI	Surface Enhanced Laser Desorption/Ionization
SIMS	Secondary Ion Mass Spectrometry
TOF	Time-Of-Flight
TOF-MS	Time-Of-Flight Mass Spectrometer

References

1 K. Deisewerd. The desorption process in MALDI. *Chem. Rev.* **2003**, *103*, 395–425.

2 M. Karas, R. Krüger. Ion formation in MALDI. *Chem. Rev.* **2003**, *103*, 427–439.

3 R. Knochenmuss, R. Zenobi. MALDI ionization: The role of in-plume processes. *Chem. Rev.* **2003**, *103*, 441–452.

4 H.D. Beckey. *Principles of Field Ionization and Field Desorption Mass Spectrometry, International Series in Analytical Chemistry*. Pergamon Press, 1977.

5 A. Benninghoven, W. Sichtermann. Detection, identification and structural investigation of biologically important compounds by secondary ion mass spectrometry. *Anal. Chem.* **1978**, *50*, 1180–1184.

6 D. Zakett, A.E. Schoen, R.G. Cooks, P.H. Hemberger. Laser-desorption mass spectrometry/mass spectrometry and the mechanism of desorption ionization. *J. Am. Chem. Soc.* **1981**, *103*, 1295–1297.

7 D.F. Torgerson, R.P. Skrowonski, R.D. Macfarlane. New approach to the mass spectroscopy of non volatile compounds. *Biochem. Biophys. Res. Commun.* **1974**, *60*, 616–619.

8 G.P. Jonsson, A.B. Hedin, P.L. Håkansson, B.U.R. Sundqvist, B.G. Save, P.F. Nielsen, P. Roepstorff, K.E. Johansson, I. Kamensky, M.S. Lindberg. Plasma desorption mass spectrometry of peptides and proteins adsorbed on nitrocellulose. *Anal. Chem.* **1986**, *58*, 1084–1087.

9 R.E. Honig, R.J. Woolston. Laser induced emission of electrons, ions, and neutral atoms from solid surfaces. *Appl. Phys. Lett.* **1963**, *2*, 138–141.

10 N.C. Fenner, N.R. Daly, Laser used for mass analysis. *Rev. Sci. Instrum.* **1966**, *37*, 1068–1072.

11 F.J. Vastola, R.O. Mumma, A.J. Pirone. Analysis of organic salts by laser ionization. *Org. Mass Spectrom.* **1970**, *3*, 101–104.

12 N.A. Posthumus, P.G. Kistemaker, H.L.C. Meuzelaar, M.C. Ten Noever de Brauw. Laser desorption-mass spectrometry of polar nonvolatile bio-organic molecules. *Anal. Chem.* **1978**, *50*, 985–991.

13 F. Hillenkamp, R. Kaufmann, R. Nitsche, E. Unsöld. Laser microprobe mass analysis of organic materials. *Nature* **1975**, *256*, 119–120.

14 M. Karas, D. Bachmann, F. Hillenkamp. Influence of the wavelength in high-irradiance ultraviolet laser desorption mass spectrometry of organic molecules. *Anal. Chem.* **1985**, *57*, 2935–2939.

15 F. Hillenkamp. Laser desorption mass spectrometry. A review. In: A. Benninghoven, R.J. Colton, D.S. Simons, H.W. Werner (Eds.), *Secondary Ion Mass Spectrometry SIMS V. Springer Series in Chemical Physics*, Vol. 44., Springer-Verlag, Berlin (1986), pp. 471–475.

16 M. Karas, F. Hillenkamp. Laser desorption ionization of proteins with molecular masses exceeding 10000 daltons. *Anal. Chem.* **1988**, *60*, 2299–2301.

17 K. Tanaka, H. Waki, Y. Ido, S. Akita, Y. Yoshida, T. Yoshida. Protein and polymer analyses up to m/z 100000 by laser ionization time-of-flight mass spectrometry. *Rapid Commun. Mass Spectrom.* **1988**, *2*, 151–153.

18 M. Schürenberg, K. Dreisewerd, F. Hillenkamp. Laser desorption/ionization mass spectrometry of peptides and proteins with particle suspension matrixes. *Anal. Chem.* **1999**, *71*, 221–229.

19 J. Sunner, E. Dratz, Y.-C. Chen. Graphite surface-assisted laser desorption/ionization time-of-flight mass spectrometry of peptides and proteins from liquid solutions. *Anal. Chem.* **1995**, *67*, 4335–4342.

20 J. Wei, J.M. Buriak, G. Siuzdak. Desorption–ionization mass

spectrometry on porous silicon. *Nature* **1999**, *399*, 243–246.

21 S.A. Trauger, E.P. Go, Z. Shen, J.V. Apon, B.J. Compton, E.S.P. Bouvier, M.G. Finn, G. Siuzdak. High sensitivity and analyte capture with desorption/ ionization mass spectrometry on silylated porous silicon. *Anal. Chem.* **2004**, *76*, 4484–4489.

22 K. Strupat, M. Karas, F. Hillenkamp. 2,5-Dihydroxybenzoic acid: a new matrix for laser desorption-ionization mass spectrometry. *Int. J. Mass Spectrom. Ion Proc.* **1991**, *111*, 89–102.

23 R.C. Beavis, J.N. Bridson. Epitaxial protein inclusion in sinapic acid crystals. *Appl. Phys.* **1993**, *26*, 442–447.

24 V. Horneffer. *Matrix-Analyt-Wechelwirkungen bei der Matrix-Unterstützten Laserdesorptions/ionisations Massenspektrometrie (MALDI-MS).* Dissertation, University of Münster, 2002.

25 J. Kampmeier, K. Dreisewerd, M. Schürenberg, F. Hillenkamp. Investigations of 2,5-DHB and succinic acid as matrices for IR and UV MALDI: Part 1: UV and IR laser ablation in the MALDI process. *Int. J. Mass Spectrom. Ion Proc.* **1997**, *169/170*, 31–41.

26 V. Horneffer, K. Dreisewerd, H.-C. Lüdememann, F. Hillenkamp, M. Lage, K. Strupat. Is the incorporation of analytes into matrix crystals a prerequisite for matrix-assisted laser desorption/ ionization mass spectrometry? A study of five positional isomers of dihydroxybenzoic acid. *Int. J. Mass Spectrom.* **1999**, *185/186/187*, 859–870.

27 V. Horneffer, A. Forsmann, K. Strupat, F. Hillenkamp, U. Kubitschek. Localization of analyte molecules in MALDI preparations by confocal laser scanning microscopy. *Anal. Chem.* **2001**, *73*, 1016–1022.

28 R. Krueger, A. Pfenninger, I. Fournier, M. Glückmann, M. Karas. Analyte incorporation and ionization in matrix-assisted laser desorption/ionization visualized by pH indicator molecular probes. *Anal. Chem.* **2001**, *73*, 5812–5821.

29 V. Horneffer, R. Reichelt, K. Strupat. Protein incorporation into MALDI-matrix

crystals investigated by high resolution field emission scanning electron microscopy. *Int. J. Mass Spectrom.* **2003**, *226*, 117–131.

30 R. Skelton, F. Dubois, R. Zenobi. A MALDI sample preparation method suitable for insoluble polymers. *Anal. Chem.* **2000**, *72*, 1707.

31 V. Horneffer, M. Glückmann, R. Krüger, M. Karas, K. Strupat, F. Hillenkamp. Matrix-analyte-interaction in MALDI-MS: Pellet and nano-electrospray preparations. *Int. J. Mass Spectrom.* **2006**, *249/250*, 426–432.

32 H.-C. Lüdemann. *Matrix-assisted laser desorption/ionization (MALDI): Optical spectroscopy of matrix molecules.* Dissertation, University of Münster, 2000, pp. 11–18.

33 H.-C. Lüdemann, R.W. Redmond, F. Hillenkamp. Singlet-singlet annihilation in UV-MALDI studied by fluorescence spectroscopy. *Rapid Commun. Mass Spectrom.* **2002**, *16*, 1287–1294.

34 R. Knochenmuss. Photoionization pathways and free electrons in UV-MALDI. *Anal. Chem.* **2004**, *76*, 3179–3184.

35 A. Rohlfing, Ch. Menzel, L.M. Kukreja, F. Hillenkamp, K. Dreisewerd. Photoacoustic analysis of matrix-assisted laser desorption/ionization processes with pulsed infrared lasers. *J. Phys. Chem. B* **2003**, *107*, 12275–12286.

36 A. Leisner, A. Rohlfing, U. Röhling, K. Dreisewerd, F. Hillenkamp. Time-resolved imaging of the plume dynamics in infrared matrix-assisted laser desorption/ionization with a glycerol matrix. *J. Phys. Chem. B* **2005**, *109*, 11661–11666.

37 Menzel, Ch., K. Dreisewerd, St. Berkenkamp, F. Hillenkamp. The role of the laser pulse duration in infrared matrix-assisted laser desorption/ ionization mass spectrometry. *J. Am. Soc. Mass Spectrom.* **2002**, *13*, 975–984.

38 A. Rohlfing. *Untersuchungen zum Desorptionsprozess in der UV-Matrix-unterstützten Laserdesorptions/-ionisation mit einer Flüssigmatrix.* Dissertation, University of Münster, 2006.

39 A. Vogel, V. Venugopalan. Mechanisms of pulsed laser ablation of biological

tissues. *Chem. Rev.* **2003**, *103*, 577–644.

40 L.V. Zhigilei, B.J. Garrison. Mechanisms of laser ablation from molecular dynamics simulation: dependence on the initial temperature and pulse duration. *Appl. Phys. A* **1999**, *69*, 75–80.

41 L.V. Zhigilei, B. Garrison. Microscopic mechanisms of laser ablation of organic solids in the thermal and stress confinement irradiation regimes. *J. Appl. Phys.* **2000**, *88*, 1281–1298.

42 R. Knochenmuss, F. Dubois, M.J. Dale, R. Zenobi. The matrix suppression effect and ionization mechanisms in matrix-assisted laser desorption/ionization. *Rapid Commun. Mass Spectrom.* **1996**, *10*, 871–877.

43 M. Karas, D. Bachmann, U. Bahr, F. Hillenkamp. Matrix-assisted ultraviolet laser desorption of non-volatile compounds. *Int. J. Mass Spectrom. Ion Proc.* **1987**, *78*, 53–68.

44 H. Ehring, M. Karas, F. Hillenkamp. Role of photoionization and photochemistry in ionization processes of organic molecules and relevance for matrix-assisted laser desorption ionization mass spectrometry. *Org. Mass Spectrom.* **1992**, *27*, 472–480.

45 M. Karas, M. Glückmann, J. Schäfer. Ionization in matrix-assisted laser desorption/ionization: singly charged molecular ions are the lucky survivors. *J. Mass Spectrom.* **2000**, *35*, 1–12.

46 R.D. Burton, C.H. Watson, J.R. Eyler, G.L. Lang, D.H. Powell, M.Y. Avery. Proton affinities of eight matrices used for matrix-assisted laser desorption/ionization. *Rapid Commun. Mass Spectrom.* **1997**, *11*, 443–446.

47 A.G. Harrison. The gas-phase basicities and proton affinities of amino acids and peptides. *Mass Spectrom. Rev.* **1997**, *17*, 201–217.

48 R. Knochenmuss. A quantitative model of ultraviolet matrix-assisted laser desorption/ionization. *J. Mass Spectrom.* **2002**, *37*, 867–877.

49 R. Knochenmuss. A quantitative model of ultraviolet matrix-assisted laser desorption/ionization including analyte ion generation. *Anal. Chem.* **2003**, *75*, 2199–2207.

50 R. Krüger, M. Karas. Formation and fate of ion pairs during MALDI analysis: anion adduct generation as an indicative tool to determine ionization processes. *J. Am. Soc. Mass Spectrom.* **2002**, *13*, 1218–1226.

51 A. Pfenninger, M. Karas, B. Finke, B. Stahl, G. Sawatzki. Matrix optimization for matrix-assisted laser desorption/ionization mass spectrometry of oligosaccharides from human milk. *J. Mass Spectrom.* **1999**, *34*, 98–104.

52 M. Karas, U. Bahr, F. Hillenkamp, A. Tsarbopoulos, B.N. Pramanik. Matrix dependence of metastable fragmentation of glycoproteins in MALDI TOF mass spectrometry. *Anal. Chem.* **1995**, *67*, 675–679.

53 B. Spengler, D. Kirsch, R. Kaufmann. Fundamental aspects of post-source decay in MALDI mass spectrometry. *J. Phys. Chem.* **1992**, *96*, 9678–9684.

54 B. Spengler, D. Kirsch, R. Kaufmann. Metastable decay of peptides and proteins in matrix-assisted laser desorption mass spectrometry. *Rapid Commun. Mass Spectrom.* **1991**, *5*, 198.

55 B. Spengler. Post-source decay analysis in matrix-assisted laser desorption/ionization mass spectrometry of biomolecules. *J. Mass Spectrom.* **1997**, *32*, 1019–1036.

56 P. Chaurand, F. Luetzenkirchen, B. Spengler. Peptide and protein identification by matrix-assisted laser desorption ionization (MALDI) and MALDI-post-source decay time-of-flight mass spectrometry. *J. Am. Soc. Mass Spectrom.* **1999**, *10*, 91–103.

57 V. Gabelica, E. Schulz, M. Karas. Internal energy build-up in matrix-assisted laser desorption/ionization. *J. Mass Spectrom.* **2004**, *39*, 579–593.

58 E. Schulz, M. Karas, F. Rosu, V. Gabelica. Influence of the matrix on analyte fragmentation in atmospheric pressure MALDI. *J. Am. Soc. Mass Spectrom.* **2006**, *17*, 1005–1013.

59 U. Bahr, J. Stahl-Zeng, E. Gleitsmann, M. Karas. Delayed extraction time-of-flight MALDI mass spectrometry of proteins above 25 000 Da. *J. Mass Spectrom.* **1997**, *32*, 1111–1116.

60 M. Glückmann, M. Karas. The initial velocity and its dependence on matrix, analyte and preparation method in ultraviolet matrix-assisted laser desorption/ionization. *J. Mass Spectrom.* **1999**, *34*, 467–477.

61 R.S. Brown, J.L. Lennon. Sequence-specific fragmentation of matrix-assisted laser-desorbed protein/peptide ions. *Anal. Chem.* **1995**, *67*, 3990–3999.

62 R.S. Brown, B.L. Carr, J.L. Lennon. Factors that influence the observed fast fragmentation of peptides in matrix-assisted laser desorption. *J. Am. Soc. Mass Spectrom.* **1996**, *7*, 225–232.

63 M.N. Takayama. N-C bond cleavage of the peptide backbone via hydrogen abstraction. *J. Am. Soc. Mass Spectrom.* **2001**, *12*, 1044–1049.

64 M. Takayama. In-source decay characteristics of peptides in MALDI TOF mass spectrometry. *J. Am. Soc. Mass Spectrom.* **2001**, *12*, 420–427.

65 R.S. Brown, J. Feng, D.C. Reiber. Further studies of in-source fragmentation of peptides in MALDI. *Int. J. Mass Spectrom. Ion Proc.* **1997**, *169/170*, 1–18.

66 T. Köcher, A. Engström, R.A. Zubarev. Fragmentation of peptides in MALDI in-source decay mediated by hydrogen radicals. *Anal. Chem.* **2005**, *77*, 172–177.

67 D.C. Reiber, T.A. Grover, R.S. Brown. Identifying proteins using matrix-assisted laser desorption/ionization in-source fragmentation data combined with database searching. *Anal. Chem.* **1998**, *70*, 673–683.

68 M. Takayama, A. Tsugita. Sequence information of peptides and proteins with in-source decay in matrix assisted laser desorption/ionization-time of flight-mass spectrometry. *Electrophoresis* **2000**, *21*, 1670–1677.

69 B. Rosinke, K. Strupat, F. Hillenkamp, J. Rosenbusch, N. Dencher, U. Krüger, H.-J. Galla. Matrix-assisted laser desorption/ionization mass spectrometry (MALDI-MS) of membrane proteins and non-covalent complexes. *J. Mass Spectrom.* **1995**, *30*, 1462–14.

70 L.H.R. Cohen, K. Strupat, F. Hillenkamp. Analysis of quarternary protein ensembles by matrix-assisted laser desorption/ionization mass spectrometry. *J. Am. Soc. Mass Spectrom.* **1997**, *8*, 1046–1052.

71 Horneffer, K. Strupat, F. Hillenkamp. Localization of noncovalent complexes in MALDI-preparations by CLSM. *J. Am. Soc. Mass Spectrom.* (in press).

72 A. Wortmann, T. Pimenova, R. Zenobi. Investigation of the first shot phenomenon in MALDI mass spectrometry of protein complexes. 54th ASMS Conference on Mass Spectrometry and Allied Topics, Seattle, WA, May 28–June 1, 2006, Poster WP 256.

73 P. Juhasz, K. Biemann. Mass spectrometric molecular weight determination of highly acidic compounds of biological significance via their complexes with basic peptides. *Proc. Natl. Acad. Sci. USA* **1994**, *91*, 4333–4337.

74 M. Moniatte, F.G. van der Goot, J.T. Buckle, F. Pattus, A. van Dorsselaer. Characterisation of the heptameric pore-forming complex of the *Aeromonas* toxin aerolysin using MALDI-TOF mass spectrometry. *FEBS Lett.* **1996**, *384*, 269–272.

75 M.O. Glocker, S.H.J. Bauer, J. Kast, J. Volz, M. Przybylski. Characterization of specific noncovalent protein complexes by UV matrix-assisted laser desorption ionization mass spectrometry. *J. Mass Spectrom.* **1996**, *31*, 1221–1227.

76 I. Gruic-Sovulj, H.C. Lüdemann, F. Hillenkamp, I. Weygand-Duraseciv, Z. Kucan, J. Peter-Katalinic. Detection of noncovalent tRNA aminoacyl-tRNA synthetase complexes by matrix-assisted laser desorption/ionization mass spectrometry. *J. Biol. Chem.* **1997**, *272*, 32084–32091.

77 T. Vogl. J. Roth, F. Hillenkamp, K. Strupat. Calcium-induced non-covalently linked tetramers of MRP8 and MRP14 detected by ultraviolet matrix-assisted laser desorption/ionization mass spectrometry. *J. Am. Soc. Mass Spectrom.* **1999**, *10*, 1124–1130.

78 J.G. Kiselar, K.M. Downard. Preservation and detection of specific antibody-peptide complexes by matrix-assisted laser desorption ionization mass spectrometry. *J. Am. Soc. Mass Spectrom.* **2000**, *11*, 746–750.

79 A.S. Woods, M.A. Huestis. A study of peptide-peptide interaction by matrix-assisted laser desorption/ionization. *J. Am. Soc. Mass Spectrom.* **2001**, *12*, 88–96.

80 A.S. Woods, J.M. Koomen, B.T. Ruotolo, K.J. Gillig, D.H. Russel, K. Fuhrer, M. Gonin, T.F. Egan, J.A. Schultz. A study of peptide-peptide interactions using MALDI ion mobility o-TOF and ESI mass spectrometry. *J. Am. Soc. Mass Spectrom.* **2002**, *13*, 166–169.

81 S.H. Luo, Y.M. Li, W. Qiang, Y.F. Zhao, H. Abe, T. Nemoto, X.R. Qin, H. Nakanishi. Detection of specific noncovalent interaction of peptide with DNA by MALDI-TOF. *J. Am. Soc. Mass Spectrom.* **2004**, *15*, 28–31.

82 M. Karas, U. Bahr, A. Ingendoh, F. Hillenkamp. Laser desorption/ionization mass spectrometry of proteins of mass 100 000 to 250 000 Dalton. *Angew. Chem. Int. Ed. Engl.* **1989**, *28*, 760–761.

83 F. Hillenkamp, M. Karas, R.C. Beavis, B.T. Chait. Matrix-assisted laser desorption ionization mass spectrometry of biopolymers. *Anal. Chem.* **1991**, *63*, 1193A–1203A.

84 F. Hillenkamp. *New methods for the study of biomolecular complexes; NATO ASI Series C: Mathematical and Physical Science 510.* Kluwer Academic Publishers: Dordrecht, The Netherlands, 1998; pp. 181–191.

85 T.B. Farmer, R.M. Caprioli. Determination of protein-protein interactions by matrix-assisted laser desorption/ionization mass spectrometry. *J. Mass Spectrom.* **1998**, *33*, 697–704.

86 M. Zehl, G. Allmaier. Instrumental parameters in MALDI-TOF mass spectrometric analysis of quaternary protein structures. *Anal. Chem.* **2005**, *77*, 103–110.

87 P. Lecchi, L.K. Pannelle. The detection of intact double-stranded DNA by MALDI. *J. Am. Soc. Mass Spectrom.* **1995**, *6*, 972–975.

88 F. Kirpekar, S. Berkenkamp, F. Hillenkamp. Detection of double-stranded DNA by IR- and UV-MALDI mass spectrometry. *Anal. Chem.* **1999**, *71*, 2334–2339.

89 J. Villanueva, O. Yanes, E. Querol, L. Serrano, F.X. Aviles. Identification of protein ligands in complex biological samples using intensity-fading MALDI-TOF mass spectrometry. *Anal. Chem.* **2003**, *75*, 3385–3395.

90 O. Yanes, J. Villanueva, E. Querol, F.X. Aviles. Functional screening of serine protease inhibitors in the medical leech *Hirudo medicinalis* monitored by intensity fading MALDI-TOF MS. *Mol. Cell. Proteomics* **2005**, *4*, 1602–1613.

91 O. Vorm, P. Roepstorff, M. Mann. Improved resolution and very high sensitivity in MALDI-TOF of matrix surfaces made by fast evaporation. *Anal. Chem.* **1994**, *66*, 3281–3287.

92 M. Schuerenberg, C. Luebbert, H. Eickhoff, M. Kalkum, H. Lehrach, E. Nordhoff. Prestructured MALDI-MS sample supports. *Anal. Chem.* **2000**, *72*, 3436–3442.

93 M. Karas, H. Ehring, E. Nordhoff, B. Stahl, K. Strupat, F. Hillenkamp, M. Grehl, B. Krebs. Matrix-assisted laser desorption/ionization mass spectrometry with additives to 2,5-dihydroxybenzoic acid, *Org. Mass Spectrom.* **1993**, *28*, 1476–1481.

94 Y.F. Zhu, C.N. Chung, N.I. Taranenko, S.L. Allman, S.A. Martin, L. Haff, C.H. Chen. The study of 2,3,4-trihydroxyacetophenone and 2,4,6-trihydroxyacetophenone as matrices for DNA detection in matrix-assisted laser desorption/ionization time-of-flight mass spectrometry. *Rapid Commun. Mass Spectrom.* **1996**, *10*, 383–388.

95 U. Pieles, W. Zürcher, M. Schär, H.E. Moser. Matrix-assisted laser desorption ionization time-of-flight mass spectrometry: a powerful tool for the mass and sequence analysis of natural and modified oligonucleotides. *Nucleic Acids Res.* **1993**, *21*, 3191–3196.

96 I. Smirnov, X. Zhu, T. Taylor, Y. Huang, P. Ross, I.A. Papayanopoulos, S.A. Martin, D.J. Pappin. Suppression of alpha-cyano-4-hydroxycinnamic acid matrix clusters and reduction of chemical noise in MALDI-TOF mass spectrometry. *Anal. Chem.* **2004**, *76*, 2958–2965.

97 S. Kjellström, O.N. Jensen. Phosphoric acid as a matrix additive for MALDI MS analysis of phosphopeptides and phosphoproteins. *Anal. Chem.* **2004**, *76*, 5109–5117.

98 A. Steensballe, O.N. Jensen. Phosphoric acid enhances the performance of

Fe(III) affinity chromatography and matrix-assisted laser desorption/ionization tandem mass spectrometry for recovery, detection and sequencing of phosphopeptides. *Rapid Commun. Mass Spectrom.* **2004**, *18*, 1721–1730.

99 M.R. Larsen, T.E. Thingholm, O.N. Jensen, P. Roepstorff, J.D. Jorgensen. Highly selective enrichment of phosphorylated peptides from peptide mixtures using titanium dioxide microcolumns. *Molec. Cell Proteomics* **2005**, *4*, 873–886.

100 C.T. Chen, Y.-C. Chen. Fe_3O_4/TiO_2 core/shell nanoparticles as affinity probes for the analysis of phosphopeptides using TiO_2 surface-assisted laser desorption/ionization

mass spectrometry. *Anal. Chem.* **2005**, *77*, 5912–5919.

101 D.W. Armstrong, L.K. Zhang, L. He, M.L. Gross. Ionic liquids as matrixes for matrix-assisted laser desorption/ionization mass spectrometry. *Anal. Chem.* **2001**, *73*, 3679–3686.

102 A. Tholey, E. Heinze. Ionic (liquid) matrices for matrix-assisted laser desorption/ionization mass spectrometry-applications and perspectives. *Anal. Bioanal. Chem.* **2006**, *386*, 24–37.

103 S. Trimpin, S. Keune, H.J. Rader, K. Mullen. Solvent-free MALDI-MS: developmental improvements in the reliability and the potential of MALDI in the analysis of synthetic polymers and giant organic molecules. *J. Am. Soc. Mass Spectrom.* **2006**, *17*, 661–671.

2
MALDI Mass Spectrometry Instrumentation

Peter B. O'Connor and Franz Hillenkamp

2.1
Introduction

MALDI ions can be mass analyzed using a number of different types of mass spectrometers. These range from the inexpensive ion trap mass spectrometers (QIT-MS), which generate fairly low-quality spectra, to the Fourier transform mass spectrometer (FT-MS), which generates extremely high-quality spectra, but at significantly greater cost. By far, the most common type of analyzer for MALDI ions, however, is the time-of-flight mass spectrometer (TOF-MS).

The choice of mass analyzer is governed by the specific properties of the MALDI ion source on the one hand, and the analytical task of the operator on the other hand. The typical MALDI source generates a pulse of ions of at most a few nanoseconds in duration at a rate of 1 to 100 pulses per second; thus, they are most compatible with mass spectrometers which are effectively pulsed-ion detectors, including instruments which trap the ions for later analysis. Mass spectrometers which operate on a continuous beam of ions, such as magnetic sector and quadrupole instruments, tend to suffer substantial problems in terms of sensitivity and are generally not suitable for a pulsed-ion source. For this reason, while several experimental instruments have been designed and constructed, there is very little utility in a MALDI-magnetic sector or MALDI quadrupole mass spectrometer, and these instrument geometries have not generally been followed up with commercial development. More recently, frequency-tripled Nd:YAG lasers with pulse repetition rates of 1 kHz or more have found increasing application in MALDI ion sources. Together with suitable trapping devices, they can be operated in a quasi continuous mode. Nonetheless, because MALDI usually generates a pulse of ions several nanoseconds wide, it is particularly suited to TOF analyzers, making TOF the dominant mass analysis technology for MALDI ions, though several variants exist. These and other instruments used for mass analysis of MALDI ions will be discussed in this chapter.

A second important feature of MALDI is that it can generate ions of very large masses of greater than 1 MDa. In contrast to electrospray, MALDI ions carry only a few charges with the singly charged ions generally predominant. Again, the TOF

MALDI MS. A Practical Guide to Instrumentation, Methods and Applications.
Edited by Franz Hillenkamp and Jasna Peter-Katalinić
Copyright © 2007 Wiley-VCH Verlag GmbH & Co. KGaA, Weinheim
ISBN: 978-3-527-31440-9

is by far the most suited analyzer for such large ions, because it has, in principle, an unlimited mass range. However, the high initial axial velocity of MALDI ions and their broad kinetic energy distribution causes problems for all mass analyzers. This limits the mass resolution in TOF instruments and makes efficient capture in storage devices such as ion traps or FTICR instruments difficult. The desorption/ionization also transfers considerable energy into internal vibrational energy of ions, which is only partly removed by the adiabatic expansion of the plume. This additional vibrational energy activates the ions and results in metastable ions that then undergo unimolecular decay over a period of microseconds to seconds depending on the relative stability of the cleaved bond (i.e., the rate of that particular fragmentation reaction and on the amount of excess internal energy deposited in the molecule on desorption). This metastable decay can be useful for obtaining structural information on the ion – such as the sequence of peptides – but it can also be a problem if the part of the molecule to be analyzed is the part which fragments (and is lost) on desorption. The latter situation occurs frequently with labile species such as sulfated peptides, phosphopeptides, and sialylated glycolipids. Metastable fragmentation of ions can also prevent the detection of the parent ion altogether in mass analyzers which need milliseconds to seconds for ion detection. The problem can be partially solved by increasing the pressure in the ion source during desorption, as discussed in Section 2.3.1. Some newer MALDI mass spectrometers have incorporated this concept into the design of the MALDI sources, and these have been shown to be very successful in stabilizing labile molecules during desorption and ionization.

The aim of this chapter is to provide an overview of the instrumentation available for the mass analysis of MALDI ions. It is not intended as an exhaustive treatise on each instrument; consequently, if more detail is required the reader is encouraged to examine the reference material in detail. The first sections of the chapter will discuss lasers for MALDI and how they are coupled to the mass spectrometer, vibrational cooling, and tandem mass spectrometry of MALDI ions; this is followed by a presentation of the different mass analyzer designs. Emphasis is placed on commercially available instrument configurations. Definitions of all acronyms and technical descriptions of terms such as peak centroid and resolving power are included at the end of the chapter.

2.2
Lasers for MALDI-MS

A variety of gas and solid-state lasers have been used successfully for MALDI-MS. With MALDI, only pulsed lasers are useful, because the energy necessary for the desorption and ionization must be transferred to the sample in a time which is short compared to the thermal diffusion time. Axial time-of-flight (TOF) analyzers also require a short laser pulse for a good time and mass resolution. The lasers are characterized by their emission (wavelength, pulse width, and pulse energy) and beam (beam diameter and divergence) parameters.

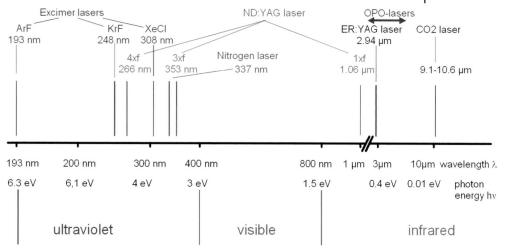

Fig. 2.1 Wavelength and photon energy of frequently used desorption lasers.

The wavelength, λ, and single photon energy, hc/λ, of the most commonly used lasers are shown in Figure 2.1. For ultraviolet-MALDI (UV-MALDI), nitrogen (N₂) gas lasers are by far the most commonly used ones because of their simplicity, small size, and relatively low cost. They emit at a wavelength of 337 nm, close to the absorption maximum of many commonly used matrices such as 2,5-DHB, HCCA, and 3-HPA (for a discussion of the different matrices, see Chapter 1). The main limitation of N₂-lasers is their limited pulse repetition frequency of typically <100 Hz and the total number of typically ≤10⁸ total emissions before the laser cartridge needs to be replaced, which correlates to between 0.5 and 2 years depending on the instrumental configuration. Both features limit their use in high-throughput analysis. For such applications, frequency-tripled Nd:YAG solid-state lasers with a wavelength of 355 nm are typically employed. For the much less common (so far) infrared MALDI (IR-MALDI), lasers emitting in the OH- or NH-stretch vibration range near 3 μm are usually employed. These are Er:YAG lasers with a wavelength of 2.94 μm (also Er:YSSG lasers, 2.79 μm) and optical parametric oscillators (OPO) which are tunable typically from about 1.7 to 2.5 μm. CO₂ lasers with a wavelength around 10 μm are much less frequently used. Depending on the laser type, the emission pulse width τ varies between about 0.5 ns and 25 ns in the UV (with 3–5 ns most frequently used) and 5–100 ns in the infrared. Picosecond and femtosecond lasers have not been used commercially to date. MALDI-MS requires relatively little energy per pulse. For UV-MALDI, 10 to 100 μJ per pulse (~1 μJ on the target) suffices, whereas for IR-MALDI 100 μJ to 1 mJ are typically required. In both cases, the fluence (energy per unit area) is the actual limitation, with the actual energy needed per pulse varying improportion to the laser spot area on the sample surface.

Besides the emission parameters of wavelength, pulse width, output energy per pulse, and maximum pulse repetition frequency, MALDI laser beam shape characteristics are also important. The laser near-field beam diameter d (measured at the laser exit port and defined typically as the full diameter at the $1/e$- intensity points; $e \approx 2.718$; Euler's number), the beam divergence θ (usually defined as the full plane angle at the $1/e$ intensity points), and the energy distribution H across the beam in the near field and the far field (measured in the focal plane of a lens focusing the beam) are important, because they determine the optimal coupling of the laser beam into the mass spectrometer and the sample. The energy per unit area, defining the energy distribution at any point along the beam is commonly called fluence, H (correctly radiant exposure), and sometimes incorrectly energy density. The power per unit area is called irradiance, I, and often incorrectly called intensity or power density. The energy absorbed, E_a, into the sample per unit volume at the surface is simply

$$E_a/V = \alpha H$$

where α is the absorption coefficient of the matrix at the laser wavelength (for details see Chapter 1.3).

Optimal MALDI performance is achieved for spot sizes on the sample of ~100–500 µm diameter and using laser spots with a fluence or irradiance distribution that is smooth, without any "hot spots". The optimal beam profile on the target should be either "bell-shaped" (near-Gaussian) or close to uniform (top hat) for fiber-optic coupling. These two requirements determine the transfer optics from the laser to the target. In the simplest case, the beam is focused onto the target by a single lens of focal length f. In this case the beam diameter on the target is given as

$$\delta = \theta \cdot f$$

where θ is the divergence angle of the beam. If the laser beam is incident onto the target at an angle ϕ other than perpendicular, the spot on the target is elliptical with the longer axis equal to $\delta/\cos\phi$. It is important to place the target precisely in the focal plane of the lens to localize the far field of the laser beam on the target. This is particularly important for N_2- or Excimer-lasers, which typically have a very inhomogeneous fluence distribution all along the beam except for the far field. Properly adjusted "axial mode only" (TEM_{oo}) solid-state lasers – and Nd:YAG lasers in particular – usually have a more benign beam profile – that is, Gaussian all along the beam. Alternatively, lasers with a less smooth beam profile can be coupled to the mass spectrometer via fiber optics, which, provided the fiber is sufficiently long, scrambles the modes and yields a flat "top hat" beam profile. Recently, Bruker Daltonics has introduced a special irradiation geometry called "smart beam", which is said to have advantages for the 355-nm wavelength and certain matrices [128].

The beam divergence θ is a fixed parameter for a given laser with a value of typically between 0.2 and 5 mrad. For Gaussian (TEM_{oo}) beams, the spot size on the target can be varied by defocusing, without sacrificing the quality of the fluence distribution. In this case the beam focus should be placed in front of the target

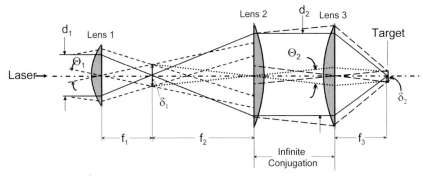

Fig. 2.2 Laser focusing optics for the adjustment of spot size on the target.

surface to prevent the maximum fluence (at the focal point) from occurring inside the sample. For laser beams with a less benign beam profile, the spot size on the target can be varied by choosing a lens (of focal length f) positioned at a distance f in front of the target. This may not always be possible, however, because of geometric constraints by other components of the ion source such as the ion lens, or because the lens is placed outside the vacuum for easy beam alignment. In this case the spot size can be varied by placing a confocal (Gaussian) telescope in between the laser and the focusing lens, as shown in Figure 2.2. Such a telescope expands (or shrinks) the beam diameter to $d_2/d_1 = f_2/f_1$ and accordingly the divergences by the inverse ratio to $\theta_2/\theta_1 = f_1/f_2$. With a focusing lens of focal length f_3 the focal spot diameter on the target will be $\delta_2 = (f_1/f_2) \cdot \theta \cdot f_3$. The distances between the two lenses and between lens 3 and the target must be adjusted carefully to $f_1 + f_2$ and f_3 (all at the laser wavelength), respectively.

The distance between the laser and lens 1, as well as that between lenses 2 and 3, can be chosen freely to accommodate the geometry of the instrument. Because laser beams are monochromatic, have small (near) diffraction-limited divergence, and are arranged in "infinite conjugation" (parallel to focused to parallel beams; see Fig. 2.2), lenses of small diameter relative to their focal length (small f-number) suffice. Therefore simple plano-convex lenses can be used without introducing measurable optical aberrations, at least as long as the focal diameter on the target stays above about 10 μm. All lenses should be oriented as shown in Figure 2.2 – that is, with the curved face towards the parallel and the plane surface towards the focused beam. For UV-MALDI, UV-grade quartz lenses must be used, whereas for IR-MALDI at wavelengths below about 3.5 μm, IR-grade quartz lenses can be chosen as these have the advantage of transmitting visible alignment laser beams if so desired. At longer wavelengths, at the 10.6 μm of the CO_2-laser in particular, ZnSe or BaF lenses must be used (depending on the wavelengths that must pass through the lenses). N_2-lasers often have a rectangular rather than a circular near field and different divergences in the two perpendicular planes. In this case, two cylinder lenses of different focal lengths may be advisable for optimal focusing, particularly when focusing into an optical fiber.

The angle of incidence of the laser beam on the target is often determined by details of the ion optics geometry. Angles between 0° (normal incidence) and about 80° have been realized in different instruments with quite comparable results. This is not too much of a surprise, because the most important parameter, the density of absorbed energy per unit volume E_a/V (for definition, see Chapter 1.3), is independent of the angle of incidence. Only the spot diameter increases and becomes elliptical with increasing angle of incidence, and the penetration depth decreases accordingly. Additionally, since matrix crystals are often oriented randomly, the facets that are angled more toward perpendicularity to the laser beam contribute more ions to the plume than those which face away from the incident laser light. This results, when averaged over all ions, in a desorption plume direction being angled slightly (~10° in extreme cases) toward the laser beam rather than normal to the surface.

Instead of coupling the laser beam directly into the mass spectrometer, optical fibers can be used for UV-wavelengths above about 300 nm [1]. Provided that long enough fibers with sufficient bending are used, the fibers will mix the beam modes so that the result is a uniformly illuminated exit face of the fiber, and the laser beam profile will have a flat-top or "top-hat" shape. The necessary fiber length should be a few meters for low-coherency lasers such as the N_2-laser, and up to 50 m for high-quality beam lasers such as TEM_{oo}-mode Nd:YAG lasers. Fiber optics are also useful in complicated coupling geometries and can simplify service of the instrument with respect to exchange of lasers and laser cartridges. In order to maintain the uniform (flat top) illumination on the sample, the fiber exit face must be imaged onto the target. This is best achieved with two plano-convex lenses in infinite conjugation, as discussed above. In this case, the fiber end face would be placed in the left side focal plane of lens 2 in Figure 2.2. The main disadvantage of fiber-optics coupling is the divergence of the (essentially non-coherent) beam as it exits the fiber. The numerical aperture of fiber optics is typically 0.1–0.2, about two orders of magnitude larger than that of the laser beam. (*Note*: numerical aperture, $NA = \sin(\Theta/2)$, where Θ is the full-angle of the divergence of the beam; thus, $NA = 0.2$ equals $\Theta = 22°$.) The imaging lenses must, therefore, have a suitably large diameter-to-focal length ratio (f-number). However, the large diameter of the beam as it converges towards the target is usually a more problematic constraint. This larger-diameter beam may interfere with structures of the ion lens; for example, if ions are being desorbed into a quadrupole, the beam must fit between the quadrupole rods. Furthermore, even though energies in the range of 1 μJ to 1 mJ suffice for most MALDI applications, because the beam must be focused into the fiber, the fluence and irradiance at the fiber entrance face can reach high values close to or above the destruction limits of most fiber materials. So-called "high-energy" or "high-power" quartz/quartz fibers of typically 100–200 μm core diameter are therefore used, and great care must be taken to image the laser beam onto the fiber front face such that the core diameter is optimally illuminated. Too small a spot, focusing into the fiber core in particular, will damage the fiber end face, while too large a spot size will lead to excessive losses. Additionally, the focal point of the coupling optics must be placed in front of the face of the fiber; if it is placed inside the fiber, the fluence inside the fiber at the focal point will be beyond the damage threshold of the fiber, and the front face of the fiber will literally explode, requiring replacement of the fiber. Most fibers have a protective polymer coating; illumination of this coating will ablate some of the material, part of which will then settle on the front face of the core, causing damage by successive laser exposures. It is, therefore, good practice to retract the protective coating by a few millimeters. No practically useful fiber optics are currently available for IR-MALDI.

Careful adjustment of the laser fluence (commonly, and incorrectly, called laser power) on the target is one of the essential prerequisites for a successful MALDI analysis. This adjustment is particularly critical for TOF analyzers with axial ion extraction and analog ion detection – by far the most commonly used MALDI mass spectrometers. Lasers for MALDI mass spectrometers should therefore have a reproducibility of the output energy from shot to shot of better than 3%. Too high a fluence will lead to saturation of the detector and/or the analog-to-digital converter (usually with a 8-bit dynamic range) and will often generate excessive fragmentation. Detector saturation is particularly critical for microchannel-plate detectors, because even partial saturation by matrix ions will decrease the detection sensitivity for the higher-mass analyte ions. TOF analyzers with orthogonal extraction, ion sources that operate with a cooling gas environment, and trapped-ion mass analyzers avoid some of these problems. Direct control of the laser output energy is not advisable, because it has adverse effects on the reproducibility of the output energy from shot to shot, and may also influence the beam parameters such as divergence and fluence distribution across the beam. Control of the transmitted fraction of the near field by a motor-driven iris for fluence adjustment, though sometimes used, is not the ideal solution. It is rather nonlinear and the dependence of the energy on the iris diameter becomes very steep below ~10% of transmitted energy. Better solutions use graded neutral density filters or angle-dependent reflection of dielectric optical coatings to control laser fluence. Laser beam attenuators should be placed into a parallel section of the beam to limit optical aberration (e.g., between lenses 2 and 3 in Fig. 2.2). Care should also be taken to avoid beam walk on the target as a function of the attenuator setting because of wedges or beam displacements of the attenuators. For many MALDI-MS applications, as well as for monitoring the analyzer performance it would also be helpful to monitor the energy incident on the target, by measuring a small fraction of the beam behind the attenuator with a photodiode and an integrating amplifier. Although this signal would only be relative, it would allow a comparison to be made from sample to sample, and from day to day. Unfortunately, such a function is not integrated in most commercial MALDI instruments.

2.3
Fragmentation of MALDI Ions

Many MALDI ions are vibrationally excited or "hot" (i.e., they are metastable on the time scale of detection in the mass spectrometer) as they emerge from the desorption plume. This is due to three separate processes. First, ions absorb some of the laser energy directly and are also heated by collisions with the matrix molecules which absorb the laser light very efficiently. Second, an electric field in the source will accelerate the ions and drag them through the matrix plume, causing collisional activation. Third, chemical reactions such as hydrogen and proton transfer occur inside the plume and have certain exothermicities associated with them. This internal energy causes many molecules – particularly labile molecules such as phospho- or sulfo- peptides and glycolipids – to dissociate before (or during) detection. Depending on when and where in the mass analyzer this fragmentation takes place, they may go undetected, be detected at their fragment mass, or get lost in the noise. The details of this will be discussed for the different mass spectrometers later in the chapter.

2.3.1
MALDI at Elevated Pressure

Elevated pressure of typically a few millibars of inert gases such as He, Ar, or N_2 in the ion source, can cool the ions and thereby increase their stability. If the MALDI ions collide with these thermal gas molecules, some of their vibrational energy gets removed; this process is known as "collisional cooling".

In 1998, Krutchinsky et al. demonstrated conclusively [2] that desorbing ions into a space of elevated pressure decreased metastable decay. This created substantial excitement, and since then ion sources using this principle have been designed for qTOF [3], oTOF [4,5], Fourier transform ion cyclotron resonance mass spectrometers (FTICRMS) [6–9], and ion trap mass spectrometers [10]. The fundamental reason for the excitement is that desorbing ions in the presence of a cooling gas solved one of the longstanding problems of MALDI, namely the ionization of labile molecules.

Vibrationally cooled ions from MALDI are much more stable than their non-cooled counterparts. This is clear from the studies of O'Connor and Costello [5,11,12] in which intact gangliosides were desorbed and detected in a Fourier transform mass spectrometer. Without cooling, the ions could barely be detected, and even then they had lost all of the labile sialic acid groups. Collisional cooling removed the excess vibrational energy from the glycolipids, allowing their intact detection. This method is potentially extensible to a number of other labile species, ranging from sulfated sugars (glycosaminoglycans) to DNA.

The vibrational cooling effect has been extended to the analysis of ions desorbed at atmospheric pressure [13–18]. This arrangement is extremely simple, merely placing the MALDI sample directly in front of an electrospray ionization source orifice. This worked well, and Cotter et al. shortly thereafter duplicated the ganglioside investigations of O'Connor et al., thus proving that the atmospheric pressure MALDI source was also a collisionally cooled ion source design [19,20]. However, compared to operation at a few millibars, ionization at atmospheric pressure requires that the ions be extracted down a small tube and through a molecular beam skimmer in order to move the ions into the vacuum where they can be analyzed. Considering the losses expected on the capillary walls and on the skimmer, it is unlikely that the AP-MALDI method will become as sensitive as the vacuum or low-pressure MALDI experiment.

2.3.2
Tandem Mass Spectrometry of MALDI Ions

Tandem mass spectrometry is the process of selecting an ion, causing it to fragment, and obtaining a mass spectrum of the resulting fragment ions. Fragment ions carry information about the ion structure. For example, the tandem mass spectrum of a peptide encodes by mass the partial or full sequence of that peptide. A simple, single-stage tandem mass spectrometry experiment is also often called an "MS/MS" experiment, but secondary (and higher-order) stages of fragmentation can also be carried out in which a fragment ion (often called a "daughter ion") is further fragmented (into a "grand-daughter ion"), and so forth. These experiments are called "MS^n" experiments.

For MALDI, because the ions are generally metastable, a "Pseudo-MS/MS" technique exists where the parent ions are fragmented, but isolation of the precursor is not performed. This method is often (and incorrectly) called tandem mass spectrometry, but is correctly called simply metastable decay and, in TOF instruments, can be classified as either "in-source" decay or "post-source" decay. Metastable decay does generate fragments, and, provided that the sample is homogeneous, these fragments are useful for structural characterization, as in tandem mass spectrometry. However, if the sample is a mixture (e.g., the tryptic peptides from a digested protein), it is difficult or impossible to determine which metastable fragment originated from which precursor ion.

Two different natural processes generate fragments of MALDI ions during desorption, chemical reactions of the ions presumably with hydrogen radicals [21] and metastable decay, as discussed above. The chemical reaction-induced fragmentation has been mostly investigated for peptides. The process generates predominantly a-, c-, y-, and z-ions, similar to electron capture dissociation (ECD) spectra of multiply charged ions in electrospray ionization. They are generated during a very short initial time of the plume development, and typically faster than the time delay for ion extraction in axial TOF instruments [22,23]. The fragment yield is very low, and these fragments are, therefore, not used for routine analysis. Unimolecular decay of ions due to excess internal energy is the method of choice to obtain structural information about MALDI ions. Often, ions carry sufficient internal energy from the desorption/ionization process to generate enough of a fragment yield on the micro- to millisecond time scale. For peptides, metastable decay leads predominantly to b- and y-fragments (following the Roepstorff nomenclature [24] as modified by Biemann [25]; see Fig. 3.5). Larger ions, and proteins in particular, typically lose small neutrals such as water or ammonia, which are not structurally useful. If the instrument used does not have sufficient resolving power, these small molecule losses can degrade mass resolution. The yield of fragments from metastable decay can be increased either by choosing a "hot" matrix

A. Collisionally Activated Dissociation

B. Photodissociation

C. Ion Electron Reactions

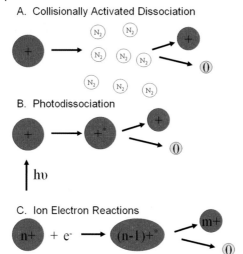

Fig. 2.3 Tandem mass spectrometry techniques.

such as alpha-cyano-4-hydroxycinnamic acid, or by prompt ion extraction in TOF mass spectrometers, which will also increase fragmentation via collisional activation by accelerating the ions through the cloud of neutral matrix molecules in the still-dense plume. Increasing the laser fluence has a surprisingly small influence on metastable fragmentation.

Often, the yield of structurally relevant fragments from metastable decay is not sufficient. In this case the internal ion energy must be increased by heating them using external means, and for this collisional activation of the molecules by an inert gas in a collision cell is by far the most common method. This process is referred to either as collisionally activated dissociation (CAD) or collision-induced dissociation (CID); the two terms are used interchangeably and are identical in meaning. In low-energy CAD the mean collisional energy of individual collisions is only a few eV/charge. In this case, the instrumental parameters are set to provide for many such collisions for any given ion, which thereby accumulates internal energy and randomizes it over the accessible internal degrees of freedom. Low-energy CAD fragments are very similar to those of metastable decay due to the internal ion energy, as discussed above. If the collision energy is >100 eV, high-energy CAD will generate additional types of fragments which can carry specific structural information such as the side-chain structure in peptides (d-, w-, and z-ions).

Ultraviolet (UV) photodissociation has been used with MALDI ions by several research groups, and has generated some rather unusual fragmentation patterns which are currently being investigated. Single-photon UV photodissociation has the advantage over CAD in that the timing of the fragmentation event can be controlled to the picosecond. If fragmentation is prompt, then the fragment ion resolution in a TOF instrument will not suffer; but if fragmentation is not prompt, then metastable lifetimes can be accurately studied because the activation time is known exactly. However, the energy needed for fragmentation is extremely high, and the most commonly used laser for this type of work is the fluorine excimer laser which emits at 157 nm (~7.9 eV/photon). Even then, a suitable chromophore must be present, which is generally the case for peptides, but may not be the case for other molecules such as oligosaccharides.

More generally, infrared (IR) photodissociation is used, commonly called infrared multiphoton photodissociation (IRMPD). This mode normally uses CO_2 lasers which emit near 10.6 μm (~0.1 eV/photon). This mode overlaps with many stretching and bending modes in organic molecules (particularly as CO_2 lasers emit in a fairly wide wavelength range from 9.1–10.6 μm). This is effectively a slow-heating method which requires the absorption of hundreds or thousands of photons prior to cleavage. Thus, it is not suitable for TOF instruments where the ion residence time inside the laser beam is very short. It is, however, a very good fragmentation method for QIT and FTICR mass spectrometers (and also hybrids; see Section 2.7) where the ions can be easily irradiated for milliseconds.

IRMPD generates fragment ion spectra similar to those of metastable decay, but with a high coverage because primary fragment ions (daughter ions) remain confined within the laser beam and continue to absorb energy, thus generating secondary (grand-daughter) and higher-order (great-grand-daughter, etc.) fragments. While this is a very simple method and does increase the number of fragments and overall coverage of fragmentation within a molecule, it also can result in extensive secondary fragmentation which yields many internal fragments that are often difficult to interpret. Thus, the laser power and duration of the laser pulse must be carefully adjusted to achieve the extent of fragmentation necessary to solve the analytical problem at hand.

Electron capture dissociation (ECD) is not usually applicable to MALDI ions as it involves the capture of a low-energy electron by a positively charged precursor ion; this reduces the charge by one. Since MALDI ions are usually singly charged, ECD would result in primarily undetectable neutral species. However, there is usually a low abundance of multiply charged ions in MALDI which increase in abundance with the mass of the precursor. Thus, MALDI-ECD has been demonstrated [26], but is usually very difficult. However, while ECD usually uses low-energy electrons in the 0–1 eV range, if higher energy electrons in the 8–12 eV range are used, then fragmentation can be induced on +1 and −1 charge state ions, respectively called electron-induced dissociation (EID) [26] and electron detachment dissociation (EDD) [27]. In this energy range electrons appear to scatter from peptide ions, causing fragmentation without partially neutralizing them. Neither method is currently used to any great degree as they are relatively difficult to adjust, though they may become useful in the future.

2.4
Mass Analyzers

Mass spectra and mass analyzers are typically characterized by a number of parameters. The most important of these are mass accuracy, mass resolution, sensitivity/limit of detection, and signal/noise (S/N) ratio.

Mass accuracy is usually the most important parameter as it defines the most important data within an experiment. Mass accuracy is usually presented in parts per million (ppm). For example, 20 ppm means that the real mass of an ion measured to 1000 Da has a predicted error margin of ±20 mDa. While it is not usually specified, the best way to define this error is at ±2 standard deviations in the errors on the calibration, which means that a measured mass is ~95% likely to be within that range. It is not critically important how this expected error is defined, provided that it is clearly stated exactly how it is calculated. Mass accuracy is limited by the quality of the calibration (how accurate it was to start with, how much it drifts with time, how much space charge has shifted the current spectrum relative to the instrument calibration, etc.), and by how accurately the center of the peak can be determined (which is determined by the number of points on a peak and how accurately the peak follows a theoretical peak shape).

Mass resolution of a given ion signal is defined as the mass spacing at which spectral features (peaks) can be clearly separated, and is reported in Daltons. More often, the term "resolution" actually refers to "resolving power", which is measured as $m/\delta m$, with m the mass of the peak center and δm the resolution (often defined as the width of a peak at half height or Full Width at Half-Maximum; FWHM). *Resolving power* is a unitless parameter. A poor mass resolution will degrade mass accuracy because the exact center (the centroid) of broad peaks cannot be found with high accuracy. Even more importantly, a poor mass resolution may fail to separate overlapping peaks or isotopic distributions, which will distort peak shapes and thus distort mass accuracy [28]. Mass resolution is determined by the instrument used and its particular quirks, and will be discussed below.

Sensitivity is defined as the slope of the intensity/moles of sample plot. It is often incorrectly used to refer to the limits of detection – which is the smallest amount of sample that can be used to achieve a detectable signal (often defined as a peak with S/N ratio >3). Sensitivity and limits of detection are universally determined by the sample, not by the instrument. Every instrument on the market can easily achieve low-femtomole or better limits of detection for clean, synthetic samples of peptides. However, samples from biological sources are extremely difficult to clean, and consequently all other biological molecules and contaminants are well below the concentration of the analyte. If the sample contains 1 fmol of peptide, but 0.1 pmol of surfactant or salt, the mass spectrometer (no matter which type) will detect surfactant and salts, even if many research groups have published data reporting that their mass spectrometer can detect attomoles of that particular peptide.

Finally, the S/N ratio is just that – the signal intensity divided by the nearby average noise value. The only trick is to define exactly how the average noise value is calculated. There are many methods available, but the most accepted calculation is to use the root-mean-square of the noise, neglecting noise spikes – provided that they do not interfere with the real sample peaks. Normally, it is very easy to identify noise spikes because almost all real ions have an isotopic peak distribution.

2.4.1
Axial TOF Mass Spectrometers

By far, the most common type of mass analyzer for MALDI applications is the axial TOF spectrometer, the principles of which are outlined in Figure 2.4. Axial TOF instruments are ideally suited to MALDI sources because they need only very short pulse of ions which then fly down the flight tube and hit a detector, such as a microchannel plate (MCP). Then, the ion current signal is simply plotted versus arrival time at the detector using a digital storage oscilloscope. The oscilloscope must be triggered carefully so that its scan is started when the ions are generated; this is usually achieved by splitting a tiny fraction of the laser light off onto a photodiode, which generates the trigger. For signal averaging and calibration, it is critical that this trigger signal has a time jitter which is very small (usually <1 ns) relative to the laser emission.

In axial TOF instruments all ions of charge q are first accelerated by a potential difference U, between the sample support and a nearby grid or open electrode, to a kinetic energy E_k before they travel down a field (and force) free path of typically about 1 m length to the detector. Thus,

$$E_k = zeU.$$

where z is the charge state and e the elementary charge. Potential differences, U, of 20 kV are typically used in modern TOF instruments. Ions with the same kinetic energy, E_k, will have different velocities, v, based on their mass, m, following the equation

$$E_k = \frac{1}{2}mv^2, \text{ or } v = \sqrt{\frac{2E_k}{m}} = \sqrt{\frac{2zeU}{m}}$$

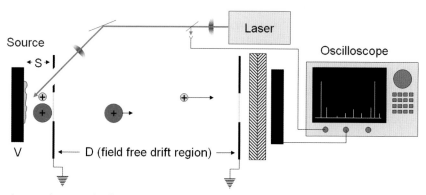

Fig. 2.4 The principle of MALDI time-of-flight mass spectrometry.

Thus, the arrival time will be:

$$t = L\sqrt{\frac{m}{2E_k}} = L\sqrt{\frac{m}{2zeU}}$$

where L is the flight tube length. This can be inverted to:

$$\frac{m}{z} = 2eU\left(\frac{t}{L}\right)^2$$

Thus, for a given experiment the calibration equation is $m/z = At^2$, where A is the calibration constant which is determined by least-squares fitting of this equation to a series of peaks, the m/z values of which are known and the arrival times measured. However, in order to account for an electronic delay of the trigger signal and other minor effects during the ion generation, a calibration equation of

$$m/z = At^2 + B$$

is normally used, adding a second calibration constant, which partially corrects for the initial velocity of the plume. For commercial instruments this works well down to the 10- to 20-ppm range of mass accuracy.

For TOF mass spectrometry instruments, there are two primary sources of error in the arrival times of ions at the detector. First, MALDI ions are extracted from their desorption plume with an initial velocity distribution which causes ions of a particular mass to arrive at slightly different times, thus degrading mass resolution. Second, if the surface is irregular (as happens with desorption from thin-layer chromatography plates or matrix crystals that are uneven), the initial position of the ions in the potential gradient will also affect their final velocity distribution. For example, a 10-μm height deviation on a crystal in a source with 20 kV acceleration potential over 5 mm will cause a 0.2% shift in the final kinetic energy, again degrading mass resolution.

The initial energy distribution of MALDI ions must be added to the energy imparted on them by the accelerating electric field, and this initial velocity distribution leads to a corresponding distribution of arrival times of ions of a given mass, thereby degrading mass resolution. To some extent this influence can be minimized by increasing the acceleration potential, but considerations of handling very high voltages in laboratory instruments limit this approach. Modern TOF mass spectrometers compensate this distribution by pulsed-ion extraction [commonly called delayed extraction (DE) or time-lag focusing] [29–34]. In this mode of operation the electric field between the sample and the first electrode is initially kept at zero or very low. After some time (typically tens of nanoseconds) this field is switched on to extract the ions into the flight tube. During the initial phase, fast ions will move farther towards the first electrode than slower ones. Upon switching on the electric field, the faster ions will experience less of a potential difference then the slower ones, compensating for their higher initial energy. As a result, the initially slow ions will catch up with the fast ones at some point down the flight tube; this point can be chosen by adjusting the pulse potential and timing. If the ion detector is placed at this "space focus", the time and mass resolution is greatly increased. In commercial TOF instruments there is usually a second electrode in front of the first one, with the total acceleration potential being split between them in order to move the space focus further down the tube and thereby increase the flight times. Unfortunately, time focusing by delayed extraction is mass-dependent, so the improved mass resolution is only achieved for a limited mass range. The instrument operator must set the switched potential as well as the delay time for the ion extraction properly in order to achieve maximum resolution for the ions of particular interest. In the peptide mass range up to a few thousand Da, mass resolutions in excess of 20000 have been achieved with relatively simple linear TOF mass spectrometers. Above about 30 kDa, delayed extraction gradually loses its power for improving mass resolution. With delayed extraction the flight time equations becomes considerably more complicated, and higher-order polynomials together with an increased number of calibrant masses are usually implemented for mass calibration [35–38].

2.4.2
Reflectron TOF Mass Spectrometers

Another option to partly compensate for the initial velocity/energy distribution of MALDI ions is the ion reflector, invented by Mamyrin and coworkers in St. Petersburg [39]. Like delayed extraction, the reflectron – or ion mirror – was also designed to re-focus the ion packets onto the detector. The basic geometry of a reflectron TOF instrument is shown in Figure 2.5. The reflectron is actually a series

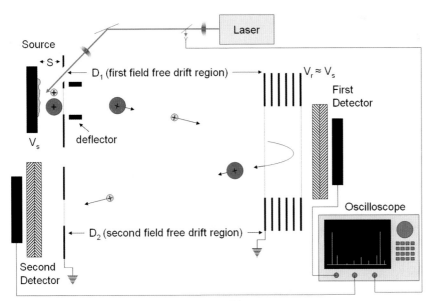

Fig. 2.5 Combined linear/reflectron MALDI time-of-flight mass spectrometer.

of ring electrodes which, near their center, ideally create a constant electric field through a linear voltage gradient that slows down the ions and turns them around (provided that the potential of the last electrode is larger than the acceleration potential of the ions) to send them back to a second detector – just like rolling a ball up a hill. For any particular ion mass, if one simply considers ions on the high-energy and low-energy tails of the kinetic energy distribution, the higher-energy ions, which arrive at the reflectron ahead of the slower ones, will roll up the voltage "hill" farther than the low-energy ones, thus increasing their flight time in the reflectron. With proper adjustment of the reflectron voltage, both the high- and low-energy ions can be focused to hit the detector at the same time, thus improving the resolving power of the instrument. A simple reflectron as shown in Figure 2.5 can correct the flight time dispersion to first order, and mass resolutions of over 10 000 have been achieved with this technique. Reflectrons with two different electric fields (two-stage reflectrons) can compensate the dispersion to second order [40]. Ion mirrors with nonlinear electric field distributions – so-called "curved field reflectrons" [41,42] – have been designed primarily for post source decay analysis (for a discussion, see below). Modern research-grade TOF mass spectrometers have both reflectrons and delayed ion extraction. The DE compensates mainly for the flight time dispersion due to the initial velocity distribution of the ions, while the reflectron will compensate for the energy dispersion, which might result, for example, from a non-flat sample morphology. Both methods together will render mass resolutions of >20 000 and a mass accuracy of 5–10 ppm, although selected and very highly tuned experiments have achieved 2–5 ppm accuracy. One important caveat for achieving these results is that the peaks must be isotopically resolved and a high S/N ratio. For example, with large proteins the mass resolution of reflectron TOF mass spectrometers is limited by small neutral losses to about 120 at mass 150 kDa of a monoclonal antibody for UV-MALDI. Infrared MALDI generates cooler ions, which increases the mass resolution for this case to 230 [43].

2.4.3
Tandem TOF Mass Spectrometers

Tandem mass spectrometry is a critical methodology used for the structural analysis of all types of molecules. The procedure involves selecting a precursor ion, fragmenting it, and generating a mass spectrum of the fragments. Structural information on the ions can be obtained in axial TOF analyzers using either metastable decay (in which ion selection is avoided, but this can complicate the analysis) or externally induced fragmentation. The classic example is that a peptide of interest (from a biological source or from an enzymatic digest of a protein) can be selected and fragmented; the fragments will then reveal the sequence of the peptide and often will allow determination of sites of deamidation, phosphorylation, or sequence variation from mutation of the peptide's parent DNA.

For the metastable decay fragments, groups conducting research into MALDI-TOF have made a rather arbitrary distinction between "in-source decay" [22,23] and "post-source decay" [44] ions. The "in-source decay ions" are generated by the desorption/ionization event on a time scale which is short compared to the transit time through the acceleration region or, in the DE mode, on a time scale which is short compared to the delay time of the ion extraction. In both cases, the fragment ions receive the full acceleration energy and will be detected in the spectrum at their correct mass. Strictly speaking, ions which fragment in the source during the rest of the acceleration time period would also generate "in-source" fragments, but they will experience a very wide range of different acceleration energies depending on where in this region the precursor ion decays. They will, therefore, appear in the spectrum distributed over a large mass range and are mostly lost in the noise (though they are occasionally observed as weird baseline drifts and humps). "In source decay" fragments are often those resulting from a chemical reaction hydrogen radicals [21,45] generating c- and z- type fragment ions. While these fragments are in principle very useful (similar to ECD, they can contain near-complete sequence information of not too-large proteins) [46] their intensity is usually too low for practical purposes.

The metastable decay of ions due to an excess internal energy can usually be described by first-order kinetics, with time constants of typically micro- to milliseconds. Early fragments during the ion acceleration are mostly lost in the noise, for the reason explained above. Precursor ions, decaying in the field-free region of the flight tube are called "post source decay" (PSD) fragment ions. Delayed extraction, in general, reduces the abundance of these PSD fragment ions because the high-pressure plume disperses somewhat before application of the ion extraction potential, thus reducing collisional activation of the ions when the electric field drags them through the dense cloud of neutrals. Conversely, the abundance of fragment ions can be increased by increasing the pressure in the source when the acceleration voltage is on.

PSD fragment ions will have the same velocity as the precursor. In a linear instrument, both the precursor and the fragments will hit the detector at the same time and thus go undetected. While the PSD precursor and its fragments have equal velocities, the total kinetic energy of the precursor ion will, in good approximation, be split between the fragments in the ratio of their masses. In the reflectron – which is an energy-dispersive element as explained above – the fragment ions will be dispersed and poorly focused onto the second detector. In general, if they are detected at all, they generate very poorly resolved peaks that are barely above the baseline noise of the instrument. However, PSD ions can be refocused onto the detector by reducing the reflectron voltage [44,47]. For each reduced reflectron voltage one obtains a small range of fragment masses which are well resolved and which, at the reduced reflector potential, retrace the same path as the precursor ion at full voltage. The valid m/z range over which these fragments will be focused on the detector is determined by the size of the detector (lower m/z ions will shift to the "inward" side of the detector) and by the ion-focusing optics (if any are used). Thus, by sequentially stepping down the reflectron voltages, thereby acquiring spectra at a series of V_r values, and stitching together the valid m/z ranges of the resulting spectra, a complete mass scan of the metastable fragment ions can be generated. This experiment is known as a "PSD scan". The calibration of the mass scale of PSD spectra is more complex compared to simple TOF spectra, and needs special attention. For this calibration the reader is referred either to the literature [35–38] or to the user's manual of their instrument.

The dispersion of fragment flight times in the simple constant electric field reflector can be partially compensated over a large mass range if the electric reflector field is chosen nonlinear ("curved") in a specific manner [41,42]. This greatly simplifies PSD scans, albeit at some expense of the overall mass resolution. Another solution to this problem is the "lift" TOF mass spectrometer [48]. In this instrumental configuration the fragment ions are "lifted up" in their potential while inside a special section of the flight tube. This adds kinetic energy so that the range of fragment energies relative to their total energy is reduced, thereby improving the fragment mass resolution over a mass range as large as 800 to 3000.

Mass spectra generated from metastable fragments would become very crowded and difficult to interpret if the fragments of all ions generated in the source were to be displayed in a single spectrum. It is, therefore, necessary to select a single mass – or at most a mass window of a few ions – for this analysis. Generally, a pair of electrodes is used to deflect unwanted ions, but a set of thin wires biased alternatively positive and negative can also achieve deflection [49]. There is usually a tradeoff between the width of the selected mass window and the transmission of the gate. For example, narrower windows, set to transmit only a monoisotopic mass signal, will have a compromised transmission of often below 50%.

Often, the metastable ion fragmentation will not suffice to extract all the structural information of interest, and consequently the internal vibrational energy of the molecules must be increased to generate more fragments. The most common method to accomplish this is by collisions with neutral gas molecules, typically in a specially designed collision cell. A typical tandem TOF instrument (often referred to as a TOF/TOF instrument) is shown in Figure 2.6 [50–53]. The first linear drift tube separates the ions into packets of different m/z. An ion gate in front of the collision cell is then switched open at a properly chosen delay time for a short period and passes only the precursor ions of interest.

The kinetic energy of the ions entering the collision cell is controlled by adjusting the offset potential. For example, if the source generates ions at 10 keV/charge, then the collision cell could be held at 9500 V to allow ions to collide at 500 eV/charge. This collision energy can be varied, but useful values are in the range of 25 to a few hundred eV/charge [54]. These higher voltages are in the so-called "high-energy CAD" range where carbon–carbon bonds can be cleaved, allowing structural analysis of branched hydrocarbons and isomeric species such as leucine/isoleucine. The collision cell with its fragment ions then acts as the ion source for the second high-resolution reflectron TOF mass spectrometer. Once the ion is dissociated, the fragment ions are accelerated to a high velocity once again and then allowed to time-separate in the second high-resolution reflectron TOF. The mass calibration of the MS/MS spectrum is more complicated than a normal

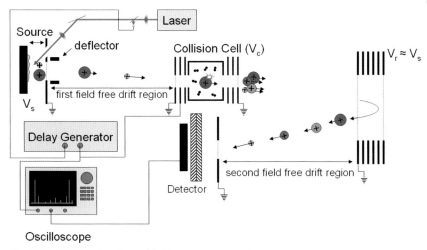

Fig. 2.6 MALDI tandem time-of-flight mass spectrometry.

MS spectrum because the fragment ions have a residual velocity from the precursor; thus, an additional constant velocity offset is added to the calibration equation to correct for this. In general, the fragment ion spectra are lower in resolving power ($\sim[1-2]\cdot10^3$) and mass accuracy (~50 ppm) than the precursor ions.

The "lift" TOF mass spectrometers [48,55] can be used for "top-down" mass spectrometry of whole proteins. Although this has been done in practice, it either yields huge numbers of very small peptide fragments (e.g., immonium ions) or it sequences the termini of the protein nicely, but it does not yield the large complementary ions needed for good top-down analyses. However, this "top-down" mass spectrometry parameter space is not yet fully explored for TOF/TOF instruments.

TOF/TOF instruments also have their limitations. First, in order to obtain sufficient initial signal for TOF/TOF experiments, the laser fluence for the MALDI source is usually increased by a factor of two. This generates many more ions, but at the expense of a wide kinetic energy distribution in the plume, and consequently the resolution suffers, both in isolation and in detection. Second, ions in the collision cell experience a statistically distributed number of collisions, and thus end up with a range of velocities on exiting the chamber. This results in wider than normal kinetic energy distributions in the second TOF, and thus greatly reduced resolution of the fragment ion signals. Third, ion positions (both parent and fragment) are generally scattered along the ion path prior to extraction, which causes them to have a distribution of ion kinetic energies in the second stage. For this reason, the resolving power for the fragment ions is often an order of magnitude or more lower than the instrument's "best case" precursor ion scan. Also, it is frequently difficult to distinguish fragments formed in the collision cell from metastable fragments that are formed between the source and the timed-ion-selector.

2.4.4
Orthogonal TOF Mass Analyzers

A more recent instrumental system for MALDI mass spectrometry is the ortho-gonal time-of-flight (oTOF) mass analyzer [2,3,56,58] (Fig. 2.7). This instrument generates MALDI ions in a plume as usual, but instead of extracting them along the axis of the plume, a voltage pulse is used to deflect them sideways so that the ion beam path is deflected perpendicular to the original direction of motion. This eliminates the high initial axial velocity distribution of the plume from the arrival times of the ions, thus simplifying the calibration equation and improving mass accuracy. In current, commercial oTOF instruments, the ions are well collimated by the ion transfer optics and have minimal radial velocity, so that they can usu-ally achieve higher resolving power and mass accuracy compared with the axial TOF instruments. Orthogonal TOF instruments can routinely achieve $(1–2) \cdot 10^4$ resolving power and <10 ppm mass accuracy on the intense peaks, provided that they are highly oversampled so that peak-shape fitting can tightly determine the peak position.

The ion-transfer optics include at least one quadrupole (or hexa- or octopole) at elevated pressure of about 1 mbar for thermalization of the ions to stabilize them for the longer analysis time. The transfer optics also essentially decouples the TOF analyzer from the ion source; for high laser pulse repetition rates the transfer

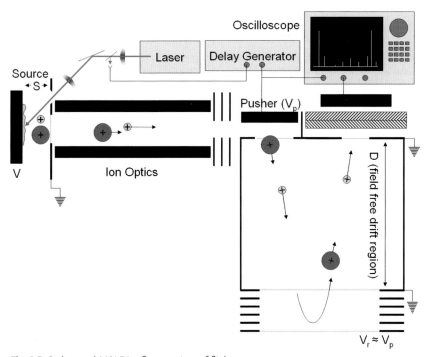

Fig. 2.7 Orthogonal MALDI reflectron time-of-flight mass spectrometry.

optics actually converts the pulsed ion source into a continuous source. Ions are either continuously transferred into the pulser region of the TOF, or accumulated and extracted discontinuously. The fast pusher then injects them into the TOF section of the instrument. The maximum rate of ion injection is limited by the transit time of the largest ion through the TOF analyzer to typically a few kHz.

However, oTOF instruments have their limitations. First, ions drifting from the source to the pusher will experience some TOF separation, so that the mass distribution observed is a function of the pulser timing relative to the laser pulse. Clearly, this timing must be adjusted carefully. Second, ions are pulsed out of the pusher as a cylinder and must travel to the detector as a sausage-shaped ion cloud. Ions that are not in the pusher region at the time of pulsing will be lost, thus limiting duty cycle and mass range. Third, the m/z range of the transfer ion optics determines the m/z range of the instrument as a whole. The best-known (and only commercially available) oTOF instruments involve quadrupole ion optics for manipulation, storage, and transmission of ions from the source to the cell.

Furthermore, the time frame of the mass spectrometry experiment is generally increased from microseconds to milliseconds, depending on the ion-transfer optics, so that most metastable decay is complete by the time the ions enter the TOF analyzer where all fragments are then recorded at their mass. Whether this is an advantage or disadvantage depends on the experiment, but with labile species such as some phosphopeptides, glycosylated peptides, oligosaccharides, or glycosylated lipids, it is critical to be able to control and/or suppress metastable decay in order to determine the position and type of modification. It is particularly the case for these analytes that MALDI-oTOF must have some type of ion cooling, as described above.

2.4.5
Tandem Mass Spectrometry in oTOF Mass Analyzers

To date, all tandem mass spectrometry in oTOF mass spectrometers has occurred in the front-end injection optics. These instruments are usually called "quadrupole time-of flight" (QTOF) instruments, because they start with a series of quadrupoles to focus and perform ion selection prior to accelerating ions (at a few tens of eV/charge) into a collision cell. The collisions are of low (a few eV) energy, so precursor ions must undergo a large number of collisions in order to acquire sufficient energy for fragmentation. If no fragmentation is desired, both quadrupoles are operated as ion guides in the radiofrequency (RF)-only mode. This arrangement resembles that of classical triple quadrupole instruments, and has the same sensitivity problems – namely that, during ion isolation, most of the ions are being rejected most of the time.

2.4.6
Ion Detectors and Data Processing in MALDI TOF analyzers

Detectors in MALDI TOF instruments are usually microchannel plates (MCP), and less frequently discrete dynode secondary electron multipliers (SEM). Both detectors convert incoming ions into electrons which are then amplified in a cascade by many orders of magnitude. MCPs are essentially glass plates with a very large number of channels, of 2–10 µm diameter, angled at 10–20° to the surface normal. Typically, these channels are very densely arranged so that the open ratio is ~60%. The surface of the plate, as well as that of the channels, are specially coated to achieve a high ion/electron conversion and electron multiplication yield. A voltage of typically 1 kV is applied between the two sides of the plate. Ions hitting the front face of the plate will elicit some electrons which will then cascade down the channels and exit the plate at the back face. The amplification is typically about 10^6 electrons for every ion. Two such plates are often arranged in series, termed a "chevron" configuration, to maximize the gain. Electrons exiting the last plate are collected by an anode, the electron current of which is converted to a voltage and amplified to the detector output. Secondary electron multipliers function by essentially the same principle, except that the amplifying electron cascade takes place between properly shaped discrete electrodes, called dynodes. Their amplification factor is comparable to that of the MCPs. Microchannel plates have the advantage of a higher response time down to less than 1 ns if correctly impedance-matched circuitry is used. Such fast time constants are required for TOF analyzers with maximum time and mass resolution, particularly in the mass range below a few thousand Daltons.

The disadvantage of MCPs is that they easily saturate. Saturation is a particular problem for samples of low analyte concentration or a high degree of impurities, both of which require higher laser fluences which in turn generate intense background signals in the mass range below ~800 Da. Because of the long recovery time, detector saturation in the low mass range will severely lower the sensitivity for the higher-mass analyte ions. Most commercial instruments have means to deflect ions in the low mass range away from the detector or to "lock" the MCP with some delay to limit the effect of saturation, but this cannot be fully avoided in all circumstances. SEMs are much less prone to saturation, particularly if the last few dynodes are boosted by fast capacitors. Their time response is, however, only about 5 ns or longer because of a more pronounced time dispersion of the electron cascade down the tube.

Because the matrix ion peaks are often very abundant compared to the analyte ion signals, MALDI-TOF instruments often operate the laser just above threshold fluence (to minimize matrix and ion signal/shot and thus prevent saturation) and then signal average a large number of exposures. Both ion detectors have limited sensitivity for large ions with an impact velocity below about $10^4 ms^{-1}$ (~30 kDa at an acceleration potential of 20 kV). Postaccelerating the ions onto the detector is sometimes used to alleviate this problem; conversion dynodes in front of the detector which convert the large ions into small ones by sputtering for an efficient secondary ion/electron conversion is another solution, though this will limit the time and mass resolution even further. Several designs for special cryodetectors at liquid helium temperatures have also been described in the literature for the detection of large ions [59–63].

The electron current signal can be captured into the computer in two ways, using either standard analog-to-digital converters (ADC) or time-to-digital converters (TDC) [64]. ADCs sample the analog detector voltage at given intervals and store the value in a memory from which the signal is then reconstructed or read out by the computer. For a high-fidelity reconstruction, at least four sampling points per peak are required for noise-free signals, and several more for noisy ones as encountered in MALDI spectra. ADCs or digital oscilloscopes for MALDI-TOFs typically operate at a sampling rate of 1–4 GHz which sample every 250 ps to 1 ns, well adapted to the fastest detector signals as discussed above. The high sampling rate limits the dynamic range of stored signal amplitudes. Currently, the practical limit of these types of board is 8 bits at GHz acquisitions speeds. This means that the maximum dynamic range (the ratio of the largest peak amplitude to that of the smallest detectable signal in a single scan) possible with such a detector is $2^8 = 256$. Furthermore, saturation of the digitizer can "lock" it for a few time bins, which causes flat-topped peaks and large errors in ion-abundance measurements. Saturation can also occur upstream of the ADC in the analog current-to-voltage amplifiers, causing them also to "lock" or return to baseline slowly.

Time-to-digital converters (TDC) instead record the arrival time of ions not as an analog signal, but as a simple digital "one" or "zero" at a particular time bin. The recording electronics simply notes the arrival time of the signals and then, over hundreds of scans, a histogram of arrival times is generated which is then converted into a mass spectrum using the normal calibration equation. There are several advantages to this mode of detection. First, if the detection threshold is properly adjusted, there is no baseline noise. Second, this method is extremely fast and can operate with relatively simple electronics at 10 GHz – that is, with many time bins for any single mass peak. And third, the data is only 1 bit deep, so that data transfer rates and signal averaging is very fast. However, the TDC is also limited in that it can only detect that ion(s) have arrived – not how many of them there are. If more than one ion hits a detector at any one time, then the abundance accuracy of the final mass spectrum is distorted. TDCs are limited to unit dynamic range. Higher dynamic range experiments are only possible by signal averaging a large number of scans. Overall, the ADC method of detection is better in terms of accuracy both in mass and abundance, but the TDC method is simpler and faster (and thus cheaper, particularly at high acquisition rates ≥10 GHz), although it cannot detect more than one ion per time bin. Very likely, there will be a shift toward ADC detection as the electronics improve.

Axial TOF analyzers typically record the spectra in the ADC mode, whereas oTOF instruments use TDC detection. In the latter case the number of accumulated runs is given by the repetition rate of the ion pusher of the TOF rather than the pulse repetition rate of the laser. Because in the TDC mode multiple ions of given arrival time must be avoided, the pusher repetition rate in oTOF instruments is typically a factor of about 10 larger than the laser repetition frequency.

Overall, the main advantages of TOF mass analyzers are the very high speed at which they can operate (often 1000 scans per second) and their effectively unlimited mass range. Additionally, they have very high sensitivity as a single ion hitting the detector is usually observable, and this is amplified by the ability to signal average a large number of scans in a short period of time (e.g., $1000 \, scans \, s^{-1}$). However, limits of detection are somewhat limited in most instruments due to a large background of chemical interference peaks – known colloquially as "chemical noise" [4]. TOF instruments suffer primarily in their limited ability for tandem mass spectrometry (several hybrids discussed below have been designed to address this) and their relatively poor mass resolution ($\sim[1–2]10^4$ FWHM in good cases, $\sim[1–2] \cdot 10^3$ with lower-quality TOFs) and accuracy (<10 ppm is possible in some modern instruments with high abundance, oversampled peaks, but 20–50 ppm is more normal [5]).

2.5
Fourier Transform Ion Cyclotron Resonance Mass Spectrometers

The best possible results in terms of resolution and mass accuracy are currently achievable by Fourier transform ion cyclotron resonance mass spectrometers (FTICRMS or FTMS) [65–67]. These instruments operate on a relatively simple principle, namely that ions in a magnetic field rotate around that field at their cyclotron frequency, $\omega_c = qB/m$. A typical open-cylindrical FTICR ion cell, with the magnetic field B aligned along the axis of the cell, is shown in Figure 2.8A, while a cross-section of the center of the cell, with the magnetic field passing into the paper, is shown in Figure 2.8B. If an excitation sine wave is applied to the excitation plates (E+/−) at the resonant frequency of an ion in the cell, the ion is excited in a spiraling orbit until it either hits the cell walls (and is neutralized) or the excitation pulse is turned off. When the pulse is turned off, the ions continue to oscillate at their final orbital radius, and they generate a very small electric field on the detection plates (D+/−) which is amplified, digitized, and stored. In

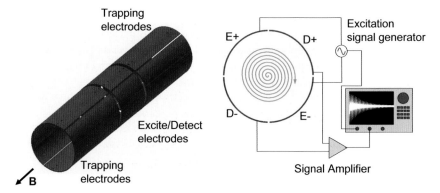

A. Typical open cylindrical cell   B. Typical Excite/Detect geometry

Fig. 2.8 The principle of FTICRMS.

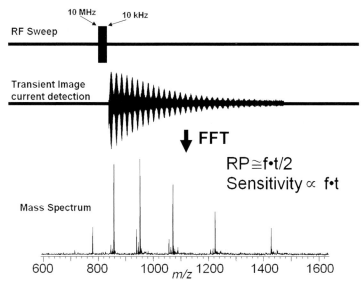

Fig. 2.9 The principle of Fourier transform ion cyclotron resonance mass spectrometers (FTICRMS).

reality, the detection of all ions is the goal, not simply the detection of one ion of known frequency (and hence mass). Therefore, instead of a resonant excitation pulse (as drawn), a frequency sweep pulse (called a "chirp") is generated that slowly shifts through all expected frequencies (and masses) and excites them all to a similarly high orbital radius which is usually approximately one-half the cell radius (see Fig. 2.9). The detected signal is a sum of decaying sine waves as:

$$c(t_n) = \sum_{i=0}^{m} A_i e^{\left(\frac{-t_n}{\tau_i} + j\omega_i t_n + \theta_i \right)}$$

where i is incremented through all the ions in the cell with A_i, and τ_i, θ_i and ω_i are the abundance, decay constant, phase, and frequency, respectively, of each ion; j is $(-1)^{1/2}$. The oscilloscope trace in Figure 2.8B is typical. The Fourier transform (FT) converts this time domain data into frequency domain data and, by peak-fitting, a mass spectrometrist can extract frequencies/masses (peak centroids), amplitude (peak area), and decay constants (peak width). The phase data (θ) are usually lost during the FT process because the latter converts "real" time domain data to "imaginary" frequency domain data of the form $a + jb$, but the FT data are usually reported as a magnitude $(a^2 + b^2)^{1/2}$ versus frequency plot. Decay constants are primarily due to collision of the ions with background gas molecules and to dephasing of the ion cloud by coulombic interactions with other ions (space charge). Higher resolution requires lower decay constants (τ_i), which means that FTICRMS instruments usually operate in ultra-high vacuum environments (typically $<1 \times 10^{-9}$ mbar) with a small number of ions.

The frequencies observed are converted to mass values using a calibration equation. There are several different equations that are functionally equivalent [68–70] (which have recently been reviewed [71]), but the most accepted equation is

$$\frac{m}{z} = \frac{A}{f} + \frac{B}{f^2} + C$$

where A, B, and C are calibration constants. The first term is usually the largest and is essentially the magnetic field term (note that if B and C are set to zero, the equation reverts to the simple cyclotron frequency equation), the second term is an electric field correction term (and is usually a small ~100 ppm correction), and the third term is essentially zero. This is a quadratic equation in $1/f$, so provided that three or more measured peak frequencies can be assigned to known masses, a simple least-squares approach will allow rapid solution for the calibration constants. With internal calibration, this equation should be accurate to ~±1 ppm for modern 7+ Tesla instruments, and deviations from this value (at least for peaks with S/N ratio >10) indicates a problem with the experiment, usually involving space charge.

Before excitation/detection of the ions, it is necessary to move them into the ICR cell and to trap them inside the magnet. If the ions can be moved into the cell with <1 eV/charge kinetic energy and the outer cylinders of the trap (Fig. 2.8A) are raised to >1 V, then the ions will be trapped along the magnetic field axis (commonly called the z-axis). Once the ions are in the cell, they are usually allowed to "cool" their translational energy via collisions with background gas and other ions for a period from tens of milliseconds to tens of seconds, depending on the experiment. Often, a pulse of collision gas is added to assist this process, but it is not required. After cooling, the ions often have low milli-eVs of residual translational kinetic energy, so the trapping potentials can be dropped substantially (e.g., to 0.1 V) while still retaining most of the ions. If there are too many ions in the trap, this also allows for some evaporative cooling of the remaining ion population via loss of the energetic ions, which can improve performance.

There are two general modes for moving ions into the trap. The first mode (Fig. 2.10) is referred to as "internal ionization", where the MALDI ions are generated inside (or very close to) the cell and are trapped immediately with higher voltages (which are mass-dependent due to the initial velocity of the plume); the voltages are then decreased to ~1 V prior to excitation/detection (this is necessary for several reasons, which will be discussed below). This method works well and achieves very high sensitivity, but the practical requirement of inserting a sample through a vacuum airlock and into an ultra-high vacuum region means that the sample will usually require several hours to outgas and achieve a sufficiently low vacuum. Furthermore, MALDI generates a large plume of neutral matrix species which coat the cell walls, and this necessitates cleaning of the cell every few months. If the cell plates become too dirty, then static charging of the surface layer will shift the potentials in the cell and either distort or prevent acquisition of good mass spectra.

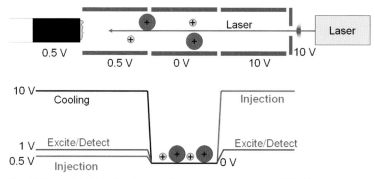

Fig. 2.10 In-cell MALDI Fourier transform mass spectrometer (FTMS).

Although the second method for moving ions into the ICR cell is generally preferred, it requires substantial ion optics to transfer the ions from an external
MALDI source into the ICR cell. Figure 2.11 shows one such instrument [9] in
which ions are desorbed into an external ion trap, thermalized/manipulated, and
then transferred as a packet down a ~1 m-long RF-only multipole into the ICR cell.
The hexapoles are able to confine ions radially, and while the long transfer hexapole is aligned and centered with the magnetic field, the ions experience little or
no magnetic forces as they enter the high magnetic field. This is the standard
method used by instruments from IonSpec and Thermo, but Bruker uses a series
of electrostatic ion optics and accelerates the ions to ~3 keV/charge to transfer the
ions. The multipole ion optics are simpler and more reliable over time and generally have higher transmission efficiency, but the static ion optics yield acceptable
performance provided that they are very well aligned with the magnetic field. Re-
gardless, the external injection method of moving ions into the ICR cell has the
advantage of separating the dirtier, low-vacuum source from the ultra-high vacuum ICR cell, but the disadvantage that ions pulsed from the source experience a
TOF separation event as they drift down the long hexapole [72]. When the ions

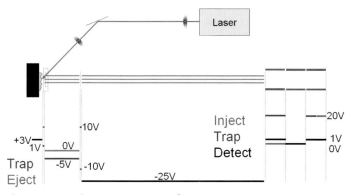

Fig. 2.11 External MALDI Fourier transform mass spectrometers (FTMS).

have entered the cell, the front trapping plate is raised from 0 to 20 V and those ions which are in the cell and have less than 20 eV/charge of kinetic energy will be trapped. However, the TOF delay from the source implies a limited trapping mass range. Furthermore, all RF ion guides transmit a defined window of masses that is roughly one order of magnitude wide for quadrupoles (e.g., 100–1000 Da e^{-1}), two orders of magnitude wide for hexapoles, and three orders of magnitude wide for octopoles [73]. While the latter problem is a fundamental property of RF ion guides, the TOF delay problem can potentially be solved by sectioning the multipole and continuously shuttling the voltages to operate the various sections as a conveyor-belt, a development that is progressing in Bruce's laboratory [74].

Once the ions are trapped, they will stay trapped for hours or days depending on the collisions. Ions in an ICR cell have three modes of motion – the axial mode, and two radial modes (cyclotron and magnetron). The first two are stable, but the magnetron mode is not. Background gas collisions and coulombic collisions with other ions scatter the ions both axially and radially. Axial motion is trapped, so the ions will slowly lose signal from that mode and return to the center of the cell. Similarly, the cyclotron mode is stable, but the magnetron mode is not. Scattering velocities in the radial direction will thus partition into these two modes, and the ions will drift slowly and radially out of the cell in the magnetron mode. Magnetron motion is almost always to be avoided in FTICRMS instruments as it leads to ion loss. Higher trapping voltages will amplify the effects of ion scatter and allow the ions to drift out of the cell more rapidly; therefore FTICRMS operators usually keep the trapping voltages as low as possible without dumping the ions out of the trap axially. One interesting exception to this rule is Bruker's "Side-Kick" [75] ion-accumulation mode, which imposes a small, radial electric field on the ion injection optics that pushes the ions radially into a magnetron orbit as they enter the trap. This greatly increases the path length of the ions within the cell, thus increasing the probability that they will have a collision with a background gas molecule and also increasing the probability that the ion will be trapped. However, the ions have a small magnetron moment in the cell, and it is particularly difficult to achieve high resolution and accuracy under these conditions.

If ions have not had sufficient time to cool their axial motion prior to excitation and detection, the inhomogeneities of the axial trapping field will cause periodic frequency shifts in the ions; this will blur out their peaks and reduce resolution. In general, since these modulations are periodic, the FT can largely average out most of these perturbations [76]. However, excess number of ions in the cell generates a time-varying electric field which can greatly perturb the frequencies in a non-periodic manner; this situation is called "space-charge", and results from coulombic repulsion among the ions. It results in many of the same types of problem that are caused by higher trapping voltages, but because its effects are often non-periodic, it can cause substantial problems with mass accuracy and resolution. Space-charge is usually the limiting factor in mass accuracy in FTICR instruments, and is first noticeable by peak shifts, but at high levels it can cause substantial peak distortion [77] and odd calibration behaviors. It is also important to remember that, while chemical noise is often unobserved in FTICR (see below), this does not necessarily mean that it is not there. For example, an excellent calibration on a standard may not be applicable to a spectrum from a biological source – even if the total ion intensity is the same – because the biological sample has many, very low abundance, chemical noise peaks which are not included in a total ion intensity calculation (because they are at or below the white noise level) but still contribute to the global space charge environment. This is a general fallacy which usually negates attempts to correct for space-charge within the calibration equations [69,78,79]. Finally, if the space charge frequency shifts are changing from scan to scan, then signal averaging will cause peaks to have shoulders which are simply artifacts and not chemically relevant. Thus, minimizing or controlling space charge levels is very important in FTICR mass spectrometers.

Overall, FTICRMS instruments provide much higher resolving power than other mass spectrometers, and if the operator is careful not to overload the cell, then 5 ppm mass accuracy with external calibration and 1 ppm mass accuracy with internal calibration are routinely achieved. A particular advantage of the current Thermo LTQ-FT instrument is that the LTQ allows automatic gain control (AGC) [80] in which a feedback mechanism is used to monitor and control the number of ions in the trap. By doing so, the space charge modulations are almost eliminated and the system is able to achieve ~2 ppm externally calibrated. Resolving powers of $(50–100) \cdot 10^3$ are routine on modern FTICRMS instruments, but if a narrow-band mode (also called heterodyne or mixer mode) experiment is used, then a resolving power of $>10^6$ can be achieved, provided that the cell is well aligned with the magnet and the base pressure is low.

The limit of detection of FTICRMS instruments is not dissimilar to that of TOF instruments which, at first glance, seems strange. FTICRMS instruments rely on the inductive detection of slow-moving ions (an inherently low-amplitude signal), while TOF instruments collide ions at many keVs/charge into a detector; this generates a cascade of electrons that amplify the signal by many orders of magnitude. However, while TOF instruments receive a single pulse of ions, the FTICRMS re-measures the ions over hundreds of thousands of cycles. This is termed the "Fellgett advantage", which notes that any FT-based detection system will constantly monitor and sum all signal channels, whereas other systems (e.g., TOFs), which only monitor one signal channel at a time, cannot signal average throughout the scan. Furthermore, MALDI-TOF mass spectrometers are subject to high levels of "chemical noise". Such noise roughly follows a $1/m$ dependence so that "chemical noise" is very bad at the low molecular mass end of the spectrum. MALDI-FTICRMS instruments are much less subject to this chemical noise for three reasons:

- The higher resolution of FTICRMS resolves many of these interference peaks so that they do not generate sufficient signal intensity for detection.
- Matrix clusters tend to be metastable and will dissociate with time; this means that, on the ~1 s timeframe of an FTICRMS, most of them have dissociated into the low m/z region.

- Many of the low m/z ions are filtered out by the RF-only multipole ion guides. That said, in "best case" comparisons [81,82] the TOF instruments have about an order of magnitude lower limit of detection than FTICRMS instruments, although in both cases it is clear that the limiting factor is sample handling, and not instrument performance.

TOF instruments are superior to FTICRMS instruments in terms of speed (on a per-shot basis) and m/z range. TOF instruments can (and sometimes do) acquire 1000 spectra per second (depending primarily on data transfer rates and the speed of the laser), while FTICRMS instruments cannot drop below one spectrum per second if the 50k resolving power range is to be preserved. However, for TOF instruments, in order to maintain $>10^4$ resolving power (particularly TOFs using time-to-digital converters for detection), these instruments must operate at threshold fluence where only a few ions are generated in each spectrum, and then hundreds to thousands of spectra are usually generated and signal-averaged. FTICRMS instruments usually use several laser shots to generate a good cloud of ions, but then use only a single excite/detect pulse. So, the speed advantage of the TOF instruments is somewhat lessened for higher-resolution or S/N ratio level experiments. MALDI-FTICRMS instruments are severely limited in the high m/z regime, and generally cannot detect above ~5–10 kDa e^{-1} without major losses in resolution [83,84]. However, this does mean that MALDI-FTICRMS instruments are ideal for generating very high-accuracy MS and MS/MS data on smaller molecules – particularly in peptide mass fingerprinting and proteomics experiments.

2.5.1
Tandem Mass Spectrometry on FTICR mass Spectrometers

FTICR mass spectrometers have the highest flexibility of all mass spectrometers in terms of methods of fragmentation. In quadrupole-FTICRMS instruments, the ions can be isolated, accumulated, and accelerated into collision cells for collisionally activated fragmentation. For ion-trap-FTICRMS hybrid instruments, ions can be resonantly fragmented in the ion trap itself. Moreover, once the ions are in the ICR trap, they can be manipulated extensively.

FTICRMS instruments operate on the principle of ion cyclotron resonance. As ions have resonant frequencies, these frequencies can be used to isolate the ions prior to further fragmentation or manipulation. For example, a resonant frequency pulse on the excite plates (E+/− in Fig. 2.8B) will eject the ions at, or near, that frequency. Furthermore, frequency sweeps, carefully defined to not excite the ion of interest, can be used to eject unwanted ions. However, the most elegant method for ion isolation is that of SWIFT (Stored Waveform Inverse Fourier Transform) [85] in which a ion-excitation pattern of interest is chosen, inverse Fourier transformed, and the resulting time domain signal stored in memory. This stored signal is then clocked-out, amplified, and sent to the excite plates when needed. The typical isolation waveform in SWIFT uses a simple excitation box with a notch at the frequencies of the ion of interest ± a few kHz.

As with all other mass spectrometers, CAD is the predominant method of generating fragments for MS/MS experiments. The easiest way to perform CAD is to resonantly excite an ion of interest and then simultaneously to trigger a pulse of gas into the cell. This works well, but has the disadvantage that fragment ions are generated off-axis (i.e., with a large magnetron moment), which means that they are often difficult to detect or that they rapidly fall out of the cell. To counter this problem, a number of different collisional activation methods were created including Multiple Excitation Collisional Activation (MECA, which involves using many, low-level collisional activation events) [86], Very Low-Energy CAD (VLE-CAD, which involves using a resonant excite, a 180% phase shift of the excitation pulse, and resonant de-excitation) [87], and Sustained Off-Resonant CAD (SORI-CAD, which excites the ions a few kHz off resonant so that they experience a "beat" pattern which alternately excites and de-excites them) [88,89]. These methods use low-level excitation and phase shifting to ensure that the ions of interest are always near the axis of the cell, such that the fragments are also generated near the center of the cell. Of these methods, SORI-CAD is generally the most used because of its simplicity and high efficiency. These methods result in many low-energy collisions with the background gas; this causes the molecule slowly to heat up, resulting in cleavage of the most labile bonds first. Moreover, because the procedure is a resonant activation method, once the parent ion dissociates, the daughter ions are not resonant and cannot be further activated, and hence primary fragmentation predominates. For peptides, this results in generally incomplete sequence coverage, with a few labile peptide cleavages dominating the spectra.

For MALDI ions, photodissociation is the next most common method for generating fragments. Here, a chromophore is required to absorb the laser light, because of which infrared multiphoton dissociation (IRMPD) [90] using a CO_2 laser is preferred because such lasers emit in a wavelength where bending and rotational modes of biomolecules absorb. This is particularly true because the CO_2 emission wavelengths are fairly broad, typically from 9.1–10.6 μm. Once the ions are in the cell and isolated, IRMPD simply requires the laser to be triggered for a few hundreds of milliseconds. The ions then absorb hundreds to thousands of photons, and dissociate through the lowest energy channels. For peptides, typical values are ~200–500 ms irradiation time at 10 W laser power, but for DNA or oligosaccharides the typical values are 25–50 ms irradiation time at 10 W. UV photodissociation has been attempted on FTICR instruments with limited success, generating predominantly immonium ions for peptides [91], but the full characterization of this method has not yet been attempted.

The new fragmentation techniques of ECD/EDD [27,92,93] and electron transfer dissociation (ETD) [94,95] have not been applied much to MALDI because they require multiply charged ions, which are generally weak with MALDI. However, the methods are occasionally applicable, and a newer method termed EID [26] has been developed which may prove to be very useful indeed.

2.6
Quadrupole Ion Trap Mass Spectrometers

MALDI ion sources have also been used with quadrupole ion trap (QIT) mass spectrometers, which are also referred to as Paul-traps as they were invented by Wolfgang Paul who won the 1989 Nobel prize in physics for these studies. Figure 2.12A illustrates the approximate geometry of a classic, hyperbolic QIT mass spectrometer. These ion traps are similar to the RF-ion guides used above, in that the radial ring (*r*-direction in Fig. 2.12A) and the end-caps (± *z*-direction electrodes) are driven 180° out of phase with an RF electric field (typically ~1 MHz and a few hundred volts peak-to-peak). As the field oscillates, ions that are in the center of the field (at the saddle point in Fig. 2.12B) undergo characteristic motion which is defined by the RF amplitude and frequency and by the mass of the ions. Some ions will be stably confined (provided that they have a low enough initial kinetic energy), while ions that are too heavy will slowly wander out of the trap and hit the electrodes, while ions that are too light will be quickly ejected. Normally, QIT mass spectrometers are operated with a few millitorr of helium as a buffer gas to collisionally cool the ions' translational kinetic energy. Newer linear ion traps (LITs) operate on the same principles, but only in the X-Y plane. The Z-direction is bounded by placing end-plate electrodes at a high potential, similar to what is done to trap ions in FTICR instruments.

With QITs, a relatively simple way of analyzing the mass range available is through the use of Mathieu stability diagrams. These stability diagrams are derived from the differential equations of motion of ions in oscillating electric fields:

$$\frac{d^2u}{d\xi^2} + (a_U - 2q_u \cos 2\xi)u = 0 \quad a_z = -2a_u = \frac{8eU_{dc}}{mr_0^2\Omega^2} \quad q_z = -2q_u = \frac{4eV_{rf}}{mr_0^2\Omega^2}$$

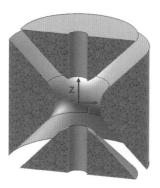

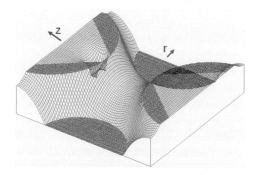

A. a cross-section of a hyperbolic quadrupole ion trap

B. a potential energy diagram of the QIT showing the saddlepoint in the electric field (generated using Simion 7.0)

Fig. 2.12 The shape of Paul ion trap mass spectrometers.

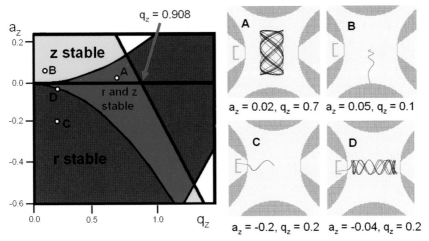

Fig. 2.13 Mathieu stability diagram with four stability points marked. Typical corresponding ion trajectories are shown on the right.

where u represents the coordinate axes (x or y), $\xi = \Omega t/2$, m = mass, e = electric charge, U_{dc} is the DC voltage difference between the ring and the endcaps, V_{rf} is the RF amplitude (zero-to-peak voltage), Ω is the applied RF frequency, and r_0 is the radius of the trap. Note that the a_z values and q_z values are inversely related to the mass/charge ratio of an ion.

The Mathieu stability diagram plots a-values versus q-values and marks the boundaries where ions will be confined to the trap. Figure 2.13 is a plot of the Mathieu stability diagram with four points A–D marked. The trajectory of ions at those four points is shown in plots A–D on the right. Point A is within the r and z stability regions of the diagram, as shown in plot A, while point B is stable in the z-direction but unstable in the r-direction and falls out of the trap radially. Point C is stable in the r-direction but unstable in the z-direction and falls out axially. Point D is just barely unstable, and this ion hangs around for a long time before eventually falling out of the trap. These four trajectories are representative, but the true trajectories are highly dependent on the ions' initial velocities, positions, and initial momentum vectors.

It must be remembered that each of the points in Figure 2.13 is defined by the a_z and q_z equations above and, thus, variance in their values can represent variance in m/z, V_{rf}, or U_{dc} (potentially also Ω or r_0, but these are usually fixed). Therefore, by slowly ramping up V_{rf} or U_{dc} (or both), ions can be shifted from stable to unstable trajectories and ejected from the trap. If done correctly, these ions can be made to impact on a conversion dynode (which ejects an electron when hit with a positive ion with sufficient kinetic energy) and secondary electron multiplier detector. When the electron current signal on the detector is plotted versus the a_z or q_z values calculated from the V_{rf} or U_{dc} ramp, a mass spectrum is produced. Normally, the conversion dynode/electron multiplier detector is placed on the z-axis, so that the $q_z = 0.908$ value as marked in Figure 2.13 is the transition point where an ion becomes unstable in the z-direction (at $a_z = 0$, or $U_{dc} = 0$).

As always in physics, the Mathieu equations and stability diagrams make certain assumptions – that is, they neglect certain effects that are deemed to be negligible or uncommon. In particular, they neglect the effect of background gas pressure and the effect of the initial trajectory of the ions, assuming that the ions start in the center of the trap at rest. They also neglect ion–ion coulombic, "space-charge" interactions. All ion modes of motion (x, y, and z) in a quadrupole ion trap are stable, which means that the ion motion in that dimension is bounded and, in the absence of other stimuli, will eventually allow the ions to cool to a resting position in the center of the cell. The addition of background gas dampens the amplitude of the ions' excursions, slowly cooling them to the center of the trap. Therefore, it is particularly easy to trap ions for indefinite periods of time in QITs. Any ion with any initially trapped trajectory will eventually end up at the exact center of the trap. However, if there are many ions, this effect increases local space charge, so that there is a balancing act in which damping pushes the ions inward while coulombic repulsion pushes them outward. Thus, a defined number of ions can be trapped, and this is known as the "space-charge" limit. Furthermore, as collisions increase, the ions are translationally cooled to the center of the trap, but vibrationally excited at the same time. For this reason, helium is usually used as a collision gas because it decreases the center-of-mass collision energy of the ions and decreases vibrational activation. However, at high ion loadings in the trap, ions will be pushed far enough outward that the vibrational activation will warm some ions enough to start fragmentation. Hofstadler et al. have termed this effect "multipole storage-assisted dissociation" (MSAD) [17].

The trapping dimensions in QITs and LITs are essentially simple harmonic oscillators, at least near the center of the traps. Because of this, in addition to trapping fields, ions have their own resonant frequencies inside the trap which can be used to excite them resonantly to high excursion trajectories. At high trajectories, their vibrational activation will cause them to fragment, allowing MS/MS experiments to be conducted. Resonant excitation is also used to selectively eject ions or to eject them using a SWIFT function [18], similar to that done in FTICRMS. Collisional activation experiments in QIT instruments have the interesting effect that, while the precursor ion is fragmenting, the fragment ions are not resonant and quickly cool back to the center of the trap, which means that the fragmentation is very selective. This effect is the same as has been observed for SORI-CAD on an FTICR mass spectrometer.

A typical MALDI-QIT instrument diagram is shown in Figure 2.14, except that a microchannel plate is drawn instead of a conversion dynode/electron multiplier detector. This instrument is similar to the design of Krutchinsky et al. [19] In this instrument, MALDI ions are generated in a plume that is directed into a transfer RF-only multipole (usually a hexapole or octopole). Ions can be either transferred directly into the ion trap or trapped and cooled in the RF-only multipole before being pulsed into the trap. Either way, in order to achieve the best sensitivity and detection limits in the instrument, it is critical that the ion pulse is synchronized

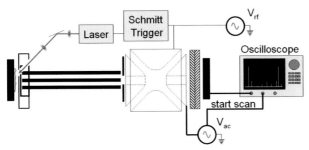

Fig. 2.14 MALDI ion trap mass spectrometry.

to the RF of the trap. If the RF voltage is high when the ion pulse reaches the trap, the ions will be repelled; if it is a large negative voltage, the ions will be accelerated and thrown across to the other side of the trap (note: this assumes positive ions; negative ions are the reverse). For highest transmission and trapping efficiencies, ions must enter the trap when the RF voltage is close to zero. Since the cycle period of the RF trapping field is about 1 µs, the window for entrance is ~10 ns, which is comparable to the pulse width of the UV lasers typically used (~3–5 ns). If the ion cloud is allowed to disperse in time, then the trapping efficiency will be low.

MALDI ions trapped in a QIT will experience a competition between activation and cooling, as discussed above. The MALDI laser generates a plume with ions of considerable internal energy, generating metastable ions. Plume expansion will result in adiabatic cooling of the ions, while collisions with neutrals upon extraction will result in activation. Transfer and trapping of the ions, which usually occurs at ~1 × 10^{-4} mbar of helium, will generally cool the ions. However, the long time frame of ion trap experiments (typically tens of milliseconds) means that metastable ions have an opportunity to fragment so that some dissociation is inevitable. Clearly, MALDI-QIT mass spectrometry experiments involve some balancing between activation and cooling, similar to many other MALDI instruments.

MALDI sources, as noted above, generate a large number of matrix cluster ions which contribute to a "chemical noise" background. In general, chemical noise is basically matrix clusters with sodium, protons, and any other contaminants (e.g., detergents, phosphate, finger-print oils) which get into the sample. It generates a hump of peaks at every *m/z* value throughout the spectrum – a "peak at every mass" with an intensity that correlates to hundreds of ions. Because QIT systems are capable of detecting single ions hitting the detector, Krutchinsky and Chait recognized that, even if an expected analyte peak is not observed at a particular *m/z*-value, MS/MS scans which eject all other peaks prior to fragmentation should be able to observe the MS/MS spectrum of the analyte because the chemical noise is essentially eliminated in these scans. Thus, these authors devised a "hypothesis-driven" scan mode [20] where expected *m/z* ranges are isolated and fragmented, even if there is no peak observed. For example, if a peptide is observed that may be partially phosphorylated at low stoichiometry, an *m/z* region 80 Da higher is quickly isolated and fragmented. If the phosphopeptide is there, its tandem mass spectrum should be seen. The same concept can be applied to any predicted (but not observed) peak.

Overall, QITs and the more recent LITs are robust, cheap mass spectrometers with high scan speed and high sensitivity. The MS/MS capability of these instruments is very advanced, due to the fact that all ion modes are stable – which means a very high percentage of fragment ions can be observed. The resolving power of these instruments is about $(1-2) \cdot 10^{-3}$ and the mass accuracy at m/z 1000 is usually about 0.1%, or ~1 Da. While the m/z range of these instruments is variable, on commercial instruments it is usually limited to between 100 and 2000 Da. Thus, provided that mass accuracy or resolution is not important, and that the m/z range to be observed is below ~2000 Da, these instruments are a cheap, efficient, and robust way to analyze samples.

However, these caveats are also their downfall. Frequently, QITs are used in proteomics experiments which require high mass accuracy (and resolution) in the mass range from ~700 to 4000 Da for tryptic peptides. The resulting data are then used to search databases in an attempt to assign identities to the proteins from which the peptide MS/MS data are generated. Because of the low mass accuracy and the depth of normal sequence databases, the false-positive rate on these database searches is often >50% (i.e., >50% of the assignments are wrong, but which 50% is not clear) [21,22]. Thus, extensive data curation and manual data interpretation is required to correct for this false-positive rate, though improved data quality would eliminate this problem.

Furthermore, using a MALDI source for proteomic samples often results in masses that are beyond the m/z range of a QIT. For example, tryptic peptides from glycoproteins routinely generate peptide and glycopeptide masses beyond 3 kDa. Overall, QITs are extremely cheap, robust mass spectrometers that generate high-efficiency MS/MS data, but they are limited to the low-molecular mass range <2 kDa, and to experiments where resolution and mass accuracy are not important.

2.6.1
RF-Only Ion Guides and LIT Mass Spectrometers

RF-only ion guides operate in essentially the same way as QITs, except that they operate with (as named) the DC offset set to zero. This corresponds to the $a_z = 0$ line in the Mathieu stability diagram (see Fig. 2.13). Ion guides are usually quadrupoles, hexapoles or octopoles, and essentially act as ion pipes, confining the ions to a small region along the axis of the multipole. All ion guides have a sharp low m/z cut-off (defined by low m/z instability, $q_z = 0.908$ for quadrupoles but hexapoles and octopoles are not easily amenable to stability diagram analysis [73]), and a very shallow high m/z cut-off (defined by the ion beam angle and kinetic energy).

LITs are simply RF-only ion guides with plates at either end. Provided that these plates have potentials which are higher than the axial kinetic energy of the ions, they act as repelling voltages, thus trapping the ions axially. The RF field traps the ions radially. Newer LITs [96] (Fig. 2.15) then allow radial resonant excitation of

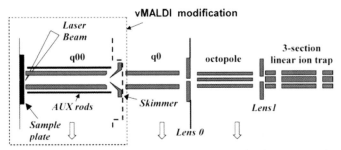

Fig. 2.15 Linear ion trap mass spectrometry. Reprinted with permission from Ref. [96].

ions into external detectors, or for ejection. RF field traps have the advantage of being substantially more sensitive than QITs, and modern electronics allows accurate and stable resonant ejection scanning on a 10 scan s^{-1} timescale.

2.6.2
Tandem Mass Spectrometry on QIT Mass Spectrometers

The trapping dimensions in QITs and LITs are essentially simple harmonic oscillators, at least near the center of the traps. Because of this, in addition to trapping fields, ions have their own resonant frequencies inside the trap which can be used to resonantly excite them to high-excursion trajectories and to manipulate the ion populations to a degree surpassed only by FTICR instruments. Because of this, QIT mass spectrometers are extremely flexible in MS/MS experiments.

Similar to FTICR mass spectrometers, ions in LITs and QITs have resonant modes whereby their trajectories can be excited by the application of low-amplitude resonant frequencies on the endcaps (for QIT-MS) or quadrupole rods (for LIT-MS). This means that the SWIFT technique can also be applied selectively to excite and/or eject ions from the trap. For a better description of the SWIFT method, see Section 2.5.1.

Most tandem mass spectrometry in QIT and LIT instruments relies on collisional activation to heat the molecule of interest until it fragments. Because these instruments usually operate in an environment of ~10^{-3} mbar of helium, collisional activation only requires that the ions be excited by the addition of a resonant excitation pulse to the endcaps or quadrupole rods. However, as with SORI-CAD (see Section 2.5.2), ions of interest are collisionally excited and dissociate, but the resulting fragment ions are not resonant and generally do not fragment – so that this mode of activation results in a strong preference for fragmentation of the weakest bonds. This is less of a problem in QIT and LIT instruments than it is in FTICR instruments, because all three ion-oscillation modes of motion are bound, which is why it is fairly easy to achieve MS^n experiments with N > 5 in QIT and LIT instruments. There is a caveat to this, however, namely that QIT and LIT instruments, being subject to the Mathieu stability diagram, can only trap a certain range of ions, with a very sharp cut-off in the trapping range at low m/z. If the fragment ions that are generated fall below this cut-off, they are not stable and will fall out of the trap. Furthermore, the addition of a resonant excitation pulse at a particular m/z further destablilizes low-m/z ions. The rule of thumb is that excitation at a particular m/z will eject all ions below one-third of that m/z value (e.g., excitation at m/z 1000 will eject all ions below m/z ~300), though this rule is being greatly weakened by some new high-voltage pulsed-excitation methods [94].

ECD (see Section 2.5.1), which was first introduced in FTICR instruments, has also been reproduced in QIT instruments [98,99], albeit with great difficulty due to the problem of achieving simultaneous stable trapping of electrons and ions. However, a derivative method, ETD [94,95], has been shown to be extremely effective. ETD involves the reaction of even-electron, multiply charged ions ($[M+nH]^{n+}$ with n ≥ 2) with low-molecular mass radical anions (M^{-}). Provided that the correct radical anion molecule is chosen, it will donate an electron to the even-electron ion, making it into a radical cation ($[M+nH]^{(n-1)+\bullet}$), which then undergoes similar free-radical chemistry that results in ECD-like fragments. Due to the high efficiency of trapping of fragment ions in QIT and LIT instruments, and due to the higher rate of reaction of multiply charged ions over singly charged ions, the reaction can be driven to generate extremely high-quality MS/MS spectra. ETD is now being widely implemented with many varied designs on most hybrid instruments which have multipole ion traps.

Although photodissociation has also been performed extensively on QIT instruments [100–108], the results have not been so widely useful that the method has been implemented on commercial systems. There are two reasons for this: First, the excursion paths of ions in QIT instruments is often wider than the laser beams, so that some ions may not be irradiated. Second, with IRMPD in particular where hundreds to thousands of photons must be absorbed in order to heat the molecules up for dissociation, ions experience a competition between heating (from the laser) and cooling (from collisions and blackbody radiation). At 10^{-3} mbar, the collisional cooling is substantial, so that higher laser powers or controlling the pressure via pulsed valves is necessary. UV photodissociation has not been extensively explored in QIT and LIT instruments.

2.7
Hybrid Mass Spectrometers

Each of the individual types of mass spectrometers, TOF, FTICR, quadrupole/multipole ion guides, and QITs have both advantages and disadvantages as discussed. However, a number of research groups have built "hybrid" MALDI mass spectrometers which combine the various components to achieve interesting combined advantages.

2.7.1
Quadrupole TOF Mass Spectrometers

MALDI-QTOF instruments basically follow an ion optical arrangement similar to that shown in Figure 2.16. The first quadrupole, Q_0, is simply used to cool the ions'

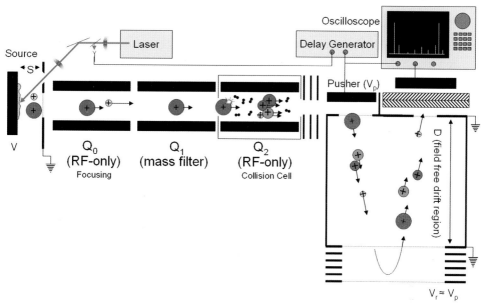

Fig. 2.16 Quadrupole time-of-flight hybrid.

kinetic energy and focus them onto the quadrupole axis so that they enter Q_1 with a well-defined initial trajectory, and is always used in RF-only mode. Q_1 is a mass filtering quadrupole which can either be used in RF-only mode to pass a wide range of mass/charge values or can be used in mass filtering mode to select particular ions. Ions leaving Q_1, either broad m/z ranges or selected ions, are then directed into Q_2. By adjusting the offset voltages of the quadrupoles, the ions can be accelerated into Q_2, where they can collide and fragment. Fragment ions are then directed into the orthogonal TOF where they can be detected.

QTOF hybrids are simple to use and reasonably flexible. The quadrupoles allow beam focusing, mass selection and dissociation, while the TOF provides high sensitivity and resolving power. The orthogonal injection geometry into the TOF decouples the surface desorption from the detection so that irregular, bumpy MALDI crystals or other surfaces (such as thin-layer chromatography plates) can be used. The plume velocity does not affect mass accuracy, so a wider range of laser fluence can be used which improves sensitivity. Moreover, the focusing effect of the quadrupoles allows ions to be injected into the oTOF in a thin beam; this improves the resolution and as a result the QTOF instruments can easily achieve $>10^4$ resolving power. Furthermore, this geometry allows fragmentation in Q_2 with rapid and sensitive detection of the products.

In addition, ions can be accumulated in Q_2 prior to dumping them into the TOF. This "selected ion accumulation" can improve detection sensitivity by as much as 100-fold, depending on the duty cycle of the instrument. The LINAC arrangement from Sciex [57] assists with these experiments by allowing improved ejection of these stored ions from Q_2.

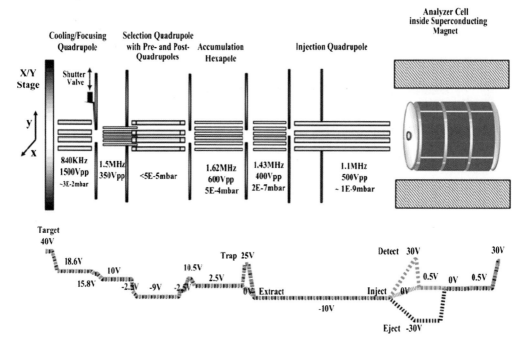

Fig. 2.17 Quadrupole Fourier transform mass spectrometer
(FTMS). Reprinted with permission from Ref. [7].

2.7.2
Quadrupole FT Mass Spectrometers

Putting a quadrupole mass filter in front of a FT mass spectrometer is also useful. The first MALDI-Q-FTICRMS was built by Brock et al. [7], and is shown schematically in Figure 2.17. This instrument can generate ions, focus them through the first quadrupole and octopole into the mass resolving quadrupole, selectively accumulate or fragment ions in the accumulation hexapole, and then transfer the accumulated ions to the ICR cell using the final octopole and quadrupole ion optics. Selected ion accumulation allows the operator to fill the ICR cell to its space charge limit with only the ion of interest, which allows higher-quality MS/MS spectra in the cell. It also, for the same reason, eliminates most chemical noise in the cell. Furthermore, mass selection in a quadrupole is much faster and simpler than selection in the ICR cell.

Bruker and Thermo have each adopted similar ion optical arrangements for their commercial MALDI-FTICRMS instruments. Bruker's instrument (Fig. 2.18) [109] uses a combined MALDI/electrospray ion (ESI) source which focuses the ions through a hexapole and into a quadrupole for mass selection. The ions are then accumulated or fragmented in a second hexapole before they are pulsed out and down their electrostatic ion guide into the ICR cell.

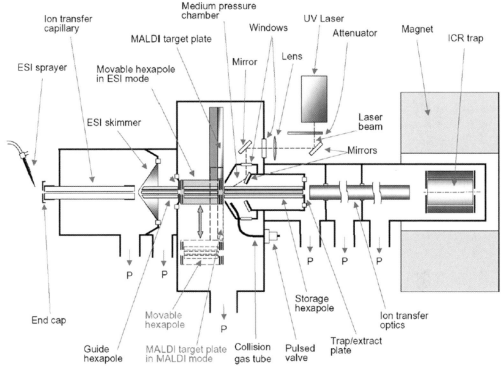

Fig. 2.18 MALDI-ESI/FTMS hybrid from Bruker. Reprinted with permission from Ref. [109].

The Thermo instrument [110] (Fig. 2.19) does not have a mass selection quadrupole prior to its linear ion trap, but it is able to use the linear ion trap as a rapid-scanning, low-resolution mass spectrometer in addition to ejecting ions into a long multipole which transports them to the ICR cell. Typically, three scans can be performed in the ion trap at the same time as performing one high-resolution scan in the FTICR. The Thermo LTQ ion trap adds substantial scan flexibility, automatic gain control, and single-ion detection limits, but without a filtering quadrupole prior to the trap there is no selected ion accumulation capability. Additionally, the Thermo LTQ software substantially limits the ability of the user to adjust pulse sequences or experimental parameters; this has the advantage of preventing inexperienced users from de-tuning the instrument, but limits experienced users from exploring and testing different pulse sequences.

2.7.3
QIT-TOF Mass Spectrometers

A more recent hybrid involves the coupling of QITs to TOF mass spectrometers. This instrument configuration originated in the laboratory of Lubman [25], but has recently been commercialized by Shimadzu (Fig. 2.20). These instruments allow

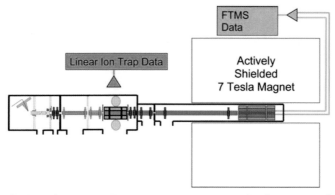

Fig. 2.19 Thermo LTQ-FT mass spectrometer. Reprinted with permission from Ref. [110].

MS^n experiments in the ion trap, followed by ~10^4 resolving power in the TOF. While these performance characteristics are reasonable, no reliable data have yet been published by independent research groups to confirm them. The difficulty is that ions, when ejected into a TOF, must be confined to a small spatial region with little or no initial kinetic energy variance in order to achieve higher resolving power. Ions oscillating in an ion trap generally have a wide range of initial velocity vectors, so the RF trapping field must be shut off just as the ions are ejected into the TOF. This can be done, but to date, the results are less than encouraging.

Nonetheless, this type of hybrid is interesting for studying ion–molecule reactions. Ion traps in general are very good for studying these reactions because ions are contained under high pressure for long periods of time. In particular, the newer ETD experiments should be possible on such an instrument which would allow the generation of both odd-electron and even-electron fragmentation, which tend to be complementary.

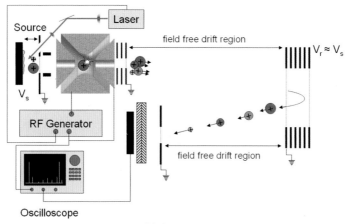

Fig. 2.20 MALDI ion trap time-of-flight mass spectrometer.

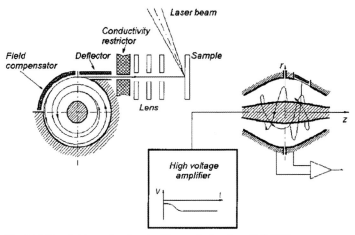

Fig. 2.21 The Orbitrap. Reprinted with permission from Ref. [111].

2.7.4
Orbitrap

A new instrument design, called the Orbitrap (Fig. 2.21), has recently been released onto the market by Thermo [111–115]. This instrument pulses ions into a
chamber which has a very special shape that allows ions of a very narrow kinetic
energy range to orbit around a central electrode. The ions also oscillate axially
along this central electrode, and this axial oscillation is roughly simple harmonic
motion. The outer electrode is sectioned into two halves, and the differential, induced electric field on these two halves is amplified, digitized, and stored. This
stored waveform is very much like the transient signals in FTICR instruments
and, provided that the ions oscillate for a sufficient duration, provides many of the
resolution and mass accuracy advantages of the FTICR. To date, however, we
are unaware of any published MALDI-Orbitrap data, though these may appear
shortly.

The limitations of the Orbitrap are not yet clear, but some expected advantages
and disadvantages of the design can be postulated. First, the instrument clearly
should suffer less from space charge limitations than the FTICR or QIT/LIT counterparts, because the central electrode essentially shields many of the ions from
each other. Second, the electrode shape and quality of the electrode surfaces is
critical, as in the QIT/LIT instruments, but unlike the FTICR instruments. Electrode cleanliness and the accuracy of the machining will limit the accuracy of the
electric fields inside the Orbitrap and thus limit coherence of the ion cloud and
hence resolution and accuracy. Third, as with the FTICR instruments, collisions
with background gas limit ion cloud coherence and, hence, resolution and accuracy, so that pressures of $<10^{-9}$ mbar are required.

2.8
Future Directions

Mass spectrometry instrumentation for MALDI ion sources has considerable room for growth. The hybrids listed above are highly effective, but very few are commercially available. However, new hybrids are very likely to continue being developed, and new fragmentation techniques are clearly being pursued.

Within the field, however, the biggest improvement needed is in sample preparation. At present, several companies provide sample preparation robots to perform everything from detecting and slicing gel electrophoresis spots to enzymatic digestion to spotting the resulting peptides onto the MALDI sample plate along with matrix. These robots greatly improve the reliability and robustness of the analysis, and are clearly helpful for achieving maximum data throughput. Some robots are also capable of depositing spots or trails of chromatographic effluent, which allows the coupling of separation methods with MALDI.

One interesting method is the coupling of thin-layer chromatography (TLC) or other planar separation techniques with MALDI-qTOF [5,116,117] and MALDI-FTICRMS [12] instruments. Although the coupling of TLC to MALDI-TOF has been previously explored [118–120], the use of such an approach was limited due to the low resolving power achievable in axial TOF instruments from irregular, bumpy surfaces. MALDI-qTOF and MALDI-FTICRMS instruments decouple the source from the analyzer, so that irregular surfaces, surface static charging, and variable kinetic energy distributions are not problematic. Consequently, this method is likely to be extended to a wide variety of surface techniques such as 2D gels or microfluidic channels.

Since the advent of MALDI, there has been a continuing exploration of the different types of lasers and matrices that can be used. One particularly interesting direction is the use of IR-MALDI [121–126] (in the 2- to 10-μm range) and liquid matrices (including water). Some use has been made of these systems, and the IR-MALDI lasers appear to generate signals with less internal vibrational energy than traditional UV lasers. Furthermore, IR-MALDI appears to generate higher charge state ions. Whilst there is substantial scope for further investigation in this realm, however, the field is currently limited by a lack of reliable IR lasers.

Definitions and Acronyms

2,5 DHB	2,5-dihydroxybenzoic acid, a common MALDI matrix
3-HPA	3-hydroxypicolinic acid
ADC	analog-to-digital converter
AGC	automatic gain control
AP-MALDI	atmospheric pressure MALDI
CAD	collisionally activated dissociation
Centroid	The center-of-mass of a peak. The best method for determining peak centroids is non-linear least-squares

fitting of the raw peak data with a function which has the proper peak shape for the instrument. Other methods include: (i) using a center-of-mass calculation (weighted average) of the raw peak data (pretty good); (ii) fitting the upper half of the peak with a parabola and reporting the zero of the second derivative (OK if there are many points on the peak); and (iii) taking the highest point of the peak and defining this to be the centroid (generally a bad method, but rapid).

CID	collision-induced dissociation
DE	delayed extraction
Dalton (Da)	A Dalton is defined as 1/12th of the mass of a ^{12}C atom. It differs from an atomic mass unit (amu), which is defined as 1/16th of the mass of a ^{16}O atom.
ECD	electron capture dissociation
EDD	electron detachment dissociation
EID	electron-induced dissociation
Er:YAG	erbium-doped yttrium aluminum garnet, a crystal used for lasers. It lases at a fundamental wavelength of 2.94 μm.
ETD	electron transfer dissociation
FFT	Fast Fourier transform [127]
Fragmentation efficiency	The ratio of the sum of the fragment ions divided by the reduction in the precursor ion abundance.
FT	Fourier transform
FTICR	Fourier transform ion cyclotron resonance
FWHM	full-width at half-maximum, the standard way of determining peak width. It is used to calculate resolving power (see below).
HCCA (also CHCA)	alpha-cyano-4-hydroxycinnamic acid
ICR	ion cyclotron resonance
IR	infrared
IRMPD	infrared multiphoton dissociation
Limit of detection	The smallest signal that can be detected reliably in the mass spectrometer. Because ion sources and transfer optics are not 100% efficient at converting analyte molecules into detectable signal, the correct way to determine this value is to measure the number of analyte molecules (in moles) that are deposited onto the MALDI target and then run a dilution series until the signal from the mass spectrometer falls below a chosen signal/noise ratio (usually ~3). It is best to report the final analyte molecule number and the signal/noise at that number. Dilution series experiments are frequently plagued by systematic errors so that

measuring detection limits can often be problematic. These systematic errors are often due to sample carry-over down the dilution series due to sample sticking to the insides of pipettes or sample target plates. Extreme care must be taken to avoid these problems in order to be able to report accurate limits of detection.

LINAC	linear accelerator
LIT	linear ion trap
MALDI	matrix-assisted laser desorption/ionization
Mass accuracy	Mass spectrometers are usually characterized by their internal and external calibration mass accuracy, or how closely the measured mass matches the theoretical mass. It is usually reported in Da or in parts-per-million, but as mass accuracy varies across the spectrum in many mass spectrometers it must also be reported at a particular m/z value. Every mass spectrometer has a characteristic mass accuracy capability, but achieving this "best-case" limit usually requires careful control of the instrumental parameters and making sure the ion source and ion optics are clean.
MCP	microchannel plate
MECA	multiple excitation collisional activation
MS	mass spectrometry
MSAD	multipole storage assisted dissociation
MS^n	nth order tandem mass spectrometry
NA	numerical aperture, $NA = \sin(\Theta/2)$, where Θ is the full-angle of the divergence of the beam; thus, $NA = 0.2$ equals $\Theta = 22°$.
Nd:YAG	neodymium-doped yttrium aluminum garnet, a crystal that is used for lasers. It lases at a fundamental wavelength of 1064 nm, but can be frequency tripled to 355 nm.
Noise	Most mass spectrometers suffer from white noise (Johnson noise and shot noise), RF interference (i.e., radio stations, fluorescent light ballast supplies, switching power supplies, quadrupoles, etc.), and chemical noise (chemical interference peaks). Generally, the noise level reported for signal/noise calculations should be reported as root-mean-square (RMS) of the white noise, but this value can sometimes be difficult to determine if the experiment has much chemical noise. RF noise is not usually dense enough to be a major problem. Regardless, if signal/noise values are reported, it is important to report exactly how noise is calculated.

OPO	optical parametric oscillator. A device for scanning or choosing wavelengths in the 1.5- to 3-μm range.
oTOF	orthogonal time-of-flight
PPM	parts per million
PSD	post source decay
QIT	quadrupole ion trap
Resolution	The lowest m/z spacing where different peaks can be distinguished. It is usually determined at FWHM, but 10% valley and 90% valley positions are also sometimes used. Resolution has units of m/z.
Resolving power	Resolving power is the most common way of discussing resolution in a mass spectrometer. It is calculated as $M/\Delta M$, where ΔM is FWHM. For example, an ion at 1000 m/z with a FWHM peak width of 0.3 m/z will have a resolving power of 3000. Resolving power is a unitless parameter.
RF	radiofrequency, anywhere from a few hundred kHz to a few hundred MHz
qTOF	quadrupole time-of-flight
TDC	time-to-digital converter
TEM$_{00}$	transverse electric field mode indicates that the electric field of the laser beam is perpendicular to its direction of propagation. TEM$_{00}$ implies a Gaussian beam profile.
TLC	thin-layer chromatography
TOF	time-of-flight
SORI-CAD	sustained off-resonance irradiation-CAD
Sensitivity	defined as the slope of a concentration versus signal intensity plot. This term is often used incorrectly to mean detection limits.
SEM	secondary electron multiplier
SWIFT	stored waveform inverse Fourier transform
UV	ultraviolet
VC-MALDI	vibrationally cooled MALDI
VLE-CAD	very low-energy-CAD

References

1 M. Schurenberg, T. Schulz, K. Dreisewered, F. Hillenkamp. Matrix-assisted laser desorption/ionization in transmission geometry – instrumental implementation and mechanistic implications. *Rapid Commun. Mass Spectrom.* **1996**, *10*, 1873–1880.

2 A.N. Krutchinsky, A.V. Loboda, V.L. Spicer, R. Dworschak, W. Ens, K.G. Standing. Orthogonal injection of matrix-assisted laser desorption/ionization ions into a time-of-flight spectrometer through a collisional damping interface. *Rapid Commun.*

Mass Spectrom. **1998**, *12*, 508–518.

3   A.V. Loboda, A.N. Krutchinsky, M. Bromirski, W. Ens, K.G. Standing. A tandem quadrupole/time-of-flight mass spectrometer with a matrix-assisted laser desorption/ionization source: Design and performance. *Rapid Commun. Mass Spectrom.* **2000**, *14*, 1047–1057.

4   A.V. Loboda, S. Ackloo, I.V. Chernushevich. A high-performance matrix-assisted laser desorption/ionization orthogonal time-of-flight mass spectrometer with collisional cooling. *Rapid Commun. Mass Spectrom.* **2003**, *17*, 2508–2516.

5   V.B. Ivleva, L.M. Sapp, P.B. O'Connor, C.E. Costello. Ganglioside analysis by thin-layer chromatography matrix-assisted laser desorption/ionization orthogonal time-of-flight mass spectrometry. *J. Am. Soc. Mass Spectrom.* **2005**, *16*, 1552–1560.

6   S.C. Moyer, B.A. Budnik, J.L. Pittman, C.E. Costello, P.B. O'Connor. Attomole peptide analysis by high pressure matrix-assisted laser desorption/ionization Fourier transform mass spectrometry. *Anal. Chem.* **2003**, *75*, 6449–6454.

7   A. Brock, D.M. Horn, E.C. Peters, C.M. Shaw, C. Ericson, Q.T. Phung, A.R. Saloma. An automated matrix-assisted laser desorption/ionization quadrupole Fourier transform ion cyclotron resonance mass spectrometer for "bottom-up" proteomics. *Anal. Chem.* **2003**, *75*, 3419–3428.

8   B.A. Budnik, S.C. Moyer, J.L. Pittman, V.B. Ivleva, U. Sommer, C.E. Costello, P.B. O'Connor. High pressure MALDI-FTMS: Implications for proteomics. *Int. J. Mass Spectrom. Ion Processes* **2004**, *234*, 203–212.

9   P.B. O'Connor, B.A. Budnik, V.B. Ivleva, P. Kaur, S.C. Moyer, J.L. Pittman, C.E. Costello. A high pressure matrix-assisted laser desorption ion source for Fourier transform mass spectrometry designed to accommodate large targets with diverse surfaces. *J. Am. Soc. Mass Spectrom.* **2004**, *15*, 128–132.

10   A.N. Krutchinsky, M. Kalkum, B.T. Chait. Automatic identification of proteins with a MALDI-quadrupole ion trap mass spectrometer. *Anal. Chem.* **2001**, *73*, 5066–5077.

11   P.B. O'Connor, E. Mirgorodskaya, C.E. Costello. High pressure matrix-assisted laser desorption/ionization Fourier transform mass spectrometry for minimization of ganglioside fragmentation. *J. Am. Soc. Mass Spectrom.* **2002**, *13*, 402–407.

12   V.B. Ivleva, Y.N. Elkin, B.A. Budnik, S.C. Moyer, P.B. O'Connor, C.E. Costello. Coupling thin layer chromatography with high pressure matrix assisted laser desorption/ionization Fourier transform mass spectrometry for the analysis of ganglioside mixtures. *Anal. Chem.* **2004**, *76*, 6484–6491.

13   V.V. Laiko, S.C. Moyer, R.J. Cotter. Atmospheric pressure MALDI/ion trap mass spectrometry. *Anal. Chem.* **2000**, *72*, 5239–5243.

14   V.V. Laiko, M.A. Baldwin, A.L. Burlingame. Atmospheric pressure matrix-assisted laser desorption/ionization mass spectrometry. *Anal. Chem.* **2000**, *72*, 652–657.

15   V.M. Doroshenko, V.V. Laiko, N.I. Taranenko, V.D. Berkout, H.S. Lee. Recent developments in atmospheric pressure MALDI mass spectrometry. *Int. J. Mass Spectrom. Ion Processes* **2002**, *221*, 39–58.

16   K.A. Kellersberger, P.V. Tan, V.V. Laiko, V.M. Doroshenko, D. Fabris. Atmospheric pressure MALDI-Fourier transform mass spectrometry. *Anal. Chem.* **2004**, *76*, 3930–3934.

17   S.C. Moyer, L.A. Marzilli, A.S. Woods, V.V. Laiko, V.M. Doroshenko, R.J. Cotter. Atmospheric pressure matrix-assisted laser desorption/ionization (AP MALDI) on a quadrupole ion trap mass spectrometer. *Int. J. Mass Spectrom. Ion Processes* **2003**, *226*, 133–150.

18   S.G. Moyer, R.J. Cotter. Atmospheric pressure MALDI. *Anal. Chem.* **2002**, *74*, 468A–476A.

19   C.E. Von Seggern, P.E. Zarek, R.J. Cotter. Fragmentation of sialylated carbohydrates using infrared atmospheric pressure MALDI ion trap mass spectrometry from cation-doped liquid matrixes. *Anal. Chem.* **2003**, *75*, 6523–6530.

20 C.E. Von Seggern, S.C. Moyer, R.J. Cotter. Liquid infrared atmospheric pressure matrix-assisted laser desorption/ionization ion trap mass spectrometry of sialylated carbohydrates. *Anal. Chem.* **2003**, *75*, 3212–3218.

21 T. Kocher, A. Engstrom, R.A. Zubarev. Fragmentation of peptides in MALDI in-source decay mediated by hydrogen radicals. *Anal. Chem.* **2005**, *77*, 172–177.

22 R.S. Brown, B.L. Carr, J.J. Lennon. Factors that influence the observed fast fragmentation of peptides in matrix-assisted laser desorption. *J. Am. Soc. Mass Spectrom.* **1996**, *7*, 225–232.

23 R.S. Brown, J.H. Feng, D.C. Reiber. Further studies of in-source fragmentation of peptides in matrix-assisted laser desorption-ionization. *Int. J. Mass Spectrom. Ion Processes* **1997**, *169*, 1–18.

24 P. Roepstorff, J. Fohlman. Proposal for a common nomenclature for sequence ions in mass spectra of peptides. *Biomed. Mass Spectrom.* **1984**, *11*, 601.

25 K. Biemann. Sequencing of peptides by tandem mass spectrometry and high-energy collision-induced dissociation. *Methods Enzymol.* **1990**, *193*, 455–479.

26 B.A. Budnik, J. Jebanathirajah, V.B. Ivleva, C.E. Costello, P.B. O'Connor. Fragmentation methods of odd and even electron ions in HP MALDI FTMS. In: *52nd ASMS*: Nashville, **2004**.

27 B.A. Budnik, K.F. Haselmann, R.A. Zubarev. Electron detachment dissociation of peptide di-anions: An electron-hole recombination phenomenon. *Chem. Phys. Lett.* **2001**, *342*, 299–302.

28 L. Sleno, D.A. Volmer, A.G. Marshall. Assigning product ions from complex MS/MS spectra: The importance of mass uncertainty and resolving power. *J. Am. Soc. Mass Spectrom.* **2005**, *16*, 183–198.

29 W.C. Wiley, I.H. McLaren. Time-of-flight mass spectrometer with improved resolution. *Rev. Sci. Inst.* **1955**, *29*, 1150–1157.

30 P. Juhasz, M.T. Roskey, I.P. Smirnov, L.A. Haff, M.L. Vestal, S.A. Martin. Applications of delayed extraction matrix-assisted laser desorption ionization time-of-flight mass spectrometry to oligonucleotide analysis. *Anal. Chem.* **1996**, *68*, 941–946.

31 M.L. Vestal, P. Juhasz, S.A Martin. Delayed extraction matrix-assisted laser-desorption time-of-flight mass-spectrometry. *Rapid Commun. Mass Spectrom.* **1995**, *9*, 1044–1050.

32 R.S. Brown, J.J. Lennon. Mass resolution improvement by incorporation of pulsed ion extraction in a matrix-assisted laser-desorption ionization linear time-of-flight mass-spectrometer. *Anal. Chem.* **1995**, *67*, 1998–2003.

33 S.M. Colby, J.P. Reilly. Space-velocity correlation focusing. *Anal. Chem.* **1996**, *68*, 1419–1428.

34 T.B. King, S.M. Colby, J.P. Reilly. High-resolution MALDI-TOF mass-spectra of 3 proteins obtained using space-velocity correlation focusing. *Int. J. Mass Spectrom. Ion Processes* **1995**, *145*, L1–L7.

35 P. Juhasz, M.L. Vestal, S.A. Martin. On the initial velocity of ions generated by matrix-assisted laser desorption ionization and its effect on the calibration of delayed extraction time-of-flight mass spectra. *J. Am. Soc. Mass Spectrom.* **1997**, *8*, 209–217.

36 N.P. Christian, R.J. Arnold, J.P. Reilly. Improved calibration of time-of-flight mass spectra by simplex optimization of electrostatic ion calculations. *Anal. Chem.* **2000**, *72*, 3327–3337.

37 J. Gobom, M. Mueller, V. Egelhofer, D. Theiss, H. Lehrach, E. Nordhoff. A calibration method that simplifies and improves accurate determination of peptide molecular masses by MALDI-TOF ms. *Anal. Chem.* **2002**, *74*, 3915–3923.

38 E. Moskovets, H.S. Chen, A. Pashkova, T. Rejtar, V. Andreev, B.L. Karger. Closely spaced external standard: A universal method of achieving 5 ppm mass accuracy over the entire MALDI plate in axial matrix-assisted laser desorption/ionization time-of-flight mass spectrometry. *Rapid Commun. Mass Spectrom.* **2003**, *17*, 2177–2187.

39 B.A. Mamyrin, V.I. Karataev, D.B. Schmikk, B.A. Zagulin. The mass-reflectron, a new nonmagnetic time-of-flight mass spectrometer with high

resolution. *Sov. Phys. JETP* **1973**, *37*, 45–48.

40* VCB.A. Mamyrin. Laser-assisted reflectron time-of-flight mass-spectrometry. *Int. J. Mass Spectrom. Ion Processes* **1994**, *131*, 1–19.

41 R.J. Cotter, S. Iltchenko, D.X. Wang. The curved-field reflectron: PSD and CID without scanning, stepping or lifting. *Int. J. Mass Spectrom. Ion Processes* **2005**, *240*, 169–182.

42 T.J. Cornish, R.J. Cotter. A curved-field reflectron for improved energy focusing of product ions in time-of-flight mass-spectrometry. *Rapid Commun. Mass Spectrom.* **1993**, *7*, 1037–1040.

43 S. Berkenkamp, F. Hillenkamp. Infrared MALDI with an orthogonal ion extraction TOF for high performance analysis of large proteins. In: *53rd American Society for Mass Spectrometry and Allied Topics*: San Antonio, Texas, **2005**.

44 R. Kaufmann, P. Chaurand, D. Kirsch, B. Spengler. Post-source decay and delayed extraction in matrix-assisted laser desorption/ionization/reflectron time-of-flight mass spectrometry – are there trade-offs. *Rapid Commun. Mass Spectrom.* **1996**, *10*, 1199–1208.

45 M. Karas, M. Gluckmann, J. Schafer. Ionization in matrix-assisted laser desorption/ionization: Singly charged molecular ions are the lucky survivors. *J. Mass Spectrom.* **2000**, *35*, 1–12.

46 D.C. Reiber, T.A. Grover, R.S. Brown. Identifying proteins using matrix-assisted laser desorption/ionization in-source fragmentation data combined with database searching. *Anal. Chem.* **1998**, *70*, 673–683.

47 B. Spengler. Post-source decay analysis in matrix-assisted laser desorption/ionization mass spectrometry of biomolecules. *J. Mass Spectrom.* **1997**, *32*, 1019–1036.

48 D. Suckau, A. Resemann, M. Schuerenberg, P. Hufnagel, J. Franzen, A. Holle. A novel MALDI lift-TOF/TOF mass spectrometer for proteomics. *Anal. Bioanal. Chem.* **2003**, *376*, 952–965.

49 A. Brock, N. Rodriguez, R.N. Zare. Hadamard transform time of flight mass

spectrometry. *Anal. Chem.* **1998**, *70*, 3735–3741.

50 R.J. Cotter, B.D. Gardner, S. Iltchenko, R.D. English. Tandem time-of-flight mass spectrometry with a curved field reflectron. *Anal. Chem.* **2004**, *76*, 1976–1981.

51 A.L. Yergey, J.R. Coorssen, P.S. Backlund, P.S. Blank, G.A. Humphrey, J. Zimmerberg, J.M. Campbell, M.L. Vestal. De novo sequencing of peptides using MALDI/TOF-TOF. *J. Am. Soc. Mass Spectrom.* **2002**, *13*, 784–791.

52 J.M. Campbell, S.E. Stein, P.S. Blank, J. Epstein, M.L. Vestal, A.L. Yergey. Fundamentals of peptide fragmentation as a function of laser fluence in a MALDI TOF-TOF mass spectrometer. *Abstracts Papers Am. Chem. Soc.* **2005**, *229*, U148–U148.

53 M.L. Vestal, J.M. Campbell. In: *Biological Mass Spectrometry*, **2005**, Vol. 402, pp. 79–108.

54 K.F. Medzihradszky, J.M. Campbell, M.A. Baldwin, A.M. Falick, P. Juhasz, M.L. Vestal, A.L. Burlingame. The characteristics of peptide collision-induced dissociation using a high-performance MALDI-TOF/TOF tandem mass spectrometer. *Anal. Chem.* **2000**, *72*, 552–558.

55 D. Suckau, D.S. Cornett. Protein sequencing by ISD and PSD MALDI-TOF ms. *Analusis* **1998**, *26*, M18–M21.

56 A. Shevchenko, A. Loboda, W. Ens, K.G. Standing. Maldi quadrupole time-of-flight mass spectrometry: A powerful tool for proteomic research. *Anal. Chem.* **2000**, *72*, 2132–2141.

57 A. Loboda, A. Krutchinsky, O. Loboda, J. McNabb, V. Spicer, W. Ens, K. Standing. Novel LINAC II electrode geometry for creating an axial field in a multipole ion guide. *Eur. J. Mass Spectrom.* **2000**, *6*, 531–536.

58 P. Verhaert, S. Uttenweiler-Joseph, M., de Vries A. Loboda, W. Ens, K.G. Standing. Matrix-assisted laser desorption/ionization quadrupole time-of-flight mass spectrometry: An elegant tool for peptidomics. *Proteomics* **2001**, *1*, 118–131.

59 H. Kraus. Cryogenic detectors and their application to mass spectrometry. *Int. J.*

Mass Spectrom. Ion Processes **2002**, *215*, 45–58.

60 S.V. Uchaikin, F. Probst, W. Seidel. Developing of a fast cryodetector read-out for mass spectrometry. *Physica C* **2001**, *350*, 177–179.

61 M. Frank. Mass spectrometry with cryogenic detectors. *Nuclear Instruments & Methods in Physics Research Section a-Accelerators Spectrometers Detectors and Associated Equipment* **2000**, *444*, 375–384.

62 P. Christ, S. Rutzinger, W. Seidel, S. Uchaikin, F. Probst, C. Koy, M.O. Glocker. High detection sensitivity achieved with cryogenic detectors in combination with matrix-assisted laser desorption/ionisation time-of-flight mass spectrometry. *Eur. J. Mass Spectrom.* **2004**, *10*, 469–476.

63 M. Frank, C.A. Mears, S.E. Labov, W.H. Benner, D. Horn, J.M. Jaklevic, A.T. Barfknecht. High-efficiency detection of 66 000 Da protein molecules using a cryogenic detector in a matrix-assisted laser desorption/ionization time-of-flight mass spectrometer. *Rapid Commun. Mass Spectrom.* **1996**, *10*, 1946–1950.

64 P.L. Ferguson. High-speed analog signal averager as oatofms data collector. *Data Acquisition and Interfacing* **2005**, *February*, 27–32.

65 M.B. Comisarow, A.G. Marshall. Fourier transform ion cyclotron resonance spectroscopy. *Chem. Phys. Lett.* **1974**, *25*, 282–283.

66 A.G. Marshall, C.L. Hendrickson, S.D. Shi. Scaling ms plateaus with high-resolution FTICRMS. *Anal. Chem.* **2002**, *74*, 252A–259A.

67 I.J. Amster. Fourier transform mass spectrometry. *J. Mass Spectrom.* **1996**, *31*, 1325–1337.

68 E.B. Ledford, Jr., D.L., Rempel M.L. Gross. Mass calibration law for quadrupolar potential. *Anal. Chem.* **1984**, *56*, 2744–2748.

69 T. Francl, M.G. Sherman, R.L. Hunter, M.J. Locke, W.D. Bowers, R.T. McIver. Experimental determination of the effects of space charge on ion cyclotron resonance frequencies. *Int. J. Mass Spectrom. Ion Processes* **1983**, *54*, 169–187.

70 S.D.H. Shi, J.J. Drader, M.A. Freitas, C.L. Hendrickson, A.G. Marshall.

Comparison and interconversion of the two most common frequency-to-mass calibration functions for Fourier transform ion cyclotron resonance mass spectrometry. *Int. J. Mass Spectrom. Ion Processes* **2000**, *196*, 591–598.

71 L.K. Zhang, D. Rempel, B.N. Pramanik, M.L. Gross. Accurate mass measurements by Fourier transform mass spectrometry. *Mass Spectrom. Rev.* **2005**, *24*, 286–309.

72 P.B. O'Connor, M.C. Duursma, G.J. Rooij, R.M.A. van Heeren, J.J. Boon. Correction of time of flight shifted polymeric molecular weight distributions in MALDI-FTMS. *Anal. Chem.* **1997**, *69*, 2751–2755.

73 D. Gerlich. In: Ng, C.-Y., Baer, M. (Eds.), *State-selected and state-to-state ion-molecule reaction dynamics*. John Wiley & Sons, Inc., New York, **1992**, Vol. LXXXII, pp. 1–176.

74 N.K. Kaiser, S. Wu, K. Zhang, D.C. Prior, M.A. Bushback, G.A. Anderson, J.E. Bruce. Development of a novel ion guide for improved ion transmission in FTICR mass spectrometry In: *53rd American Society for Mass Spectrometry and Allied Topics*: San Antonio, Texas, **2005**.

75 P. Caravatti. US Patent no. 4,924,089, issued 8 May, **1990**.

76 S.D.H. Shi, J.J. Drader, C.L. Hendrickson, A.G. Marshall. Fourier transform ion cyclotron resonance mass spectrometry in a high homogeneity 25 Tesla resistive magnet. *J. Am. Soc. Mass Spectrom.* **1999**, *10*, 265–268.

77 S. Guan, M.C. Wahl, A.G. Marshall. Elimination of frequency drift from FTICR mass spectra by digital quadrature heterodyning: Ultrahigh mass resolving power for laser-desorbed molecules. *Anal. Chem.* **1993**, *65*, 3647–3653.

78 M.L. Easterling, T.H. Mize, I.J. Amster. Routine part-per-million mass accuracy for high-mass ions: Space-charge effects in MALDI FTICR. *Anal. Chem.* **1999**, *71*, 624–632.

79 E.V. Nikolaev, N.V. Miluchihin, M. Inoue. Evolution of an ion cloud in a FTICRMS during signal detection: Its influence on spectral line shape and position. *Int. J.*

Mass Spectrom. Ion Processes **1995**, *148*, 145–157.

80 M.E. Belov, R. Zhang, E.F. Strittmatter, D.C. Prior, K. Tang, R.D. Smith. Automated gain control and internal calibration with external ion accumulation capillary liquid chromatography-electrospray ionization-Fourier transform ion cyclotron resonance. *Anal. Chem.* **2003**, *75*, 4195–4205.

81 B.O. Keller, L. Li. Detection of 25,0000 molecules of substance p by MALDI-TOF mass spectrometry and investigations into the fundamental limits of detection in MALDI. *J. Am. Soc. Mass Spectrom.* **2001**, *12*, 1055–1063.

82 S.C. Moyer, B.A. Budnik, J.L. Pittman, C.E. Costello, P.B. O'Connor. Attomole peptide analysis by high-pressure matrix-assisted laser desorption/ionization Fourier transform mass spectrometry. *Anal. Chem.* **2003**, *75*, 6449–6454.

83 P.B. O'Connor, C.A. Costello, W.E. Earle. A high voltage rf oscillator for driving multipole ion guides. *J. Am. Soc. Mass Spectrom.* **2002**, *13*, 1370–1375.

84 A.J. Jaber, J. Kaufman, R. Liyanage, E. Akhmetova, S. Marney, C.L. Wilkins. Trapping of wide range mass-to-charge ions and dependence on matrix amount in internal source maldi-ftms. *J. Am. Soc. Mass Spectrom.* **2005**, *16*, 1772–1780.

85 A.G. Marshall, T.-C.L. Wang, T.L. Ricca. Tailored excitation for fourier transform ion cyclotron resonance mass spectrometry. *J. Am. Chem. Soc.* **1985**, *107*, 7893–7897.

86 S.A. Lee, C.Q. Jiao, Y. Huang, B.S. Freiser. Multiple excitation collisional activation in FTMS. *Rapid Commun. Mass Spectrom.* **1993**, *7*, 819–821.

87 K.A. Boering, J. Rolfe, J.I. Brauman. Control of ion kinetic energy in ion cyclotron resonance spectrometry: Very-low-energy collision-induced dissociation. *Rapid Commun. Mass Spectrom.* **1992**, *6*, 303–305.

88 J.W. Gauthier, T.R. Trautman, D.B. Jacobson. Sustained off-resonance irradiation for CAD involving FTMS. CAD technique that emulates infrared multiphoton dissociation. *Anal. Chim. Acta* **1991**, *246*, 211–225.

89 E. Mirgorodskaya, P.B. O'Connor, C.E. Costello. A general method for precalculation of parameters for sustained off resonance irradiation/collision-induced dissociation. *J. Am. Soc. Mass Spectrom.* **2002**, *13*, 318–324.

90 D.P. Little, J.P. Speir, M.W. Senko, P.B. O'Connor, F.W. McLafferty. Infrared multiphoton dissociation of large multiply-charged ions for biomolecule sequencing. *Anal. Chem.* **1994**, *66*, 2809–2815.

91 E.R. Williams, F.W. McLafferty. 193-nm laser photoionization and photodissociation for isomer differentiation in Fourier-transform mass-spectrometry. *J. Am. Soc. Mass Spectrom.* **1990**, *1*, 361–365.

92 R.A. Zubarev, N.L. Kelleher, F.W. McLafferty. Electron capture dissociation of multiply charged protein cations – a nonergodic process. *J. Am. Chem. Soc.* **1998**, *120*, 3265–3266.

93 R.A. Zubarev, N.A. Kruger, E.K. Fridriksson, M.A. Lewis, D.M. Horn, B.K. Carpenter, F.W. McLafferty. Electron capture dissociation of gaseous multiply-charged proteins is favored at disulfide bonds and other sites of high hydrogen atom affinity. *J. Am. Chem. Soc.* **1999**, *121*, 2857–2862.

94 J.J. Coon, J.E.P. Syka, J.C. Schwartz, J. Shabanowitz, D.F. Hunt. Anion dependence in the partitioning between proton and electron transfer in ion/ion reactions. *Int. J. Mass Spectrom.* **2004**, *236*, 33–42.

95 J.E.P. Syka, J.J. Coon, M.J. Schroeder, J. Shabanowitz, D.F. Hunt. Peptide and protein sequence analysis by electron transfer dissociation mass spectrometry. *Proc. Natl. Acad. Sci. USA* **2004**, *101*, 9528–9533.

96 T.J. Garrett, R.A. Yost. Analysis of intact tissue by intermediate-pressure MALDI on a linear ion trap mass spectrometer. *Anal. Chem.* **2006**, *78*, 2465–2469.

97 J.C. Schwartz, J.E.P. Syka, S.T. Quarmby. Improving the fundamentals of msn on 2d linear ion traps: New ion activation and isolation techniques. In: *53rd American Society for Mass Spectrometry*

and Allied Topics: San Antonio, Texas, **2005**.

98 T. Baba, Y. Hashimoto, H. Hasegawa, A. Hirabayashi, I. Waki. Electron capture dissociation in a radio frequency ion trap. *Anal. Chem.* **2004**, *76*, 4263–4266.

99 O.A. Silivra, I.A. Ivonin, F. Kjeldsen, R.A. Zubarev. Ion-electron reactions in a quadrupole ion trap: Realization, first results and prospects. In: *52nd ASMS Conference on Mass Spectrometry*: Nashville, TN, **2004**.

100 T.Y. Kim, M.S. Thompson, J.P. Reilly. Peptide photodissociation at 157 nm in a linear ion trap mass spectrometer. *Rapid Commun. Mass Spectrom.* **2005**, *19*, 1657–1665.

101 A.H. Payne, G.L. Glish. Thermally assisted infrared multiphoton photodissociation in a quadrupole ion trap. *Anal. Chem.* **2001**, *73*, 3542–3548.

102 W.D. Cui, B. Hadas, B.P. Cao, C. Lifshitz. Time-resolved photodissociation (TRPD) of the naphthalene and azulene cations in an ion trap/reflectron. *J. Phys. Chem. A* **2000**, *104*, 7160–7160.

103 C.S. Creaser, K.E. O'Neill. Photodissociation and collisionally activated dissociation tandem mass spectrometric studies of difluoro[triazol-1-ylmethyl]benzhydrols and related compounds in a quadruple ion trap. *Int. J. Mass Spectrom. Ion Processes* **1997**, *165*, 13–23.

104 A. Colorado, J.X.X. Shen, V.H. Vartanian, J. Brodbelt. Use of infrared multiphoton photodissociation with swift for electrospray ionization and laser desorption applications in a quadrupole ion trap mass spectrometer. *Anal. Chem.* **1996**, *68*, 4033–4043.

105 M. Welling, R.I. Thompson, H. Walther. Photodissociation of MgC60+ complexes generated and stored in a linear ion trap. *Chem. Phys. Lett.* **1996**, *253*, 37–42.

106 J.L. Stephenson, M.M. Booth, S.M. Boue, J.R. Eyler, R.A. Yost. In: *Biochemical and biotechnological applications of electrospray ionization mass spectrometry*, **1996**, Vol. 619, pp. 512–564.

107 J.D. Williams, R.G. Cooks, J.E.P. Syka, P.H. Hemberger, N.S. Nogar. Determination of positions, velocities, and kinetic energies of resonantly excited ions in the quadrupole ion-trap mass-spectrometer by laser photodissociation. *J. Am. Soc. Mass Spectrom.* **1993**, *4*, 792–797.

108 J.N. Louris, J.S. Brodbelt, R.G. Cooks. Photodissociation in a quadrupole ion trap mass-spectrometer using a fiber optic interface. *Int. J. Mass Spectrom. Ion Processes* **1987**, *75*, 345–352.

109 G. Baykut, J. Fuchser, M. Witt, G. Weiss, C. Gostell. A combined ion source for fast switching between electrospray and matrix-assisted laser desorption/ionization in Fourier transform ion cyclotron resonance mass spectrometry. *Rapid Commun. Mass Spectrom.* **2002**, *16*, 1631–1641.

110 C.W. Diehnelt, S.M. Peterman, W.L. Budde. Liquid chromatography-tandem mass spectrometry and accurate m/z measurements of cyclic peptide cyanobacteria toxins. *Trends Anal. Chem.* **2005**, *24*, 622–635.

111 A. Makarov. Electrostatic axially harmonic orbital trapping: A high-performance technique of mass analysis. *Anal. Chem.* **2000**, *72*, 1156–1162.

112 J.V. Olsen, L.M.F. de Godoy, G.Q. Li, B. Macek, P. Mortensen, R. Pesch, A. Makarov, O. Lange, S. Horning, M. Mann. Parts per million mass accuracy on an orbitrap mass spectrometer via lock mass injection into a c-trap. *Mol. Cell. Proteomics* **2005**, *4*, 2010–2021.

113 M. Thevis, A.A. Makarov, S. Horning, W. Schanzer. Mass spectrometry of stanozolol and its analogues using electrospray ionization and collision-induced dissociation with quadrupole-linear ion trap and linear ion trap-orbitrap hybrid mass analyzers. *Rapid Commun. Mass Spectrom.* **2005**, *19*, 3369–3378.

114 Q.Z. Hu, R.J. Noll, H.Y. Li, A. Makarov, M. Hardman, R.G. Cooks. The orbitrap: A new mass spectrometer. *J. Mass Spectrom.* **2005**, *40*, 430–443.

115 M. Hardman, A.A. Makarov. Interfacing the orbitrap mass analyzer to an electrospray ion source. *Anal. Chem.* **2003**, *75*, 1699–1705.

116 K. Dreisewerd, S. Kolbl, J. Peter-Katalinic, S. Berkenkamp, G. Pohlentz.

Analysis of native milk oligosaccharides directly from thin-layer chromatography plates by matrix-assisted laser desorption/ionization orthogonal-time-of-flight mass spectrometry with a glycerol matrix. *J. Am. Soc. Mass Spectrom.* **2006**, *17*, 139–150.

117 K. Dreisewerd, J. Muthing, A. Rohlfing, I. Meisen, Z. Vukelic, J. Peter-Katalinic, F. Hillenkamp, S. Berkenkamp. Analysis of gangliosides directly from thin-layer chromatography plates by infrared matrix-assisted laser desorption/ionization orthogonal time-of-flight mass spectrometry with a glycerol matrix. *Anal. Chem.* **2005**, *77*, 4098–4107.

118 J. Guittard, X.P.L. Hronowski, C.E. Costello. Direct matrix-assisted laser desorption/ionization mass spectrometric analysis of glycosphingolipids on thin layer chromatographic plates and transfer membranes. *Rapid Commun. Mass Spectrom.* **1999**, *13*, 1838–1849.

119 A.I. Gusev, A. Proctor, Y.I. Rabinovich, D.M. Hercules. Thin-layer chromatography combined with matrix-assisted laser desorption ionization mass spectrometry. *Anal. Chem.* **1995**, *67*, 1805–1814.

120 A.I. Gusev, O.J. Vasseur, A. Proctor, A.G. Sharkey, D.M. Hercules. Imaging of thin-layer chromatograms using matrix/assisted laser desorption/ionization mass spectrometry. *Anal. Chem.* **1995**, *67*, 4565–4570.

121 K. Dreisewerd, S. Berkenkamp, A. Leisner, A. Rohlfing, C. Menzel. Fundamentals of matrix-assisted laser desorption/ionization mass spectrometry with pulsed infrared lasers. *Int. J. Mass Spectrom. Ion Processes* **2003**, *226*, 189–209.

122 C. Menzel, K. Dreisewerd, S. Berkenkamp, F. Hillenkamp. The role of the laser pulse duration in infrared matrix-assisted laser desorption/ionization mass spectrometry. *J. Am. Soc. Mass Spectrom.* **2002**, *13*, 975–984.

123 C. Menzel, K. Dreisewerd, S. Berkenkamp, F. Hillenkamp. Mechanisms of energy deposition in infrared matrix-assisted laser desorption/ionization mass spectrometry. *Int. J. Mass Spectrom. Ion Processes* **2001**, *207*, 73–96.

124 C. Menzel, S. Berkenkamp, F. Hillenkamp. Infrared matrix-assisted laser desorption/ionization mass spectrometry with a transversely excited atmospheric pressure carbon dioxide laser at 10.6 µm wavelength with static and delayed ion extraction. *Rapid Commun. Mass Spectrom.* **1999**, *13*, 26–32.

125 S. Berkenkamp, C. Menzel, M. Karas, F. Hillenkamp. Performance of infrared matrix-assisted laser desorption/ionization mass spectrometry with lasers emitting in the 3 µm wavelength range. *Rapid Commun. Mass Spectrom.* **1997**, *11*, 1399–1406.

126 S. Berkenkamp, M. Karas, F. Hillenkamp. Ice as a matrix for IR-matrix-assisted laser desorption/ionization: Mass spectra from a protein single crystal. *Proc. Natl. Acad. Sci. USA* **1996**, *93*, 7003–7007.

127 J.W. Cooley, J.W. Tukey. Fast Fourier transform algorithm. *Math. Comput.* **1965**, *19*, 297–301.

128 A. Holle, A. Haase, M. Kayser, J. Höhndorf. Optimizing UV-laser focus profiles for improved MALDI performance. *J. Mass Spectrom.* **2006**, *41*, 705–716.

3
MALDI-MS in Protein Chemistry and Proteomics

Karin Hjernø and Ole N. Jensen

3.1
Introduction

Proteins are a major constituent of living cells, and mediate the majority of biological processes in all organisms, from microbes to mammals. During the past
five decades, research into protein biochemistry and molecular cell biology has led
to the elucidation of the functions and structures of many proteins. However, the
overall architecture and dynamics of subcellular structures and the mechanisms
of molecular signaling networks remain to be established. Proteins have been –
and continue to be – explored for use in biotechnology, pharmacology, and biomedical applications. Consequently, a range of protein-based drugs are now
available commercially, and many proteins are targets for small-molecule
drugs. Recombinant enzymes are widely used for catalysis in bioprocessing, ranging from the production of food and food ingredients to the manufacture of
textiles and the creation of biofuels. Thus, protein research is at center stage
in many cell biology and biotechnology laboratories, both in academia and in
industry.

Within the cell, proteins are synthesized in a multi-step process which includes
the transcription of DNA into RNA, the processing of RNA into mature mRNA,
and finally translation of the mRNA into protein. Each of these steps is prone
to molecular events that lead to alterations of the protein product (Fig. 3.1). For
example, the gene may contain base substitutions and mutations that eventually
alter the amino acid sequence of the protein. Likewise, the processing and matu-
ration of RNA may lead to the elimination of distinct exons, thereby creating vari-
ant gene products, the so-called splice variants. Both co- and post-translational
modification of the polypeptide backbone lead to further diversity and heterogeneity of the gene products – that is, the proteins. Many proteins are substrates
for dynamic reversible modifications (e.g., phosphorylation and acetylation) that
regulate their biological activity and interactions, depending upon environmental
cues and the metabolic status of the cell. Thus, the elucidation of protein primary
structure and post-translational modifications is crucial to studies of biological
processes.

MALDI MS. A Practical Guide to Instrumentation, Methods and Applications.
Edited by Franz Hillenkamp and Jasna Peter-Katalinić
Copyright © 2007 Wiley-VCH Verlag GmbH & Co. KGaA, Weinheim
ISBN: 978-3-527-31440-9

DNA	➡	RNA	➡	Protein	➡	Analytical artifacts
Allelic variation *Mutation* *Damage* *Polymorphisms*		*Processing* *Splicing*		*Co‑translational modification* *Post‑translational modification* *Processing*		*Chemical modification* *Oxidation* *Degradation* *Cross‑linking* *Precipitation*

Fig. 3.1 A range of events may influence protein structure and integrity. Molecular events at the DNA and RNA levels generate heterogeneity at the protein level. Co- and post-translational modifications and processing leads to heterogeneity of proteins. Analytical artifacts during protein extraction, purification and analysis produces further heterogeneity. Mass spectrometry can detect and locate chemical changes to proteins via determination of the accurate molecular mass of protein and the derived peptides.

Modern analytical technologies are applied to elucidate the intricate protein machinery of the cell and to discover and develop new protein-based biotechnology products. Proteomic studies – that is, the systematic analysis of proteins from cells, tissues or whole organisms – calls for highly efficient and sensitive analytical methods for the identification, characterization, and quantification of proteins. Today, mass spectrometry (MS) is the most useful analytical technique for protein and proteome analysis, as it provides a relatively simple platform for determining one of the fundamental properties of biological molecules, namely the molecular mass. In this respect several strategies can be taken, depending on the complexity of the sample and the information sought. The basic steps in such MS-based strategies for protein analysis are summarized in Figure 3.2, and include: (i) the biochemical characterization of purified proteins; and (ii) two parallel strategies for the analysis of proteomes or sub-proteomes – that is, the multitude of proteins within a given cell or cell organelle at any one time.

The first approach (Fig. 3.2A) is very useful for the validation of native or recombinant proteins destined for structure analysis by nuclear magnetic resonance (NMR) or X-ray diffraction, as well as in the quality control of proteins for biotechnological applications. The method is based on an ability to obtain pure protein in soluble form, and in a solvent that is suitable for matrix-assisted laser desorption/ionization (MALDI) MS analysis. In most cases, the identity of the protein is known *a priori*, and the amino acid sequence is available either from protein databases or by translation from the corresponding gene sequences. The mass determination of a purified intact protein will reveal any discrepancies from the molecular mass of a "naked" protein – that is, the calculated molecular mass as determined by the amino acid sequence predicted from the cognate gene. Similarly, MS analysis of peptides derived from a protein by enzymatic digestion will not only confirm the amino acid composition of the individual peptides but also reveal the presence of any chemical modifications by a mass increment or a mass deficit relative to the expected masses of the unmodified peptides. This type of

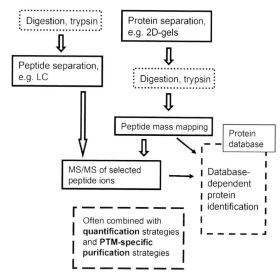

A Protein biochemistry
(known, intact purified protein)

B Two proteomics strategies
(unknown, complex protein sample)

Fig. 3.2 MALDI-MS is a very powerful tool in many kinds of analysis involving proteins, from in-depth analysis of individual proteins to studies of large complex protein samples, as in proteomic studies. Here, the outlines of two different strategies are shown: the analysis of a purified protein for which the sequence is known (A), and the analysis of all proteins in, for example, a given cell (i.e., the proteome of the cell) (B). Both the LC- and gel-based strategies are given, and these can be combined with quantification strategies and studies of modified proteins. As can be seen, many of the steps involved, such as protein digestion, peptide mass mapping and peptide fragmentation, are identical and used in all strategies shown.

analysis is referred to as "peptide mass mapping". When used for proteomics analysis and the analysis of large protein complexes (Fig. 3.2B), MS must be combined with protein or peptide separation strategies, in order to deal with the high complexity of such samples. These are typically one- or two-dimensional (1- or 2D) gel electrophoresis or liquid chromatography (LC) approaches. The individual proteins/peptides are then identified by searching of the obtained MS-data against *in-silico*-digested proteins from databases of known or predicted sequences. Correlations between the experimental and theoretical data are used for scoring of the proteins and thereby determining the identity of the proteins in the sample being analyzed. Further details on the different approaches available to achieve such identity are provided later in the chapter.

Before being analyzed by MS, soluble biomolecules must be converted into gas-phase ions. Today, MALDI and electrospray ionization (ESI) are the two main techniques used to produce peptide and protein ions for MS analysis. ESI mainly produces multiply protonated peptide ions $[M + nH]^{n+}$, whereas MALDI generates mainly singly protonated peptide ions [M + H⁺], although multiply charged species are sometimes observed. Proteins may generate both singly and multiply charged ions in MALDI.

Tandem mass spectrometry (MS/MS) is very useful for amino acid sequencing of peptides, and has been used widely in both protein biochemistry and proteomics to identify proteins, to deduce the sequence of a peptide, and to detect and locate post-translational modifications. Until a few years ago, the concept of amino acid sequencing by MS-technologies was synonymous with ESI-MS/MS, but today MALDI-MS/MS techniques are implemented in high-performance instruments such that the quality of MALDI tandem mass spectra is comparable with that of ESI-MS/MS spectra. Currently, MALDI tandem mass spectrometers exist in a number of geometries, including TOF-TOF, Q-TOF, and ion trap analyzers that each provide unique analytical features for the sequencing of peptides and proteins by MS/MS (details of the instrumentation for different types of MS/MS are provided in Chapter 2).

Here, we will describe a range of applications of MALDI-MS, from the concepts of in-depth analysis of purified proteins to applications of MALDI-MS in a broader, proteomics-based research where proteins are identified, characterized, and quantified. In addition, issues of sample preparation, protein characterization and identification strategies and bioinformatic tools for data interpretation will be discussed. The concepts of peptide fragmentation, sequencing and derivatization, analysis of post-translational modifications and the clinical applications of MALDI-MS are also briefly outlined.

3.2
Sample Preparation for Protein and Peptide Analysis by MALDI-MS

In this section, we describe some of the general sample preparation issues that influence protein analysis by MALDI-MS experiments.

Often, the intact proteins need to be purified to near homogeneity by biochemical or immunological methods, for example chromatography, affinity purification or immunoprecipitation prior to MS analysis. Volatile buffers, such as ammonium bicarbonate and ammonium acetate should be used for the final stages of purification of proteins, if possible. The pH should be adjusted to <3 by addition of trifluoroacetic acid (TFA) prior to mixing the protein sample with matrix for MALDI-MS analysis. This is done in order to obtain a good crystallization of matrix and analyte. It is important to avoid ionic detergents, such as sodium dodecyl sulfate (SODS) and Triton, which will interfere with the formation of analyte-matrix crystals during sample preparation. Low levels of non-ionic detergents, such as *N*-octyl-glucoside, can be tolerated (Katayama et al., 2001; Zhang and Li, 2004).

The analysis of proteins by MALDI-MS peptide mass mapping involves proteolytic degradation of proteins into peptides. Prior to such proteolytic digestion it is advisable to reduce and S-alkylate the cysteine residues to reduce S–S bridges and the resulting secondary structure. This blocks their chemical reactivity, increases the accessibility to digesting proteases, and improves the detection efficiency of Cys-containing peptides in MALDI-MS. Iodoacetamide/iodoacetate and 4-vinylpyridine are good S-alkylating reagents. If the proteins are not reduced and alkylated prior to separation on acrylamide-containing gels, one should consider that the cysteines might react with unpolymerized acrylamide, resulting in propionamide-modified peptides (Jensen et al., 1998).

Trypsin is often used as the proteolytic reagent, as it is a highly active and specific protease, cleaving C-terminal to Lys and Arg residues to generate peptides that fall within the mass range of 500 to 5000 Da, which is suitable for peptide mass mapping. The endoproteases Lys-C, Asp-N and Glu-C are also highly useful enzymes. The latter two proteases generate fragments that often complement those generated by trypsin. Less-specific proteases are chymotrypsin, proteinase K, and subtilisin, which are helpful for proteolytic cleavage of very compact or stable protein structures. Cyanogen bromide is a useful chemical cleavage reagent that cleaves at Met residues; however, it should be used with care because of its toxicity. For gel-separated proteins, the digestion is traditionally performed while still in the gel (*in-gel* digestion) and subsequently extracted into the surrounding solvent, either spontaneously or by extraction procedure using acetonitrile and TFA (Shevchenko et al., 1996).

It is advisable to desalt and concentrate protein and peptide samples using miniaturized solid-phase extraction methods prior to MALDI-MS analysis (Fig. 3.3),

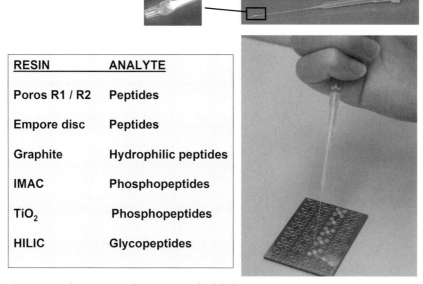

RESIN	ANALYTE
Poros R1 / R2	Peptides
Empore disc	Peptides
Graphite	Hydrophilic peptides
IMAC	Phosphopeptides
TiO$_2$	Phosphopeptides
HILIC	Glycopeptides

Fig. 3.3 Sample preparation by miniaturized solid-phase extraction or affinity enrichment. The resin is used to trap the analyte and serves to concentrate and desalt the sample prior to MALDI-MS analysis. IMAC: Immobilized metal affinity chromato-graphy; HILIC: Hydrophilic interaction chromato-graphy.

Table 3.1 MALDI-MS matrices and methods for the analysis of purified proteins.

Analyte	Matrix	Solvent/additives	Comment
Soluble proteins	SA HCCA DHB	0.1% TFA, 50% ACN	Dried droplet method or sandwich methods, rinsing
Hydrophobic proteins	HCCA HCCA	1–30% formic acid Ethyl acetate	Liquid-liquid extraction method. Hydrophobic species
Glycoproteins	DHB THAP DHAP	0.1% TFA, 30% ACN Diammonium acetate	
Protein mixtures	HCCA DHAP	0.1% TFA, 30% ACN Diammonium acetate	

ACN: acetonitrile; DHAP: 2,6-dihydroxyacetophenone; DHB: 2,5-dihydroxybenzoic acid; HCCA: alpha-cyano-4-hydroxycinnamic acid; SA: Sinapinic acid; THAP: 2,4,6-Trihydroxyacetophenone.

especially if only small amount of material are available. Miniaturized solid-phase extraction columns can be custom-made using Eppendorff gelloader tips packed with resin (Gobom et al., 1999; Rappsilber et al., 2003) or purchased from various vendors. The selectivity of the method can be customized for various applications: Poros R1/R2 resin material is useful for analysis of protein and peptide samples. Phosphopeptides are recovered by IMAC or TiO_2 (Stensballe et al., 2001; Stensballe and Jensen, 2004; Larsen et al., 2005b), whereas glycopeptides are retained by HILIC and graphite (Hagglund et al., 2004; Larsen et al., 2005a) see Section 3.3.3. In most cases, the sample is eluted using MALDI matrix solution and deposited directly onto the MALDI probe. This increases overall sensitivity of the MALDI-MS experiment. Functionalized MALDI-MS probes that facilitate sample preconcentration are also available from various manufacturers (Schuerenberg et al., 2000).

For MALDI-MS analysis of intact proteins HCCA (alpha-cyano-4-hydroxycinnamic acid), SA (sinapinic acid) or DHB (2,5-dihydroxybenzoic acid) matrices and the "dried droplet" deposition method for sample preparation are typically used (Strupat et al., 1991; Beavis and Chait, 1996) (Table 3.1). Depending on the properties of the protein, it is often necessary to test a series of solvents and matrices to optimize the outcome of the MALDI-MS experiment. Peptides and small proteins below molecular weight 20000 Da are often amenable to analysis using HCCA matrix and reflector TOF-MS mode. Larger proteins may produce better results with SA or DHB matrix in the linear TOF-MS mode. Hydrophobic proteins can be analyzed using HCCA matrix dissolved in high concentrations of formic acid (up to 30%) (Cohen and Chait, 1996). When using cinnamic acid

Table 3.2 Matrices and sample preparation methods for peptide analysis by MALDI-MS.

Analyte	Matrix	Solvent/additives	Comment
Peptides	HCCA	0.1% TFA, 70% ACN	Dried droplet method
		Acetone : water : nitrocellulose	Fast evaporation method
	DHB	0.1% TFA, 50% ACN	
			Dried droplet method
Phosphopeptides	DHB	0.1–0.5% PA, 50% ACN	Dried droplet method
	DHAP	Diammonium acetate	
Glycopeptides	DHB	0.1% TFA, 50% ACN	Dried droplet method
	sDHB		
	THAP	Diammonium citrate	
	DHAP		
Acylated peptides	HCCA	Ethyl acetate	Liquid-liquid extraction
Sulfated peptides	DHAP	Diammonium citrate	Negative ion mode

For abbreviations, see Table 3.1.

sDHB: superDHB (DHB and 2-hydroxy-5-methoxybenzoic acid at 10% v/v; Tsarbopoulos et al., 1994).

matrices, SA and HCCA, and the dried-droplet method for deposition, then it is often beneficial to perform subsequent rinsing of the dried matrix/analyte deposit using cold water, 0.1% TFA or volatile solvents/buffers (Beavis and Chait, 1990; Smirnov et al., 2004).

If the dried-droplet method does not produce good results, then various other sample deposition techniques should be attempted, including various "sandwich" methods (Kussmann and Roepstorff, 2000; Zhang and Li, 2004) and the fast evaporation method, also referred to as the "thin-layer" method (Vorm et al., 1994).

Matrix mixtures have also been used for improved MALDI-MS performance for both proteins and peptides. For example, a mixture of DHB and HCCA was found to improve spot-to-spot reproducibility for peptide mass mapping and to increase the protein sequence coverage (Laugesen and Roepstorff, 2003). For examples of other matrices useful for peptide analysis, see Table 3.2. Membrane proteins can be efficiently analyzed using optimized sample preparation methods and detergents, as described by Cadene and Chait (2000). Also, *in situ* liquid-liquid extraction using ethyl acetate and HCCA may enable analysis of integral membrane proteins by MALDI-MS (Kjellstrom and Jensen, 2003).

3.3
Strategies for Using MALDI-MS in Protein Biochemistry

In this section, we describe approaches to analyze purified proteins by using MALDI-MS in more details (see Fig. 3.2A). Using the methods described here it is also possible to analyze proteins for which the amino acid sequence is unknown or is only partially characterized. The MS analysis will then help to elucidate the complete primary structure of the proteins; the process can be very troublesome and time-consuming, however, and should be combined with other technologies such as cloning.

Depending on the size and properties of the protein, 2 to 50 pmol of purified protein is needed for complete and comprehensive molecular analysis by multiple MALDI-MS experiments, including intact mass determination, several peptide mass mapping experiments and MS/MS analysis of individual peptides (see below).

Intact mass determination is a very useful first step towards full characterization of proteins (Fig. 3.4), and the purified protein should therefore be analyzed by MALDI-MS for determination of the molecular mass. Gel electrophoresis will

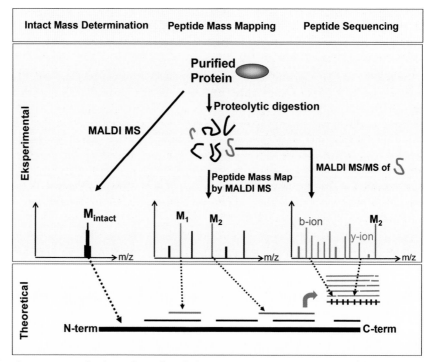

Fig. 3.4 Strategy for the analysis of purified proteins by MALDI-MS. The initial stage is intact mass determination by MALDI-MS, followed by proteolytic cleavage using one or more sequence-specific enzymes. The resulting peptide mix is analyzed by MALDI-MS. Finally, the individual peptides can be subjected to amino acid sequencing by MALDI-MS/MS. Comparison of the experimental and calculated mass values will reveal discrepancies that might originate from natural processing of the protein or from chemical artifacts. The theoretical molecular mass of a protein and the derived peptides is calculated from the amino acid sequence using computer tools.

often be used as the final stage of protein purification, depending on the purification procedure used. If not, it is advantageous to monitor the protein purification by SDS-PAGE to establish the purity and integrity of the protein sample and to estimate protein abundance. The intact proteins can sometimes be successfully extracted from SDS-PAGE gels or 2D electrophoresis gels by electroelution (Haebel et al., 1995) or by detergent-mediated passive elution (Cohen and Chait, 1997). However, this requires significant amounts of purified protein, often more than 20 pmol, in the gel. Alternatively, the proteins are analyzed directly from solution. Usually, 1–10 pmol of protein loaded onto the MALDI probe will ensure a good signal, but this will depend on protein size and sample solubility/complexity and solvent composition.

3.3.1
Peptide Mass Mapping of Purified Proteins

The next stage in the analysis includes proteolytic cleavage of the protein to generate peptides that are amenable to MALDI-MS analysis. MALDI-MS is conveniently used for the initial analysis of the crude peptide mixture in order to "read out" the generated peptide fragments and for analysis of individual peptides after their separation and isolation by LC.

The cleavage can be performed either *in-solution* or *in-gel* (Shevchenko et al., 1996). Knowledge of the cleavage specificity allows prediction of the peptide products from the original amino acid sequence of the protein, provided that its sequence is known a priori. The peptide mass map will help to validate the primary structure by experimental observation of peptides that were predicted based on knowledge of the amino acid sequence of the protein and enzyme cleavage specificity. The peptide mass map produced by specific cleavage reagents or enzymes is unique for the protein and its post-translational modifications. It is often useful to perform two or three different proteolytic digestion experiments of a protein by using proteases with complementary cleavage specificity, for example trypsin and GluC. The peptide mass maps obtained from the different experiments will provide information on complementary regions of the protein and generate partially overlapping coverage of its amino acid sequence.

In most cases it is necessary to separate peptides by HPLC prior to MALDI-MS in order to detect and characterize all components in a mixture. In theory, all peptides should be present in equal concentration in a protease digest. However, this is not reflected in MALDI spectra for several reasons. First, the ionization efficiency of different peptides can vary over several orders of magnitude, and certain peptides are not detected when present in a peptide mix. The reasons for this ionization efficiency bias of peptides (suppression effects) remain to be established. Arginine-containing peptides are found to ionize very well as compared to arginine-deficient peptides (Krause et al., 1999). Partial covalent modification of peptides leads to heterogeneity, where the peptide exists in several different versions having different concentrations. Minor components are often not detected. The detection efficiency of a peptide as a single component in a preparation compared to being a component in a mixture varies dramatically, and separation of peptides is therefore a necessity for in-depth analysis of proteins.

3.3.2
Peptide Sequencing by MALDI-MS/MS

Traditionally, proteins were initially characterized by *de-novo* sequencing using automated Edman degradation and amino acid composition analysis. Today, these techniques tend to be replaced by MS, which not only provides more flexibility and sensitivity but is also amenable to the analysis of protein and peptide mixtures. Tandem mass spectrometry (MS/MS) is used for amino acid sequencing of peptides. MALDI-MS/MS is very powerful for peptide characterization and identification via sequencing and sequence database searching.

As mentioned previously, MALDI produces mainly singly protonated peptide ions $[M + H]^+$. Upon low-energy collision-induced dissociation (CID) of the peptides in a MALDI-MS/MS instrument, the dominating ions-series will be a, b and y fragments (according to the Roepstorff/Fohlman nomenclature; Roepstorff and Fohlman, 1984) (Fig. 3.5). Diagnostic immonium ions can be observed in the low-mass region; the mass of these indicates from which amino acid residues they arise, and hence an indication is also provided of which residues the peptide is composed. These are assigned by the 1-letter code of the residue from which they originate (e.g., H and P). Moreover, fragments arising from CID involving the residue side chains can be observed when high-energy collision experiments are performed. These are denoted w-, v- and d-ions (not shown) (Johnson et al., 1988).

In contrast to ESI, many MALDI peptide ions produce rather few fragment ions upon CID. Intense ion signals are often observed, however, due to gas-phase cleavage at acidic amino acid residues (Qin and Chait, 1995). These carboxylic group-induced fragmentation pathways are called charge-remote fragmentation

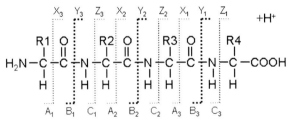

Fig. 3.5 Nomenclature for peptide sequencing. Fragments resulting from backbone cleavages are mainly observed when MALDI peptide ions are fragmented in low-energy collision-induced dissociation (CID) tandem mass spectrometers. Of these, the cleavage at the amide bond will be most efficiently cleaved, resulting in y- and b-ions, depending on which part of the peptide will retain the charge. If the charge is retained on the C-terminal fragment, the ions are called y_n-ions, where n is the number of residues in the fragment. When the charge is on the N-terminal part, the ions are called b_n-ions. The mass difference of y_n and y_{n+1} corresponds to the mass of the amino acid residues and can therefore be used to deduce the peptide sequence.

pathways, as they do not directly involve the positive charge of the peptide. This is in contrast to charge-induced fragmentation pathways that rely on "mobile protons" (Wysocki et al., 2000; Wattenberg et al., 2002). The acid-induced fragments, which result from cleavage of the amide bond at the C-terminal side of Asp (and to a lesser extent Glu), are prominent in MALDI-MS/MS spectra of arginine-containing peptides; this highly basic residue sequesters the proton and makes it less available for charge-induced fragmentation such that acid-induced fragmentation dominates the spectra instead (Wattenberg et al., 2002). This also means that it is possible to make a simple prediction of which ions should dominate the spectra if an arginine-containing peptide with one or more internal acidic residues is analyzed (see example in Fig. 3.6). Such simple predictions cannot be performed for ESI-CID experiments as these generally involve mobile protons.

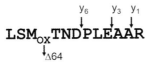

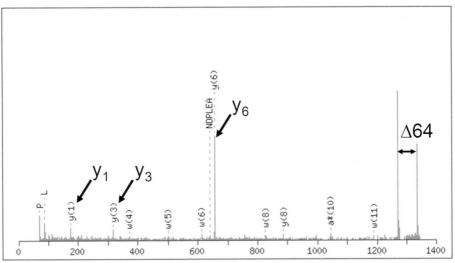

Fig. 3.6 Prediction of intense acid-induced fragments. Here, an example of an identification of the peptide LSM$_{ox}$TNDPLEAAR is given (the methionine is oxidized). The MALDI TOF/TOF spectrum is shown as annotated by the search engine MASCOT (see later). As cleavage C-terminal to acidic residues is expected to give rise to intense ions (especially C-terminal to aspartic acid), the y$_6$ and y$_3$ ions are predicted to be intense in the spectra. For peptides with arginines in the C-terminal end, the y$_1$ at 175 is also expected to be relatively intense. Furthermore, an oxidized methionine is expected to cause a neutral loss of 64 from the original peptide ion (Lagerwerf et al., 1996). The dominating ions signals can be explained by these predictions, as well as by small immonium ions (indicated by P and L in this case) and the signal from the original ion to the outer right of the spectra. Some of the weak signals (indicated by w) are assigned as side-chain fragments (Johnson et al., 1988). For a table of amino acid abbreviations, see the Appendix to this chapter.

Because of this sequestration of the proton to the basic sites in the peptides, MALDI often gives rise to poor MS/MS fragmentation efficiency of peptide ions. Therefore, the applicability of MALDI-MS/MS for *de-novo* sequencing of peptides was initially limited. This kindled interest in the development of alternative sample preparation methods for enhancing MALDI-MS/MS peptide sequencing performance. Today, several robust techniques for chemical derivatization of peptides are available, that in turn generate extensive peptide fragment ion series in MALDI-MS/MS. The CAF and SPITC reagents are used to introduce a sulfonic acid moiety at the N-terminus of peptides (Keough et al., 1999, 2003; Hellman and Bhikhabhai, 2002; Wang et al., 2004). The sulfonic acid group provides a mobile proton that can migrate to the polypeptide backbone and promote charge-induced cleavage of amide bonds. These types of reagent generate extensive y-ion series in MS/MS that often allow complete readout of the amino acid sequence of a peptide. At the same time, the b-ions are neutralized/"silenced", which reduces the complexity of the MS/MS spectra and thereby makes data interpretation straightforward. An example of SPITC derivatization of a peptide along with an example of a MS/MS analysis is shown in Figure 3.7.

3.3.3
Analysis of Post-Translational Modifications

Post-translational modification (PTM) of protein plays a pivotal role in dynamic biological processes. Protein conformation, interaction, sequestration and turnover is regulated by covalent processing of the polypeptide chain, for example by the addition of phosphate groups to specific serine or tyrosine residues (O-phosphorylation), or the conjugation of glycans to distinct asparagine residues (N-glycosylation) (see Chapter 6, Glycoconjugates). Other modifications include acetylation (Lys), palmitoylation (Cys), disulfide formation (Cys) and nitration (Tyr). More than 200 different PTMs are currently known. Such PTMs give rise to a mass increase of the amino acid residue and thereby to the digest peptide and the intact protein (Mann and Jensen, 2003; Jensen, 2006). Intact protein mass determination and subsequent peptide mass mapping by MALDI-MS is a convenient method for initial analysis of PTMs. MALDI-MS/MS analysis of the modified peptides is then used to identify the conjugated amino acids in individual peptides.

Several affinity-based sample preparation techniques are available for modification-specific enrichment of post-translationally modified peptides prior to MALDI-MS. Immobilized metal affinity chromatography (IMAC) and TiO_2 resin is useful for the analysis of phosphopeptides (Posewitz and Tempst, 1999; Stensballe et al., 2001; Pinkse et al., 2004; Larsen et al., 2005b). Hydrophilic interaction chromatography (HILIC) recovers glycopeptides, including N-linked glycans and glyco-phosphoinositol (GPI)-anchors (Hagglund et al., 2004; Omaetxebarria et al., 2006) (Fig. 3.8). Very hydrophilic species such as small phosphopeptide and glycans are retained by graphite powder and can be eluted with MALDI matrix solution (Larsen et al., 2004, 2005a).

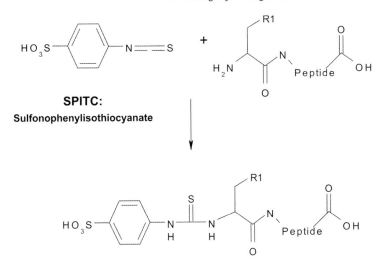

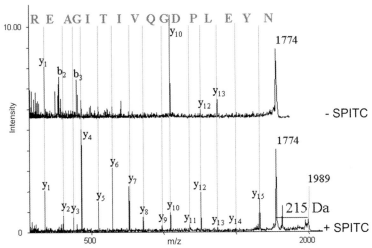

Fig. 3.7 Derivatization of peptides by sulfonic acid reagents such as sulfonophenylisothiocyanate (SPITC) enhances peptide fragmentation efficiency in MS/MS. MALDI TOF-TOF MS/MS of a singly protonated peptide NYELPDQGVITIGAER revealed only few fragment ions (top panel). In contrast, the same peptide derivatized by SPITC generated a strong y-ion series that allowed near-complete interpretation of the amino acid sequence (*de-novo* sequencing) (lower panel). The peptide sequence is shown above the spectra (written from the C-terminal to the N-terminal to reflect the y-ion series). In the upper spectrum the y_{10}-ion is dominating due to the presence of an acidic amino acid residue. The loss of SPITC is observed as a neutral loss of 215 Da.

Post-translationally modified proteins, such as glycoproteins and phosphoproteins, have the problem that they are prone to decomposition during the preparation and/or MALDI process and in the mass analyzer. Labile molecules should be analyzed using the DHB matrix (Tsarbopoulos et al., 1994) or the THAP or DHAP

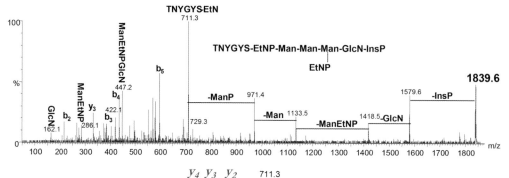

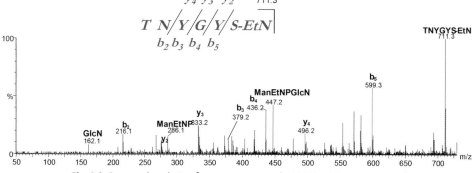

Fig. 3.8 Structural analysis of a post-translationally modified peptide by MALDI MS/MS. Porcine membrane dipeptidase, a GPI-anchored glycoprotein, was in-gel-digested by trypsin. The C-terminal GPI-anchored peptide was recovered from the peptide mixture by hydrophilic interaction chromatography (HILIC) and then analyzed by MALDI Q-TOF MS/MS. The MS/MS spectrum exhibits fragment ions that allow structural characterization of the carbohydrate structure of the GPI-anchor and amino acid sequencing of the modified peptide. Adapted from Omaetxebarria et al. (2006). For the glycan nomenclature, see Chapter 6.

matrices (Gorman et al., 1996) using the linear TOF-MS mode. Post-translationally modified peptides should be analyzed in both linear TOF mode and reflector TOF mode for observation of labile modifications (Annan and Carr, 1996) (differences between linear and reflectron TOFs in the detection of metastable ions are detailed in Chapter 2). It may also be useful to try negative and positive ion modes, as certain species ionize more efficiently in the negative ion mode (Xu et al., 2005).

Structural analysis of glycopeptides and glycans is conveniently performed by using glycosidase digestion combined with MALDI-MS. A range of specific exo- and endo-glycosidases is commercially available. Treatment of glycopeptides with various glycosidases will lead to the removal of specific sugar units by enzyme-specific cleavage of glycosidic bonds. MALDI-MS analysis of the products helps determine the type and number of glycan units in the glycopeptides and their chemical association (Mortz et al., 1996; Kuster et al., 1997; Harvey, 2005).

Similarly, phosphatases are used to detect and identify phosphopeptides in MALDI-MS. Phosphatase treatment leads to a mass decrease of 80 Da for phosphopeptides (Liao et al., 1994; Zhang et al., 1998; Larsen et al., 2001; Stensballe et al., 2001).

3.4
Applications of MALDI-MS in Proteomics

In proteome studies, one usually has to start with a highly complex mixture of proteins, typically hundreds of unknown species. This normally requires one or several protein fractionation steps such as 1- or 2-D gel electrophoresis or liquid chromatography separation of the proteins and peptides prior to MALDI-MS analysis. The different types of fractionation methods can be combined in several ways; for example, SDS-PAGE followed by liquid chromatography (LC) separation of *in-gel*-digested peptides. In contrast to ESI, LC-separation techniques cannot be directly coupled to the MALDI instruments. Instead, the LC-separated peptide or protein fractions are spotted onto the MALDI-target, and then analyzed in an "offline" approach.

MALDI-MS and MS/MS are then used to generate protein specific fingerprints and amino acid sequence information of peptides. These data are used for the identification of proteins by sequence database searching. MALDI-MS has been used in a variety of proteomics projects, from the identification of proteins from 2-D gels, and components of purified protein complexes to large-scale analysis of the yeast interactome (Shevchenko et al., 1996; Wigge et al., 1998; Gavin et al., 2002). In the following sections, we describe the concepts of protein identification by peptide mass mapping and sequencing.

3.4.1
Protein Identification by MALDI-MS Peptide Mass Mapping

In 1993, five research groups proposed the concept of peptide mass mapping for protein identification by combinations of MS and protein sequence database searching (Henzel et al., 1993; James et al., 1993; Mann et al., 1993; Pappin et al., 1993; Yates et al., 1993). The principles of peptide mass mapping are outlined in Figure 3.9.

The concept relies on the presence of only a single or few proteins in a sample (it is therefore ideal for 2-D-gel separated proteins), the use of a sequence-specific protease for producing peptides (usually trypsin), and the assumption that the corresponding protein sequence is available in a biological sequence database. The principle behind protein identification by MALDI-MS peptide mass mapping

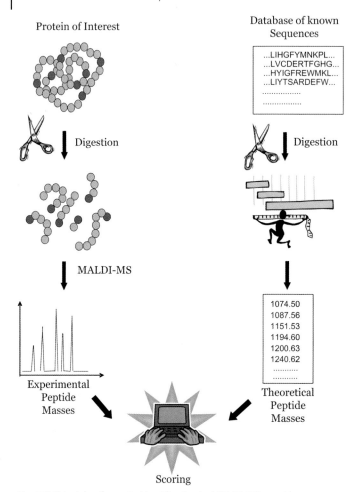

Fig. 3.9 Principle of protein identification by MALDI-MS peptide mass mapping. The experimentally determined MALDI-MS peptide mass map is compared to protein sequences in a biological database using computer-based search algorithms. A scoring scheme is used to retrieve the best match. See the main text for details.

is very simple; the protein is identified as the sequence that generates the best match between the experimental peptide mass map (a list of tryptic peptide masses, the "fingerprint") and the computationally calculated list of tryptic peptide masses from each individual protein sequence entry in a database. This type of MS-based protein identification strategy is known as peptide mass fingerprinting (see example in Fig. 3.10). In practice, these identifications are made by optimized search-algorithms and scoring schemes build into simple-to-use programs (see below).

```
Top Score       : 151 for BAB22149, AK002501 NID:  - Mus musculus
```

Probability Based Mowse Score

Protein score is -10*Log(P), where P is the probability that the observed match is a random event.
Protein scores greater than 76 are significant (p<0.05).

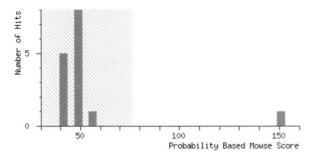

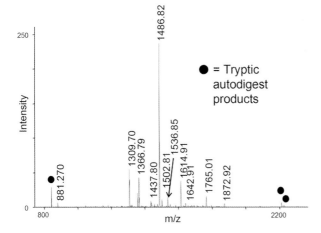

● = Tryptic autodigest products

```
1.  BAB22149          Mass: 30189    Score: 151    Expect: 1.9e-09  Queries matched: 9
    AK002501 NID: - Mus musculus
```

Start - End	Observed	Mr(expt)	Mr(calc)	Delta	Miss	Sequence
8 - 21	1502.8093	1501.8020	1501.7823	0.0197	1	R.VWCRGLLGAASVDR.G
109 - 121	1357.7598	1356.7525	1356.7249	0.0277	0	K.SLADLTAVDVPTR.Q
122 - 135	1765.0086	1764.0013	1763.9682	0.0331	1	R.QNRFEIVYNLLSLR.F
125 - 135	1366.7945	1365.7872	1365.7656	0.0216	0	R.FEIVYNLLSLR.F
199 - 210	1437.8032	1436.7959	1436.7663	0.0296	1	R.KDFPLTGYVELR.Y
200 - 210	1309.6983	1308.6910	1308.6714	0.0197	0	K.DFPLTGYVELR.Y
217 - 230	1642.9143	1641.9070	1641.8838	0.0232	1	K.RVVAEPVELAQEFR.K
218 - 230	1486.8164	1485.8091	1485.7827	0.0265	0	R.VVAEPVELAQEFR.K
218 - 231	1614.9095	1613.9022	1613.8776	0.0246	1	R.VVAEPVELAQEFRK.F

Fig. 3.10 Protein identification by MALDI-MS peptide mass mapping. The MALDI-MS peptide mass map was used to search a database using the Mascot search engine. The highest scoring protein sequence (here a mouse protein) allows the assignment of nine peptide ion masses to the masses of peptides generated from a theoretical digestion of the protein. In this case, the top hit is given a probability based Mowse score of 151.

The introduction of delayed ion extraction in MALDI-TOF instruments in 1995 resulted in a tremendous improvement in analytical performance, mass resolution and accuracy. Using this technique, peptide mass mapping gained high specificity for protein identification and an ability to identify individual components among simple protein mixtures (Jensen et al., 1996, 1997b). Improved sample preparation methods also provided enhanced sensitivity to enable analysis of femtomolar levels of protein (Vorm et al., 1994; Gobom et al., 1999). Today, the automation of MALDI-MS data acquisition enables the analysis of hundreds of samples each day in large-scale protein identification experiments (Jensen et al., 1997a).

MALDI-MS is the preferred method for peptide mass mapping for several reasons, including the direct correlation between the *m/z* value of the singly charged ions and the mass of a peptide, the high mass accuracy and sensitivity, and tolerance towards contaminants. In addition, the simplicity, good sensitivity and high speed of MALDI-MS make it the method of choice for large-scale analysis of proteins from 2-D gels and of defined protein bands from SDS-PAGE gels.

Protein identification by amino acid sequencing is achieved by searching a database using MS/MS data as query. The MS/MS fragment ions are then used to score the spectrum against protein sequences in a database. The best match is the peptide that fits the measured mass, the protease cleavage specificity, and the fragment ion pattern in the MS/MS spectrum. Usually, the peptide mass map of a protein and a series of peptide MS/MS spectra are combined to increase the specificity of the database search.

3.4.2
Quantitation of Proteins by MALDI-MS

MALDI-MS allows absolute or relative quantitation of proteins and peptides by use of internal standards. However, it should be borne in mind that the dynamic range of the MALDI-MS measurement is limited, typically less than two orders of magnitude. The method is therefore only suited for well-defined problems where the results fall within a limited peptide or protein abundance range, or where a low precision is acceptable. The heterogeneity of analyte/matrix deposits makes reproducible data acquisition a challenge. Nevertheless, with proper sample handling and analytical controls it is feasible to perform relative quantitation of peptides by MALDI-MS.

The relative quantitation of peptides is typically achieved by using stable isotope-labeled peptide analogues. For this, ^{2}H-, ^{13}C-, ^{15}N- or ^{18}O-encoded peptides are frequently used (Aebersold and Mann, 2003). Typical applications are comparison of proteins from healthy versus diseased cells, samples before and after treatment, or cells in different states of metabolism. In each of such cases the mixture analysis will generate pairs of signals that differ by a fixed mass which is easily recognized by automated spectral processing. Stable isotope-encoded peptide species are generated by *in-vivo* metabolic labeling of protein, by using stable isotope-encoded alkylating agents, or by enzymatic catalysis. Stable isotope-labeled peptides can also be synthesized using solid-phase chemistry and added to samples for absolute quantitation of specific proteins and peptide (Kirkpatrick et al., 2005).

In-vivo metabolic labeling of protein is achieved by growing cells in the presence of ^{15}N-precursors (Oda et al., 1999; Krijgsveld et al., 2003), or by the addition of isotope-labeled amino acids to the medium. The latter method is known as Stable Isotope Labeling by Amino Acids in Cell Culture (SILAC) (Zhu et al., 2002; Ong et al., 2003), and it provides the means for comparative analysis of protein abundance and post-translational modifications by MALDI-MS (Trester-Zedlitz et al., 2005). Commercially available alkylating reagents include ICAT (Cys) (Gygi et al., 1999), MassTag (Lys) (Peters et al., 2001) and iTRAQ (Lys) (Ross et al., 2004), all of which are suitable for MALDI-MS and MS/MS. A simple method, namely trypsin-catalyzed incorporation of ^{18}O into carboxyl groups, facilitates the comparative analysis of protein and peptide samples by MALDI-MS (Schnolzer et al., 1996; Mirgorodskaya et al., 2000).

3.5
Computational Tools for Protein Analysis by MALDI-MS

Mass spectrometry frequently produces very complicated datasets, and it is therefore necessary to apply computational tools to data analysis and interpretation. A range of publicly available or commercial databases, software packages and web services are available to the protein mass spectrometrist (Table 3.3). These computational resources provide functions for calculating the intact protein molecular mass and the molecular masses of peptides generated by digestion with sequence-specific proteases. The automated assignment and alignment of peptide mass and sequence to the cognate protein sequence aids in elucidating protein primary structure. Tools for protein identification by MALDI-MS peptide mass fingerprinting and MALDI-MS/MS sequencing by sequence database searching are also available, as well as tools for assisting in quantitation. For an analysis of post-translational modifications, it is often an advantage to combine and/or compare the obtained MS-based results with *in-silico* results obtained from computational tools optimized for the prediction of sites likely to be modified. For this, software packages are available that focus on the handling of large sets of data and results obtained in proteomic studies. An example is ProteinCenter, which offers the opportunity to extract various information on the identified proteins such as GO-annotation, biological functions and links to public available databases. For links to more MS-related bioinformatics tools, the reader should refer to ExPASy.org.

Table 3.3 Computational services and tools for protein mass spectrometry.

Name	Website	Purpose	References
ExPASy	www.expasy.org	Resource/software for protein sequence and mass analysis	Public
EBI	www.ebi.ac.uk	Biological databases and query tools	Public
NCBI	www.ncbi.nih.gov	Biological databases and query tools	Public
GPMAW	www.gpmaw.com	Protein sequence and mass analysis	Commercial
Mascot	www.matrixscience.com	Computational tool for protein identification by MS and MS/MS data. For links to more search engines, see ExPASy	Public/ Commercial
X! Tandem	www.thegpm.org/TANDEM/	Computational tool for protein identification by MS/MS data	Public/ Commercial
Phenyx	www.phenyx-ms.com/	Computational tool for protein identification by MS/MS data, quantitation	Commercial
VEMS	yass.sdu.dk	Computational tool for protein identification, quantitation and annotation of PTMs.	Freeware
MSQuant	msquant.sourceforge.net	Tool for protein quantitation	Freeware
Protein-Prospector	prospector.ucsf.edu/	Various tools for protein analysis and identification by MS and MS/MS data	Public
ProteinCenter	www.proxeon.com	Computational tool for analysis of large proteomic data sets	Commercial
NetPhos	Cbs.dtu.dk/services/NetPhos	Predictions for phosphorylation sites. For links to other predictors, see ExPASy	Public

3.6
Clinical Applications of MALDI-MS

The automation and high-throughput capabilities of modern instruments make MALDI-MS a potentially useful technique for large-scale clinical research and diagnostics. The underlying idea is that MALDI-MS analysis of body fluids, such as blood, urine, saliva or cerebrospinal fluid or of intact tissue samples (MALDI imaging) might reveal disease-specific protein patterns and reflect progression of the disease or the effect of medical treatment (Petricoin et al., 2002; Reyzer and Capridi 2005). Thus, MALDI-MS "protein profiling" is a very active field of research worldwide. As for protein chemistry and proteomics studies, sample preparation is a critical issue in clinical applications of MALDI-MS. In this context, sample preparation involves every step from collection of the body fluid or tissue from a person to the deposition of proteins onto the MALDI probe. Study design is crucial as it is important to have case and control groups of adequate size, while body fluid sample collection requires standardization. Sample collection and storage procedure and the consumables used should be strictly controlled in order to minimize biological and analytical variations, as has been shown for MALDI-MS protein profiling of serum samples (Villanueva et al., 2005). Several methods for sample preparation from serum have been reported, including the use of solid-phase extraction (Callesen et al., 2005) and functionalized magnetic beads (Villanueva et al., 2004). Protein profiling has been used to analyze cancers and neurological disease. The statistical analysis of large datasets might classify samples as belonging to either healthy or diseased persons (Diamandis et al., 2003). However, despite a number of encouraging results, no MALDI-MS protein profiling method has at the present time been validated for clinical studies.

3.7
Conclusions

Today, MALDI-MS is finding applications in all areas of protein biochemistry and proteomics. The acceptance of MALDI-MS in the biological community is largely due to the simplicity of sample preparation for routine analysis of proteins and peptides, and the high sensitivity of modern MALDI-MS instruments. Advanced technologies and new applications of MALDI-MS in protein analysis and proteomics are being continuously introduced. Sample handling is crucial for successful MALDI-MS analysis, and the development of increasingly robust and scalable approaches to protein and peptide preparation for MALDI-MS is a very active field of research. The relative quantification of proteins by MALDI-MS is emerging as a powerful method to study the molecular details of complex biological processes. The high performance of modern MALDI-MS/MS instruments presents a range of challenges to the computational analysis of large-scale datasets. Clearly, the integration of biological, mass spectrometric and computational techniques is a prerequisite for effective and sensitive analysis of protein samples.

Acknowledgments

The authors thank present and past members of the Protein Research Group for their contributions to the development of MALDI-MS for protein analysis. K.H. was funded by Danish Technical Sciences Research Council (BAMSE research consortia). The Protein Research Group is supported by grants from the Danish Research Councils. O.N.J. is a Lundbeck Foundation Professor.

Appendix

Table A1 Name, abbreviations and side-chain composition of the 20 common amino acid residues.

Amino acid	Abbreviation (3- and 1-letter)	Side chain
Alanine	Ala, A	$-CH_3$
Cysteine	Cys, C	$-CH_2SH$
Aspartate	Asp, D	$-CH_2COOH$
Glutamate	Glu, E	$-CH_2CH_2COOH$
Phenylalanine	Phe, F	$-CH_2C_6H_5$
Glycine	Gly, G	$-H$
Histidine	His, H	$-CH_2-C_3H_3N_2$
Isoleucine	Ile, I	$-CH(CH_3)CH_2CH_3$
Lysine	Lys, K	$-(CH_2)_4NH_2$
Leucine	Leu, L	$-CH_2CH(CH_3)_2$
Methionine	Met, M	$-CH_2CH_2SCH_3$
Asparagine	Asn, N	$-CH_2CONH_2$
Proline	Pro, P	$-CH_2CH_2CH_2-$
Glutamine	Gln, Q	$-CH_2CH_2CONH_2$
Arginine	Arg, R	$-(CH_2)_3NH-C(NH)NH_2$
Serine	Ser, S	$-CH_2OH$
Threonine	Thr, T	$-CH(OH)CH_3$
Valine	Val, V	$-CH(CH_3)_2$
Tryptophan	Trp, W	$-CH_2C_8H_6N$
Tyrosine	Tyr, Y	$-CH_2-C_6H_4OH$

References

R. Aebersold, M. Mann. Mass spectrometry-based proteomics. *Nature.* **2003**, *422*, 198–207.

R.S. Annan, S.A. Carr. Phosphopeptide analysis by matrix-assisted laser desorption time-of-flight mass spectrometry. *Anal. Chem.* **1996**, *68*, 3413–3421.

R.C. Beavis, B.T. Chait. Rapid, sensitive analysis of protein mixtures by mass spectrometry. *Proc. Natl. Acad. Sci. USA.* **1990**, *87*, 6873–6877.

R.C. Beavis, B.T. Chait. Matrix-assisted laser desorption ionization mass-spectrometry of proteins. *Methods Enzymol.* **1996**, *270*, 519–551.

M. Cadene, B.T. Chait. A robust, detergent-friendly method for mass spectrometric analysis of integral membrane proteins. *Anal. Chem.* **2000**, *72*, 5655–5658.

A.K. Callesen, S. Mohammed, J. Bunkenborg, T.A. Kruse, S. Cold, O. Mogensen, R. Christensen, W. Vach, P.E. Jorgensen, O.N. Jensen. Serum protein profiling by miniaturized solid-phase extraction and matrix-assisted laser desorption/ionization mass spectrometry. *Rapid Commun. Mass Spectrom.* **2005**, *19*, 1578–1586.

S.L. Cohen, B.T. Chait. Influence of matrix solution conditions on the MALDI-MS analysis of peptides and proteins. *Anal. Chem.* **1996**, *68*, 31–37.

S.L. Cohen, B.T. Chait. Mass spectrometry of whole proteins eluted from sodium dodecyl sulfate-polyacrylamide gel electrophoresis gels. *Anal. Biochem.* **1997**, *247*, 257–267.

E.P. Diamandis, A. Scorilas, S. Fracchioli, M. Van Gramberen, H. De Bruijn, A. Henrik,

A. Soosaipillai, L. Grass, G.M. Yousef, U.H. Stenman, M. Massobrio, A.G. Van Der Zee, I. Vergote, D. Katsaros. Human kallikrein 6 (hK6): a new potential serum biomarker for diagnosis and prognosis of ovarian carcinoma. *J. Clin. Oncol.* **2003**, *21*, 1035–1043.

A.C. Gavin, M. Bosche, R. Krause, P. Grandi, M. Marzioch, A. Bauer, J. Schultz, J.M. Rick, A.M. Michon, C.M. Cruciat, M. Remor, C. Hofert, M. Schelder, M. Brajenovic, H. Ruffner, A. Merino, K. Klein, M. Hudak, D. Dickson, T. Rudi, V. Gnau, A. Bauch, S. Bastuck, B. Huhse, C. Leutwein, M.A. Heurtier, R.R. Copley, A. Edelmann, E. Querfurth, V. Rybin, G. Drewes, M. Raida, T. Bouwmeester, P. Bork, B. Seraphin, B. Kuster, G. Neubauer, G. Superti-Furga. Functional organization of the yeast proteome by systematic analysis of protein complexes. *Nature* **2002**, *415*, 141–147.

J. Gobom, E. Nordhoff, E. Mirgorodskaya, R. Ekman, P. Roepstorff. Sample purification and preparation technique based on nano-scale reversed-phase columns for the sensitive analysis of complex peptide mixtures by matrix-assisted laser desorption/ionization mass spectrometry. *J. Mass Spectrom.* **1999**, *34*, 105–116.

J.J. Gorman, B.L. Ferguson, T.B. Nguyen. Use of 2,6-dihydroxyacetophenone for analysis of fragile peptides, disulphide bonding and small proteins by matrix-assisted laser desorption/ionization. *Rapid Commun. Mass Spectrom.* **1996**, *10*, 529–536.

S.P. Gygi, B. Rist, S.A. Gerber, F. Turecek, M.H. Gelb, R. Aebersold. Quantitative analysis of complex protein mixtures using isotope-coded affinity tags. *Nat. Biotechnol.* **1999**, *17*, 994–999.

S. Haebel, C. Jensen, S.O. Andersen, P. Roepstorff. Isoforms of a cuticular protein from larvae of the meal beetle, *Tenebrio molitor*, studied by mass spectrometry in combination with Edman degradation and two-dimensional polyacrylamide gel electrophoresis. *Protein Sci.* **1995**, *4*, 394–404.

P. Hagglund, J. Bunkenborg, F. Elortza, O.N. Jensen, P. Roepstorff. A new strategy for identification of N-glycosylated proteins and unambiguous assignment of their glycosylation sites using HILIC enrichment and partial deglycosylation. *J. Proteome Res.* **2004**, *3*, 556–566.

D.J. Harvey. Proteomic analysis of glycosylation: structural determination of N- and O-linked glycans by mass spectrometry. *Expert Rev. Proteomics* **2005**, *2*, 87–101.

U. Hellman, R. Bhikhabhai. Easy amino acid sequencing of sulfonated peptides using post-source decay on a matrix-assisted laser desorption/ionization time-of-flight mass spectrometer equipped with a variable voltage reflector. *Rapid Commun. Mass Spectrom.* **2002**, *16*, 1851–1859.

W.J. Henzel, T.M. Billeci, J.T. Stults, S.C. Wong, C. Grimley, C. Watanabe. Identifying proteins from two-dimensional gels by molecular mass searching of peptide fragments in protein sequence databases. *Proc. Natl. Acad. Sci. USA* **1993**, *90*, 5011–5015.

P. James, M. Quadroni, E. Carafoli, G. Gonnet. Protein identification by mass profile fingerprinting. *Biochem. Biophys. Res. Commun.* **1993**, *195*, 58–64.

O.N. Jensen. Interpreting the protein language using proteomics. *Nature Rev. Mol. Cell. Biol.* **2006**, *7*, 391–403.

O.N. Jensen, A. Podtelejnikov, M. Mann. Delayed extraction improves specificity in database searches by matrix-assisted laser desorption/ionization peptide maps. *Rapid Commun. Mass Spectrom.* **1996**, *10*, 1371–1378.

O.N. Jensen, P. Mortensen, O. Vorm, M. Mann. Automation of matrix-assisted laser desorption/ionization mass spectrometry using fuzzy logic feedback control. *Anal. Chem.* **1997a**, *69*, 1706–1714.

O.N. Jensen, A.V. Podtelejnikov, M. Mann. Identification of the components of simple protein mixtures by high-accuracy peptide mass mapping and database searching. *Anal. Chem.* **1997b**, *69*, 4741–4750.

O.N. Jensen, M.R. Larsen, P. Roepstorff. Mass spectrometric identification and microcharacterization of proteins from electrophoretic gels: Strategies and applications. *Proteins-Structure Function Genet.* **1998**, *Suppl. 2*, 74–89.

R.S. Johnson, S.A. Martin, K. Biemann. Collision-induced fragmentation of (M+H)+ ions of peptides – side-chain specific sequence ions. *Int. J. Mass Spectrom. Ion Processes* **1988**, *86*, 137–154.

H. Katayama, T. Nagasu, Y. Oda. Improvement of in-gel digestion protocol for peptide mass fingerprinting by matrix-assisted laser desorption/ionization time-of-flight mass spectrometry. *Rapid Commun. Mass Spectrom.* **2001**, *15*, 1416–1421.

T. Keough, R.S. Youngquist, M.P. Lacey. A method for high-sensitivity peptide sequencing using postsource decay matrix-assisted laser desorption ionization mass spectrometry. *Proc. Natl. Acad. Sci. USA* **1999**, *96*, 7131–7136.

T. Keough, R.S. Youngquist, M.P. Lacey. Sulfonic acid derivatives for peptide sequencing. *Anal. Chem.* **2003**, *75*, 156A–165A.

D.S. Kirkpatrick, S.A. Gerber, S.P. Gygi. The absolute quantification strategy: a general procedure for the quantification of proteins and post-translational modifications. *Methods* 2005, 35, 265–273.

S. Kjellstrom, O.N. Jensen. In situ liquid-liquid extraction as a sample preparation method for matrix-assisted laser desorption/ionization MS analysis of polypeptide mixtures. *Anal. Chem.* **2003**, *75*, 2362–2369.

E. Krause, H. Wenschuh, P.R. Jungblut. The dominance of arginine-containing peptides in MALDI-derived tryptic mass fingerprints of proteins. *Anal. Chem.* **1999**, *71*, 4160–4165.

J. Krijgsveld, R.F. Ketting, T. Mahmoudi, J. Johansen, M. Artal-Sanz, C.P. Verrijzer, R.H. Plasterk, A.J. Heck. Metabolic labeling of *C. elegans* and *D. melanogaster* for quantitative proteomics. *Nat. Biotechnol.* **2003**, *21*, 927–931.

M. Kussmann, P. Roepstorff, Sample preparation techniques for peptides and proteins analyzed by MALDI-MS. *Methods Mol. Biol.* **2000**, *146*, 405–424.

B. Kuster, S.F. Wheeler, A.P. Hunter, R.A. Dwek, D.J. Harvey. Sequencing of N-linked oligosaccharides directly from protein gels: in-gel deglycosylation followed by matrix-assisted laser desorption/ionization mass spectrometry and normal-phase high-performance liquid chromatography. *Anal. Biochem.* **1997**, *250*, 82–101.

F.M. Lagerwerf, M. Van de Weert, W. Heerma, J. Haverkamp. Identification of oxidized methionine in peptides. *Rapid Commun. Mass Spectrom.* **1996**, *10*, 1905–1910.

M.R. Larsen, G.L. Sorensen, S.J. Fey, P.M. Larsen, P. Roepstorff. Phospho-proteomics: evaluation of the use of enzymatic dephosphorylation and differential mass spectrometric peptide mass mapping for site specific phosphorylation assignment in proteins separated by gel electrophoresis. *Proteomics* 2001, 1, 223–238.

M.R. Larsen, M.E. Graham, P.J. Robinson, P. Roepstorff. Improved detection of hydrophilic phosphopeptides using graphite powder microcolumns and mass spectrometry: evidence for in vivo doubly phosphorylated dynamin I and dynamin III. *Mol. Cell. Proteomics* 2004, 3, 456–465.

M.R. Larsen, P. Hojrup, P. Roepstorff. Characterization of gel-separated glycoproteins using two-step proteolytic digestion combined with sequential microcolumns and mass spectrometry. *Mol. Cell. Proteomics* 2005a, 4, 107–119.

M.R. Larsen, T.E. Thingholm, O.N. Jensen, P. Roepstorff, T.J. Jorgensen. Highly selective enrichment of phosphorylated peptides from peptide mixtures using titanium dioxide microcolumns. *Mol. Cell. Proteomics* 2005b, 4, 873–886.

S. Laugesen, P. Roepstorff. Combination of two matrices results in improved performance of MALDI-MS for peptide mass mapping and protein analysis. *J. Am. Soc. Mass Spectrom.* **2003**, *14*, 992–1002.

P.C. Liao, J. Leykam, P.C. Andrews, D.A. Gage, J. Allison. An approach to locate phosphorylation sites in a phosphoprotein: mass mapping by combining specific enzymatic degradation with matrix-assisted laser desorption/ionization mass spectrometry. *Anal. Biochem.* **1994**, *219*, 9–20.

M. Mann, O.N. Jensen. Proteomic analysis of post-translational modifications. *Nature Biotechnol.* **2003**, *21*, 255–261.

M. Mann, P. Hojrup, P. Roepstorff. Use of mass spectrometric molecular weight information to identify proteins in sequence databases. *Biol. Mass Spectrom.* **1993**, *22*, 338–345.

O.A. Mirgorodskaya, Y.P. Kozmin, M.I. Titov, R. Korner, C.P. Sonksen, P. Roepstorff. Quantitation of peptides and proteins by matrix-assisted laser desorption/ionization mass spectrometry using (18)O-labeled internal standards. *Rapid Commun. Mass Spectrom.* **2000**, *14*, 1226–1232.

E. Mortz, T. Sareneva, I. Julkunen, P. Roepstorff. Does matrix-assisted laser desorption/ionization mass spectrometry allow analysis of carbohydrate heterogeneity in glycoproteins? A study of natural human interferon-gamma. *J. Mass Spectrom.* **1996**, *31*, 1109–1118.

Y. Oda, K. Huang, F.R. Cross, D. Cowburn, B.T. Chait. Accurate quantitation of protein expression and site-specific phosphorylation. *Proc. Natl. Acad. Sci. USA* **1999**, *96*, 6591–6596.

M.J. Omaetxebarria, P. Hagglund, F. Elortza, N.M. Hooper, J.M. Arizmendi, O.N. Jensen. Isolation and characterization of glycosylphosphatidylinositol-anchored peptides by hydrophilic interaction chromatography and MALDI tandem mass spectrometry. *Anal. Chem.* **2006**, *78*, 3335–3341.

S.E. Oong, L.J. Foster, M. Mann. Mass spectrometric-based approaches in quantitative proteomics. *Methods* **2003**, *29*, 124–130.

D.J. Pappin, P. Hojrup, A.J. Bleasby. Rapid identification of proteins by peptide-mass fingerprinting. *Curr. Biol.* **1993**, *3*, 327–332.

E.C. Peters, D.M. Horn, D.C. Tully, A. Brock. A novel multifunctional labeling reagent for enhanced protein characterization with mass spectrometry. *Rapid Commun. Mass Spectrom.* **2001**, *15*, 2387–2392.

E.F. Petricoin, A.M. Ardekani, B.A. Hitt, P.J. Levine, V.A. Fusaro, S.M. Steinberg, G.B. Mills, C. Simone, D.A. Fishman, E.C. Kohn, L.A. Liotta. Use of proteomic patterns in serum to identify ovarian cancer. *Lancet* **2002**, *359*, 572–577.

M.W. Pinkse, P.M. Uitto, M.J. Hilhorst, B. Ooms, A.J. Heck. Selective isolation at the femtomole level of phosphopeptides from proteolytic digests using 2D-NanoLC-ESI-MS/MS and titanium oxide precolumns. *Anal. Chem.* **2004**, *76*, 3935–3943.

M.C. Posewitz, P. Tempst. Immobilized gallium(III) affinity chromatography of phosphopeptides. *Anal. Chem.* **1999**, *71*, 2883–2892.

J. Qin, B.T. Chait. Preferential fragmentation of protonated gas-phase peptide ions adjacent to acidic amino-acid-residues. *J. Am. Chem. Soc.* **1995**, *117*, 5411–5412.

J. Rappsilber, Y. Ishihama, M. Mann. Stop and go extraction tips for matrix-assisted laser desorption/ionization, nanoelectrospray, and LC/MS sample pretreatment in proteomics. *Anal. Chem.* **2003**, *75*, 663–670.

M.L. Reyzer, R.M. Capridi. MALDI Mass Spectrometry for Direct Tissue Analysis: A New Tool for Biomarker Discovery. *J. Proteome Res.* **2005**, *4*, 1138–1142.

P. Roepstorff, J. Fohlman. Proposal for a common nomenclature for sequence ions in mass-spectra of peptides. *Biomed. Mass Spectrom.* **1984**, *11*, 601.

P.L. Ross, Y.N. Huang, J.N. Marchese, B. Williamson, K. Parker, S. Hattan, N. Khainovski, S. Pillai, S. Dey, S. Daniels, S. Purkayastha, P. Juhasz, S. Martin, M. Bartlet-Jones, F. He, A. Jacobson, D.J. Pappin. Multiplexed protein quantitation in *Saccharomyces cerevisiae* using amine-reactive isobaric tagging reagents. *Mol. Cell. Proteomics* **2004**, *3*, 1154–1169.

M. Schnolzer, P. Jedrzejewski, W.D. Lehmann. Protease-catalyzed incorporation of 18O into peptide fragments and its application for protein sequencing by electrospray and matrix-assisted laser desorption/ionization mass spectrometry. *Electrophoresis* **1996**, *17*, 945–953.

M. Schuerenberg, C. Luebbert, H. Eickhoff, M. Kalkum, H. Lehrach, E. Nordhoff. Prestructured MALDI-MS sample supports. *Anal. Chem.* **2000**, *72*, 3436–3442.

A. Shevchenko, O.N. Jensen, A.V. Podtelejnikov, F. Sagliocco, M. Wilm, O. Vorm, P. Mortensen, H. Boucherie, M. Mann. Linking genome and proteome by mass spectrometry: Large-scale identification of yeast proteins from two dimensional gels. *Proc. Natl. Acad. Sci. USA* **1996**, *93*, 14440–14445.

I.P. Smirnov, X. Zhu, T. Taylor, Y. Huang, P. Ross, I.A. Papayanopoulos, S.A. Martin, D.J. Pappin. Suppression of alpha-cyano-4-hydroxycinnamic acid matrix clusters and reduction of chemical noise in MALDI-TOF mass spectrometry. *Anal. Chem.* **2004**, *76*, 2958–2965.

A. Stensballe, O.N. Jensen. Phosphoric acid enhances the performance of Fe(III) affinity chromatography and matrix-assisted laser desorption/ionization tandem mass spectrometry for recovery, detection and sequencing of phosphopeptides. *Rapid Commun. Mass Spectrom.* **2004**, *18*, 1721–1730.

A. Stensballe, S. Andersen, O.N. Jensen. Characterization of phosphoproteins from electrophoretic gels by nanoscale Fe(III) affinity chromatography with off-line mass spectrometry analysis. *Proteomics* **2001**, *1*, 207–222.

K. Strupat, M. Karas, F. Hillenkamp. 2,5-Dihydroxybenzoic acid: a new matrix for laser desorption-ionization mass spectrometry. *Int. J. Mass Spectrom. Ion Processes* **1991**, *111*, 89–102.

M. Trester-Zedlitz, A. Burlingame, B. Kobilka, M. Von Zastrow. Mass spectrometric analysis of agonist effects on posttranslational modifications of the beta-2 adrenoceptor in mammalian cells. *Biochemistry* **2005**, *44*, 6133–6143.

A. Tsarbopoulos, M. Karas, K. Strupat, B.N. Pramanik, T.L. Nagabhushan, F. Hillenkamp. Comparative mapping of recombinant proteins and glycoproteins by plasma desorption and matrix-assisted laser desorption/ionization mass spectrometry. *Anal. Chem.* **1994**, *66*, 2062–2070.

J. Villanueva, J. Philip, D. Entenberg, C.A. Chaparro, M.K. Tanwar, E.C. Holland, P. Tempst. Serum peptide profiling by magnetic particle-assisted, automated sample processing and MALDI-TOF mass spectrometry. *Anal. Chem.* **2004**, *76*, 1560–1570.

J. Villanueva, J. Philip, C.A. Chaparro, Y. Li, R. Toledo-Crow, L. Denoyer, M. Fleisher, R.J. Robbins, P. Tempst. Correcting common errors in identifying cancer-specific serum peptide signatures. *J. Proteome Res.* **2005**, *4*, 1060–1072.

O. Vorm, P. Roepstorff, M. Mann. Improved resolution and very high sensitivity in MALDI TOF of matrix surfaces made by fast evaporation. *Anal. Chem*, **1994**, *66*, 3281–3287.

D. Wang, S.R. Kalb, R.J. Cotter. Improved procedures for N-terminal sulfonation

of peptides for matrix-assisted laser desorption/ionization post-source decay peptide sequencing. *Rapid Commun. Mass Spectrom.* **2004**, *18*, 96–102.

A. Wattenberg, A.J. Organ, K. Schneider, R. Tyldesley, R. Bordoli, R.H. Bateman. Sequence dependent fragmentation of peptides generated by MALDI quadrupole time-of-flight (MALDI Q-TOF) mass spectrometry and its implications for protein identification. *J. Am. Soc. Mass Spectrom.* **2002**, *13*, 772–783.

P.A. Wigge, O.N. Jensen, S. Holmes, S. Soues, M. Mann, J.V. Kilmartin. Analysis of the *Saccharomyces* spindle pole by matrix-assisted laser desorption/ionization (MALDI) mass spectrometry. *J. Cell Biol.* **1998**, *141*, 967–977.

V.H. Wysocki, G. Tsaprailis, L.L. Smith, L.A. Breci. Mobile and localized protons: a framework for understanding peptide dissociation. *J. Mass Spectrom.* **2000**, *35*, 1399–1406.

C.F. Xu, Y. Lu, J. Ma, M. Mohammadi, T.A. Neubert. Identification of phosphopeptides by MALDI Q-TOF-MS in positive and negative ion modes after methyl esterification. *Mol. Cell. Proteomics* **2005**, *4*, 809–818.

J.R. Yates, III, S. Speicher, P.R. Griffin, T. Hunkapiller. Peptide mass maps: a highly informative approach to protein identification. *Anal. Biochem.* **1993**, *214*, 397–408.

N. Zhang, L. Li. Effects of common surfactants on protein digestion and matrix-assisted laser desorption/ionization mass spectrometric analysis of the digested peptides using two-layer sample preparation. *Rapid Commun. Mass Spectrom.* **2004**, *18*, 889–896.

X. Zhang, C.J. Herring, P.R. Romano, J. Szczepanowska, H. Brzeska, A.G. Hinnebusch, J. Qin. Identification of phosphorylation sites in proteins separated by polyacrylamide gel electrophoresis. *Anal. Chem.* **1998**, *70*, 2050–2059.

H. Zhu, S. Pan, S. Gu, E.M. Bradbury, X. Chen. Amino acid residue specific stable isotope labeling for quantitative proteomics. *Rapid Commun. Mass Spectrom.* **2002**, *16*, 2115–2123.

4
Microprobing and Imaging MALDI for Biomarker Detection

Bernhard Spengler

4.1
Introduction

Over the past decade a new analytical method for examining biological samples and for visualizing analytical information acquired from such samples has evolved [1]. Matrix-assisted laser desorption/ionization (MALDI) imaging, as it is generally called, allows one to visualize the two-dimensional concentration profiles of bio-molecules in or on the surface of biological samples, and is expected to have a great impact on, for example, proteomics and metabolomics research in the future. Correlation of histological phenomena with molecular images will provide for a new quality of medical diagnostics and molecular understanding of disease. The data source for creating such images is a set of mass spectra acquired stepwise, with a certain lateral resolution and a defined lateral assignment for each spectrum.

The connection between the detection of molecules and medical diagnostics is often established through the word "biomarker" – one of the most frequently used (and abused) terms in the current scientific glossary. The meaning of "biomarker" can vary greatly, depending on the scientific context in which it is used. For example, in geology and astrobiology biomarkers are understood to be indicators of the presence of biological organisms. However, the medical sciences have a very different, but still variable understanding of the term. Sometimes, biomarkers are thought of as labels (e.g., radioactive) that are introduced into a biological system for diagnostic purposes. For example, they may be substances (e.g., antibodies) the detection of which indicates a certain disease state, or they may be indicators of the exposure of an organism to environmental substances. Sometimes they may simply be a "meaningful" substance or measurement quantity obtained from an individual under observation.

Biomarkers can be native to the investigated organism, or they can be external substances or derivatives (metabolites) of external substances. They can be either specific or integral, and from a chemical point of view they can be almost any type of substance class. In the context of proteomics, biomarkers are mostly understood to be proteins, glycoproteins or lipids, and to be indicators for a disease state through their presence, quantity, or modification state.

MALDI MS. A Practical Guide to Instrumentation, Methods and Applications.
Edited by Franz Hillenkamp and Jasna Peter-Katalinić
Copyright © 2007 Wiley-VCH Verlag GmbH & Co. KGaA, Weinheim
ISBN: 978-3-527-31440-9

Whether under this definition a substance is a *specific* biomarker is mostly a question of correlation rather than of a detailed understanding of biochemical mechanisms of a disease. The number of detectable parameters available for such a correlation is of significant importance for the resulting validity of the diagnostic conclusion. It has been found to be quite problematic, for example, to take a poorly resolved mass spectrometric peak acquired from an affinity capturing surface as a valid indicator for the presence of an expected protein having the observed mass. Further validation is certainly necessary in such experiments, for example by enzymatic digestion and/or fragment ion analysis. Another possible parameter for correlation that would increase validity is the localization of a suspected biomarker within the tissue. This is why biomarker discovery, in combination with molecular imaging, is gaining interest rapidly. It must be borne in mind, however, that even a reasonable localization pattern of a low-resolution mass spectrometric signal is still not sufficient proof of the presence of an expected substance that was identified beforehand from a bulk proteomics experiment.

It is an often-stressed truism that "a picture is worth a thousand words". The underlying psychological or physiological effect that pictures are perceived in a much more direct and more efficient way than words or data has a tremendous impact on many aspects of analytical imaging. The directness of visual information transfer and the pleasant feeling of looking at colorful pictures inevitably lead to an acceptance and consent level that is not necessarily justified by the scientific content of the displayed information. Needless to say, on the other hand, that coding analytical information in one meaningful picture has enormous advantages over the manual evaluation of pages of data or stacks of individual mass spectra. Today, MALDI imaging is in only the preliminary stages of its development – there is much to be done, especially in the field of quality assurance and the validation of information.

While a picture might be *worth* a thousand words, it definitely *costs* more than a thousand words. It is not only the storage of high-resolution images on computers that is much more demanding than handling single mass spectra. Rather, it is the entire process of image formation, manipulation and evaluation that implies a high level of computer power, intelligent algorithms and last, but not least, the high-quality instrumental apparatus, that is far beyond the scope of regular mass spectrometric bioanalysis. Nonetheless, such methods and instruments have been developed during the past decade, and now is the time to develop the application and validation protocols that will turn the colorful pictures into powerful information and knowledge.

The technical principle of MALDI imaging is summarized in Figure 4.1. A pulsed laser beam is focused to the size of the aspired lateral resolution. To date, mainly lasers with ultraviolet wavelengths (337 nm, 355 nm, 266 nm) and pulse lengths of a few nanoseconds have been used. The focused laser beam is directed to the surface of the sample, which is then moved in steps in order to scan the sample according to the intended lateral resolution of the system. Before analysis, the sample must be prepared in a way that the biomolecular ions can be

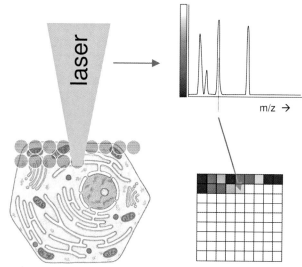

Fig. 4.1 Scheme of the technical principle of MALDI imaging. A pulsed laser scans the surface (e.g., a biological sample), producing a mass spectrum for each spot. The intensity of a selected mass signal (a molecular species or "biomarker") is transformed into a grayscale value and drawn to the according pixel map.

desorbed and ionized by the laser beam, as in regular MALDI analyses. For that, the sample (e.g., a biological tissue sample) must be covered with a suitable matrix, such as 2,5-dihydroxybenzoic acid, sinapinic acid or α-cyano-4-hydroxycinnamic acid. The method of preparation must take into account that any lateral analyte migration caused by the addition of matrix solution should be smaller than the intended lateral resolution of the analysis, with the understanding that any inclusion of analyte molecules into the matrix crystals on top of the tissue section requires a minimum axial diffusion.

The laser pulses desorb and ionize part of the material located within the laser focus. Ions corresponding to a certain pixel of the sample are mass-analyzed typically by a time-of-flight (TOF) analyzer, as described in Chapter 2. The resulting mass spectra are then evaluated for creating distribution images. The intensity of a selected mass signal formed by a molecular species or a "biomarker" of interest (i.e., a physiologically or pathologically meaningful, detectable substance), is transformed into a grayscale value and is drawn to the according pixel map at the position currently interrogated by the laser beam. Many such images can be formed from one and the same sample scan if more than one mass signal is evaluated. The primary images can then be combined to multicolor images by dedicated image-processing algorithms.

4.2
History of Mass Spectrometry Imaging and Microprobing Techniques

The bulk composition of a homogenized sample of an organ, a tissue, of cells or cell organelles can provide important information on biochemical and biological mechanisms, especially if not only qualitative information is deduced but also quantitative or dynamic data are determined. Even so, the understanding of biological processes is limited, as long as the localization of target compounds in living systems is not determined. Such targets of interest can be inorganic ions (Ca^{2+}, K^+, Na^+, etc.), trace elements (Se, Zn), drugs, metabolites, or biomolecules such as lipids, carbohydrates, peptides or proteins.

The aim of investigating biochemical processes not only on a qualitative or quantitative level but also topologically has a relatively long history. Fluorescence microscopy, for example, allows one to localize and image target compounds by employing immunochemical labeling techniques [2]. The most prominent disadvantage of these imaging techniques, however, is that the labeling of a target compound requires that the nature of the target is known in advance, and that only a limited number of targets can be investigated with reasonable effort and at the same time. Unknown targets cannot be investigated this way; for example, fluorescently labeled antibodies must be defined or green fluorescent protein (GFP) markers expressed recombinantly prior to analysis. This limitation of imaging techniques does not hold for those methods capable of determining the localized chemical composition of a biological sample without precognition or substance-specific preparation steps.

Such a method, in principle, is mass spectrometry (MS). About 30 years ago, soon after powerful UV lasers first became commercially available, a new method was developed that permitted the analysis of biological material mass spectrometrically, by using a highly focused laser beam [3]. Today, the so-called "laser microprobe mass analysis" (LAMMA) technique is still in service in many laboratories, and allows the analysis of micro-quantities of biological samples, especially with respect to inorganic compounds [4,5]. Similarly, secondary ion mass spectrometry (SIMS) has been used to detect elemental and molecular species in micro-localized areas on the surfaces of various types of sample. Initially, both LAMMA and SIMS had lateral resolutions in the range of 1 µm, and were later further developed to scan samples automatically and to create images. Nowadays, with the development of more sensitive primary ion sources, SIMS is capable of imaging smaller biomolecules such as lipids in biological samples at sub-micrometer resolution [6].

With the development of MALDI [7] and electrospray ionization (ESI) [8] during the 1980s, the focus of bioanalytical research was first directed towards the bulk analysis of large molecules, and later towards identifying and characterizing proteins, before returning to the question of the bioanalytical imaging of native samples. MALDI imaging was first described in 1994 [9] and subsequently found broad application in many laboratories. Today, it is used in a low lateral resolution mode rather broadly, on a qualitative, non-validated information level. In high lateral resolution mode, MALDI imaging remains at an early developmental stage.

4.3
Visualization of Mass Spectrometric Information

A number of data evaluation and processing steps are required to form micro-analytical images from mass spectrometric data. The basic procedure is first to store all mass spectra as formed by laser desorption ionization of material from each precisely positioned sample spot. The focus diameter of the laser beam can be taken as a first approximation of the achievable lateral resolution of the final image. Typically, the step width of the scan is equal to the laser focus diameter, leading to adjacent laser spots with no oversampling or undersampling. Over-sampling can be useful under certain conditions as an alternative approach, if the instrumental sensitivity is sufficiently high [10,13].

When the scan is finished the mass spectra are evaluated with respect to signals of interest. In the simplest case these signals are chosen manually, while more sophisticated imaging programs allow the signals to be listed and chosen auto-matically, for example according to their frequency of appearance or their local-ization over the sample area [11]. Once the mass signals (as markers for compounds of interest) are chosen, the stored mass spectra are evaluated by trans-lating the relative intensities of the signals of each position-correlated mass spec-trum into grayscale values of the pixels in the mass-specific images.

As a second step the raw grayscale images can be further processed, for exam-ple by color-coding or by the combination of several images. Using colors instead of gray values is useful for enhancing the visibility of concentration differences. Due to so-called color invariance, the human eye is able to differentiate only 32 to 64 gray levels, but about 350000 different colors. By using a color lookup table for matching gray values to color representations, the scale of visibility of information can be largely extended. Figure 4.2 shows the effect of using different color lookup tables for the same raw data.

Another powerful tool is to code three images of different compounds of a sam-ple in the three native colors of red, green and blue (RGB), and to use the addi-tive color model of light (Fig. 4.3). As a result, the presence of both the red-coded and the green-coded substance in a certain sample spot, for example, is repre-sented as a yellow pixel in the according RGB image. Equal mass signal intensi-ties of all three coded substances result in white pixels. Using the 24-bit RGB color model is a lossless representation of three 8-bit grayscale images.

A more evaluative color-coding method is to include arithmetic operations such as intensity differences of two grayscale images. Beyond these processing steps, standard image processing software offers an immense number of possibilities to modify or clarify the informational contents and the visualization quality.

All of these manipulations have in common that the validity of the representa-tion with respect to its analytical value, its reliability and truth can easily be lost.

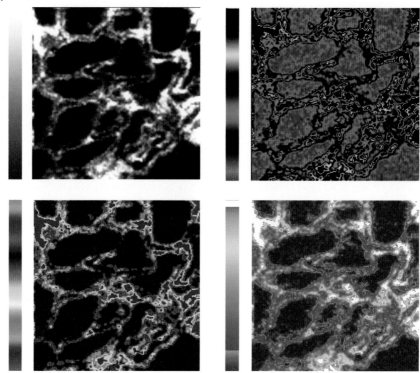

Fig. 4.2 Different color look-up tables (CLUT) applied to the same data set allow the enhancement of specific analytical information and extension of the dynamic range of visibility of intensity gradients. Using s split CLUT (upper right) is useful for samples with two (or more) particularly abundant ion species in differing signal intensity ranges. Rapidly changing colors are advantageous to enhance the visibility for a certain small intensity range (lower left). An intuitive color representation is created by using a "topographic" color map (lower right) that includes a "waterline" at low intensities, "greenland", "rocks" and "snow" for the highest intensities.

It is therefore of major importance that image-processing steps are performed under controlled and traceable conditions. Already at the level of grayscale transformation of mass spectral intensities, data processing methods must be scrutinized in detail. A visualized representation of a two-dimensional array of mass spectra is, by definition, a quantitative statement, even if the individual mass spectra originally were supposed to be only qualitatively meaningful. Although peak intensities in regular mass spectral analyses are not accepted as being directly proportional to a substance concentration, they are often nevertheless taken uncritically as concentration values in imaging mass spectrometry. A structure visible in the resulting image, however, is only analytically meaningful, if the underlying signal intensity gradient corresponds to a substance concentration gradient and not, for example, to a morphologically induced drop of the total ion current. At this

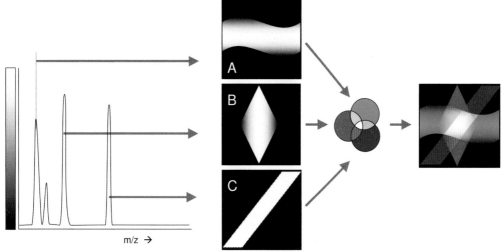

Fig. 4.3 Coding of mass spectrometric information in a red, green, blue (RGB) color image. The spatial distributions of three components A, B, and C initially are visualized in grayscale images, where white codes high signal intensity and black codes low signal intensity. Up to three components can then be combined to a RGB image in which each color channel corresponds to one grayscale image (one component). Coincidences of component localizations then lead to mixed colors, with white corresponding to presence of all three components in maximum intensity.

point, there is much to be done in future to turn colorful pictures into scientifically sustainable and exploitable results. Fundamental studies describing the nature of the best internal standard for MALDI tissue imaging analysis remain to be conducted. Candidates for this are the total ion current, the signal intensity of the MALDI matrix ion (or a fragment thereof), or the signal intensity of another analyte that is supposed to be equally dispersed throughout the sample. Unfortunately, the methods of isotopic dilution and isotopic coding (which are routinely employed in bulk quantitative analyses) cannot be used in tissue analysis, as a homogeneous distribution of the added standard cannot be guaranteed.

On a second level, image processing can even destroy analytical validity if manipulations such as smoothing, gamma correction, contrast and brightness adjustment or filtering are not documented and described in detail. Therefore, acceptable analytical image-processing programs must be capable of documenting each manipulation step, displaying the resulting correlation of the color table or grayscale with the original signal intensities or with a calibrated concentration curve, and providing the possibility of easily tracing back a certain pixel in a final manipulated image to the original mass spectral data [11].

4.4

MALDI in Micro Dimensions: Instruments and Mechanistic Differences

Images in general can be created in two different ways, as is known from electron microscopy. In transmission electron microscopy (TEM), the sample is directly imaged by visualizing the transmitted electrons onto a screen or a position-sensitive, two-dimensional detector. By contrast, in scanning electron microscopy (SEM) the image is formed by correlating pixel-by-pixel a secondary or back-scattered electron signal (detected not positionally sensitive but globally) with the sample position of a highly focused primary electron beam. In analogy to these two principles, MALDI imaging can be performed by irradiating either the whole sample or partial areas thereof with each single laser pulse, and directly imaging analyte ions onto a position-sensitive detector. An alternative is to scan a focused laser beam across the sample and to detect and store all ions for every single position. While the latter mode can be realized in principle with any MALDI mass spectrometer under certain conditions, the former approach (called "microscope mode") is possible only with a special instrumental arrangement, and as yet has not been widely employed. In the following section we will first describe the common scanning mode, before returning to the microscope mode.

Scanning MALDI imaging differs from normal MALDI-MS in many aspects. Instrumentally, the optical set-up for focusing the laser beam, the mechanical set-up to scan the sample area and the software for translating MS data into pictures are the main components that are fundamentally different in an imaging mass spectrometer compared to a regular MALDI instrument.

The original set-up of a first MALDI imaging mass spectrometer was built to create analytical images in the micrometer or sub-micrometer resolution range [9]. To obtain such lateral resolution, the laser beam must be focused to a spot size as small as physically possible. This limit is defined by

$$d = 1.22 \cdot \frac{\lambda}{A}$$

where d is the focal diameter (or more exactly the diameter of the central disc of the diffraction pattern), λ is the wavelength of the laser beam, and A is the numerical aperture of the lens [12]. With a dedicated lens system having a numerical aperture A of 0.6, a lateral resolution of 0.6 µm was obtained at 337 nm wavelength [9,13] (Fig. 4.4). The high lateral resolution mode of MALDI imaging is called SMALDI-MS, an acronym for "Scanning Microprobe MALDI-MS".

Without special optical arrangements for highest lateral resolution, the focus diameter of a commercial (non-imaging) MALDI mass spectrometer is typically equal or greater than 30 to 50 µm under optimized conditions.

A different approach aiming at high lateral resolution but without the need to use a high-quality optical lens system is described under the term "MALDI imaging in microscope mode" [14], in contrast to the above-mentioned "microprobe mode" of scanning microfocus mass spectrometers. In microscope mode,

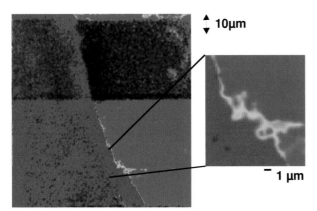

Fig. 4.4 Demonstration of analytical lateral resolution. Scanning laser desorption/ ionization-mass spectrometry (LDI-MS) of the edge between a red and a green felt-tip pen coating. Red channel: $m/z = 122\,u$, originating from red dye; green channel: $m/z = 372\,u$, originating from green dye. The scanning area was $100 \times 100\,\mu m$, with a step size of $0.5\,\mu m$; the acquisition time was 15 min. The effective lateral resolution of analytical information was found to be about $0.5\,\mu m$.

the whole area of interest is irradiated by each laser shot, and imaged onto the ion detector by a dedicated electrostatic lens systems. The lateral resolution of the final image is determined by the quality of the ion optics and the lateral resolution of the position-sensitive detector. Thus, the resulting image is not a pixel image formed by digital image processing but rather is an analog image formed on a position-sensitive ion detector (in the simplest case, a microchannel plate detector with a phosphor screen). Such an electrostatical set-up is available commercially only in one instrument family (TRIFT TOF-SIMS; Physical Electronics, Ismaning, Germany) for SIMS. The advantage of the microscope mode over the (scanning) microprobe mode is that an ion image can, in principle, be formed with a single laser shot, whereas microprobe scanning requires acquisition times of up to several hours. Another fundamental difference from microprobe instruments is that the effective lateral resolution in microscope mode is defined not by the resolution (or focus diameter) of the laser beam but by the quality and lateral resolution of the ion optical system. This means that the lateral resolution can be better than the diffraction limit of the chosen laser wavelength. It has been shown that a lateral resolution a factor of 3 below the diffraction limit is possible in infrared MALDI microscope mode imaging [15]. One considerable drawback of microscope mode MALDI imaging, however, is that instruments in this mode of operation are usually not mass spectrometric analyzers but rather mass range selectors. This is because images are obtained from two-dimensional microchannel plate detectors either by a photographic process or by electronic signal processing. A mass spectrometric resolution of these images can be achieved by fast video acquisition, fast electronic processing, or by ion gating. To date, all of these methods have resulted in rather limited mass resolution. Hence, the development of new detector designs is necessary to make the microscope mode a superior method of image formation. The effects of ion formation and interaction in the dense ion cloud formed during the desorption event are other issues of possible limitation of image quality in microscope mode.

Besides TOF instruments, other mass analyzers such as quadrupole-orthogonal TOF instruments, ion traps and Fourier transform (FT) ion cyclotron resonance mass spectrometers have been recently employed for MALDI imaging (see below). These instrumental configurations are discussed in detail in Chapter 2.

4.5
The Matrix Deposition Problem in High-Resolution Imaging

Aiming at high lateral resolution in MALDI imaging is not only a question of improving the laser optical set-up (or the ion optical and detector quality in microscope mode, respectively). An even more demanding aspect is the integrity of the sample under investigation. MALDI imaging follows the same principal requirements of ion formation processes as normal MALDI analysis. Thus, the necessary interaction of matrix and analyte must be maintained in order to achieve optimal conditions for ion formation with respect to analytical sensitivity and analytical information. One major requirement for the highly sensitive detection at least of peptides and proteins by MALDI-MS appears to be that analyte molecules are incorporated into matrix crystals upon solvent evaporation. Second, it is observed that for some matrices (e.g., 2,5-dihydroxybenzoic acid), larger crystals result in a better signal-to-noise (S/N) ratio and thus higher analytical sensitivity than smaller crystals. Both requirements are highly detrimental to the expected quality of high-resolution MALDI imaging. The incorporation of analyte molecules into large matrix crystals means that the effective resolution is limited to the size of the matrix crystals, and is probably even worse than that, due to migration during liquid phase interaction of matrix solution with tissue samples. Published attempts to employ MALDI imaging for biological questions were therefore mostly deliberately limited to a resolution of 100 or 200 μm, where such effects play a minor role. To aim at the highest possible resolution requires an understanding and optimization of the sample preparation step in microscopic detail. The fundamental and general properties of MALDI samples are illustrated in Figure 4.5 [13]. Peptide molecules are incorporated into the growing matrix crystals upon solvent evaporation, as can be inferred from the local distribution of analyte ion signals. Even more important for the ion formation process, however, might be that alkali ion impurities which are known considerably to hinder the ion-formation process (especially in other ionization methods such as ESI) are clearly excluded from the matrix crystals and thus are separated from the interaction region of analyte ion formation. The distribution images of analyte and alkali ions appear to be almost exactly inverted to each other in Figure 4.5. It can be assumed that this phenomenon of the separation of ionic components is one of the most important aspects of the MALDI process in general. Thus, it can be concluded that analyte molecules which are incorporated predominantly into matrix crystals (especially peptides and proteins) are locally separated from alkali ions and are thus

Matrix:
2,5-Dihydroxybenzoic acid

Peptide:
Substance P

Salt impurities:
Potassium

Fig. 4.5 High-resolution MALDI image showing distribution of the peptide substance P (center image) in a regular dried-droplet preparation using the matrix 2,5-dihydroxybenzoic acid (left image). The image corresponding to the distribution of alkali ion impurities (right image) indicates that the MALDI preparation procedure is in fact a purification step with analyte (peptide) molecules incorporated into the matrix crystals and inorganic ions segregated outside of the crystals.

ionized by protonation with greater efficiency if their distance is high (with larger crystals). Analyte molecules which are predominantly ionized by alkali ion attachment, on the other hand (e.g., carbohydrates) would not be expected to be affected by crystal size to the same extent. The high tolerance of MALDI with regard to ionic impurities, when compared to other ionization methods, is most likely due to the fact that dried droplet preparation is in reality a purification step in its own right. In Figure 4.5, it is suggested that lateral analytical resolution is limited to the size of the matrix crystals. Figure 4.6 shows that, furthermore, even

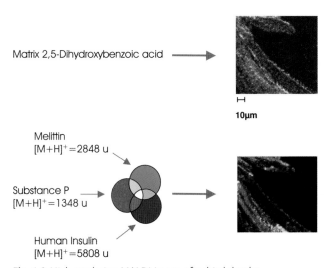

Matrix 2,5-Dihydroxybenzoic acid

10µm

Melittin
$[M+H]^+ = 2848$ u

Substance P
$[M+H]^+ = 1348$ u

Human Insulin
$[M+H]^+ = 5808$ u

Fig. 4.6 High-resolution MALDI image of a dried droplet preparation of a homogeneous solution of three peptides, substance P, melittin and insulin using 2,5-dihydroxybenzoic acid as a matrix. The distribution image indicates that analyte ions segregate during solvent evaporation in different areas of the matrix crystals.

within the crystals and among crystal communities, originally homogeneous sample mixtures are separated and segregated into specific localization areas for different components. Both observations imply that a preparation leading to small matrix crystals is preferable for imaging MALDI if analytical sensitivity is to be conserved. Thus, it can be assumed that the liquid state during sample preparation will always lead to considerable lateral transportation of analyte material over large distances, resulting in a strong reduction of effective analytical resolution in tissue analysis.

It became, therefore, the goal of several groups to identify a matrix preparation procedure that would allow highly sensitive mass analysis in imaging studies, while retaining a maximum lateral analytical resolution. The electrospray deposition of matrices (2,5-dihydroxybenzoic acid, sinapinic acid or α-cyano-4-hydroxycinnamic acid) onto tissue samples, though resulting in very small crystals, was found to be virtually unusable under these conditions for biomolecular ion formation. The reason for this behavior was thought to be that nanodroplets of matrix impinge on the tissue surface already in a desolvated state. As a result, mixing between the analyte and matrix is prevented and peptide or protein molecules are not incorporated into the matrix crystals. It was found, however, that resolubilization and recrystallization of the matrix-covered tissue sample in a wet atmosphere after electrospray or vacuum deposition led to a favorable sensitivity for biomolecule detection under still very high lateral analytical resolution [16]. In this way, an effective resolution of ca. 2 µm was obtained using this procedure.

4.6
Organisms, Organs, Tissues and Cells: Imaging MALDI at Various Lateral Resolutions

Following the demonstration of the principle of MALDI imaging in 1994 [9], the first applications of MALDI imaging in the biomedical field were published in 1997 by the group of Richard Caprioli [17]. Proteins in rat pituitary tissue and several neuropeptides were detected in the anterior, intermediate and posterior regions, and signals were correlated to the lateral position [17,18]. A major breakthrough in the recognition of MALDI imaging by the scientific community was achieved in 2001 by the group's publication of studies on cancer tissue [19]. Human glioblastoma slices were mounted onto a MALDI target and coated with sinapinic acid matrix for analysis. In this way it was possible to distinguish in the resulting images the proliferating area of the tumor from the ischemic and necrotic areas of the lesion, based on the localization of different proteins (Fig. 4.7). The imaged proteins were identified off-line by standard protein identification from tissue extracts as thymosin β.4 in the proliferating area and β Actin and S100A4, respectively, in the ischemic and necrotic areas.

The general procedure for preparation of the tissue samples in these and the later published investigations was based on cryo-sectioning the fresh-frozen tissue and air-drying the sections. A cryostat at −15 °C was used to cut 12-µm sections

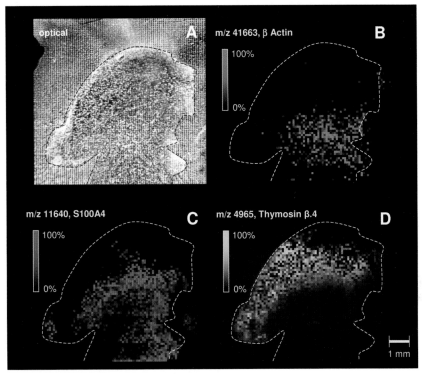

Fig. 4.7 Optical image (a) and mass spectrometric images (b–d) of human glioblastoma sections. Image (d) shows the distribution of a protein correlating with the proliferating area of the tumor, while protein images (b) and (c) correlate more to the ischemic and necrotic areas of the tumor. (Reproduced from Ref. [19], with permission.)

from the frozen material, which were immediately placed on a metallic sample target. In a cold environment at 4 °C the samples were covered with 10 μL of matrix solution (10 mg mL⁻¹ sinapinic acid in 1 : 1 acetonitrile/0.05% trifluoroacetic acid) by quickly spreading the solution over the tissue with a spatula [19]. After 45 min of air-drying and 2 h drying in a desiccator, the samples were analyzed in a high-vacuum MALDI TOF mass spectrometer.

Many applications followed the first publications of biomolecular MALDI imaging, with most to date remaining on a rather qualitative and unvalidated information level. A few more examples will be described here to demonstrate the broad applicability of the method, and its potential.

The investigation of Alzheimer's disease was one target of MALDI imaging in a pharmaceutical research environment [20], with brain tissue being examined with respect to its content of amyloid peptides. The approach has also been intensely employed to identify and optimize drug leads in the pharmaceutical industry. Phospholipid analysis is another important area of mass spectrometric imaging, addressable by both MALDI and SIMS imaging [21,22]. Here, a change in the lipid composition of cells could be characterized by MALDI imaging, and this approach was used to investigate Duchenne muscular dystrophy, a neuro-muscular disease linked to a defect of lipid metabolism. It was for the first time that slight changes in fatty acid compositions could be monitored directly from biological tissue by using MS. Another area of MALDI imaging use is in phar-maceutical research and metabolomics, where skin treatment by a xenobiotic sub-stance was monitored by imaging a skin cross-section, airspray-covered with α-cyano-4-hydroxycinnamic acid (CHCA) matrix [23]. By superimposing the images with histological images of the sample, the degree of drug permeation into the tissue could be assessed.

4.7
Whole-Cell and Single-Cell Analysis: Prospects and Limitations

Matrix crystallization is not the only problem to be faced when increasing lateral resolution. With smaller dimensions, another principle limitation enters the game, namely the available number of analyte molecules per investigated area (per pixel). Independent of the ion formation method (i.e., for MALDI imaging as well as for SIMS imaging or other imaging techniques), the accessible lateral resolu-tion is fundamentally limited by the natural concentration of analyte molecules. Assuming a monolayer of peptides (i.e., an extremely high surface concentration of an individual analyte), a laser focus of 200 μm in diameter would cover about six billion molecules. A microfocus (1 μm focus diameter) under the same condi-tions would cover only 200 000 molecules. Taking into account that a mass spec-trometer, depending on its detection technique and scanning principle, requires between 1000 and 100 000 molecules per sample to form a reasonable mass spec-trum, it is clear that only highly abundant substances can be analyzed from the surface of cells or tissues by microfocus instruments, and that less-abundant species can only be detectable with micrometer lateral resolution when averaged over extended numbers of layers (not from the surface alone) of a tissue. This prin-cipal limitation is eased slightly by the fact that the quality of a mass spectrum is more defined by the S/N ratio rather than by the absolute signal intensity. A decrease of the analyzed area of a sample thus results in a decrease in signal intensity, and at the same time in a decrease of chemical noise. A lower signal in-tensity might therefore still be detectable which, in a regular MALDI instrument, would already be covered by noise. Nevertheless, signal statistics at small num-bers of ions definitely limit the detectability of low-abundant species from biolog-ical samples at micrometer lateral resolution.

The analysis of individual cells, either by averaging analytical data from a number of individual cells ("whole-cell analysis") or by analyzing only one cell ("single-cell analysis") often has enormous advantages over tissue analysis. Tissues are normally not composed homogeneously of equal cells but rather exhibit com-plex communities of various cell types. The investigation of cultured cancer cell lines, for example, is far from examining the "real" native conditions of a cancer, although it does allow one to standardize and fix the analytical situation by focusing on only one type of cell. Single-cell analysis is even more promising than whole-cell analysis, in that it provides a snapshot-biochemistry analysis. Averaging analytical data over more than one cell means arbitrarily averaging all sorts of cell states, whereas single cells express an individual cell status in a snapshot view. Collecting data from many single-cell analyses might even allow one to describe biochemical processes time-resolved.

The complexity of mass spectra acquired directly from cells without chromatographic separation steps represents a clear analytical problem. Obtaining analytical information from complex mass spectrometric data in this case typically means employing fingerprint or cluster analysis, pattern recognition, or molecular labeling techniques. A high mass resolution is mostly required, in order to provide sufficient ability to distinguish between a large number of substance signals in a mass spectrum.

4.7.1
Cellular Analysis

Protein and peptide profiling from a small number of individual cells was first described by Jimenez et al. in 1993 [24,25]. The identification and characterization of peptides was first performed on a small number of cells from the pituitary pars

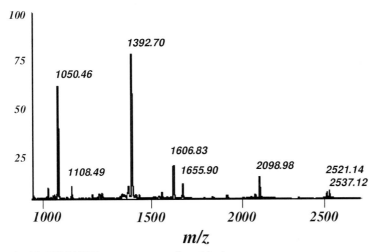

Fig. 4.8 MALDI-TOF mass spectrum of peptides in pars intermedia tissue from the amphibian *Xenopus laevis*. The spectrum indicates the presence of several expected and unexpected neuropeptides, including vasotocin (CYIQNCPRG-NH2) at 1050.46 u, vasotocin precursor (CYIQNCPRGGKR-OH) at 1392.70 u, and signals derived from propiomelanocortin POMCA and POMCB precursors. (Adapted from Ref. [26].)

intermedia of *Xenopus laevis* by MALDI postsource decay mass spectrometry [26], and later from a single cell of *Aplysia californica* [27]. The neural cells of frogs or mollusks are favorable targets for whole-cell or single-cell peptide profiling due to their relatively large size and their high concentrations of neuropeptides. More demanding is the protein profiling of individual bacteria and other microorganisms [28]. While fingerprint analyses of bacteria by the LAMMA method were demonstrated rather at an early stage [29], the profiling of large biomolecules of such species by fingerprinting (still on a large number of cells) was described much later [30–32]. Fingerprinting by MALDI-MS allowed differentiation to be made between closely related strains of bacteria based on the detection of specific biomarkers [33]. The latest developments focus on the direct classification of airborne bacterial cells by aerosol MALDI [34,35].

4.7.2
Individually Isolated Cells

Single isolated neurons were investigated with respect to differential peptide profiling [36,37]. The spreading of individual neuronal cells onto a MALDI target by mechanical elongation allows one to differentiate subcellular areas, even with an instrument that has a lateral resolution (a laser focus diameter) of ca. 100 µm. Mechanical stretching provides an efficient approach to circumvent instrumental limitations regarding lateral resolution. Neuronal cells of *Aplysia californica* were pinned down onto a formvar membrane and later stretched, together with the membrane, to separate the cell organelles one from another.

4.7.3
Direct Cellular and Subcellular Imaging

The direct analysis of individual normal-sized cells or cell organelles is only achievable by using a high instrumental lateral resolution. The first results of micrometer-resolved SMALDI analysis of cultured human renal cancer cells were presented recently [38]. In differing from high-resolution SIMS imaging, the analytical sensitivity of SMALDI imaging seems not to deteriorate considerably with increasing analyte mass, and signals up to 15 000 Da could be detected from cancer cells at a pixel resolution of 1 µm. The acquired images allowed differentiation to be made between substance-specific concentration profiles at the subcellular level.

4.8
Cell Sorting and Capturing

Laser capture micro-dissection was combined with MALDI-MS by several groups [39–42]. In one study, MALDI analysis of microdissected cells from human breast cancer in comparison to normal breast epithelial cells was reported [39]. More than

40 peaks in the MALDI mass spectra were found to differ in intensity between cancer and normal tissue. This approach represents another way of achieving laterally resolved analytical information. In this case, it was not the focus size of the desorbing laser beam in the mass spectrometer but rater the size of the captured entities which defined the resulting lateral resolution. Thus, microdissection may represent another approach to circumvent limitations in achievable lateral resolution, by first isolating the sample microarea of interest. Since single cells can be individually dissected under microscope control, the resulting lateral resolution is considerably better than that of standard MALDI imaging instruments. Although this method shows great promise, several problems have limited its applicability to date. Among these is that mass spectrometric sensitivity is generally not high enough for the analysis of individual cells, and protein or peptide identification must be performed in parallel by conventional proteome studies. Sample selection and capturing can be rather laborious processes, and tissue preparation for MALDI analysis of dissected cells is less straightforward than for direct MALDI tissue-imaging analysis. In general, the tissues must be stained for selecting cells of interest, but staining often prevents favorable results in MALDI analyses. Matrices used for MALDI preparation are mostly opaque (due to light scattering of the polycrystalline cover layers), and this causes problems in locating the underlying dissected tissue areas. The laser dissection process, furthermore, has been found to affect the neighboring tissue material – a problem that becomes especially prominent if very small tissue areas or individual cells are to be analyzed.

Opaque matrices are a methodological problem not only in the analysis of captured tissue cells but also in normal MALDI tissue imaging, when areas of interest must be positioned correctly after matrix application. Sweedler and coworkers have sought transparent matrices layers as substitutes for regular MALDI matrix substances [43]. The addition of triethylamine to solutions of standard MALDI matrices led to the formation of ionic liquids [44], and this resulted in favorable MALDI imaging performance.

4.9
Future Areas of Application

In addition to the direct analysis of biomarkers, biomolecule inventories or differential tissue characterization, MALDI imaging is about to find new applications in a rapidly growing new field of bioanalytical research, which is based on high-density protein or peptide microarrays [45]. With increasing density of protein chips, fast and highly resolving MS imaging techniques are required. One important approach in this respect is the so-called molecular scanner technique [46], which combines protein separation by gel electrophoresis with peptide mass fingerprinting techniques in an automated set-up. Key to the technique is that the digestion step following gel separation takes place with spatial resolution, thus retaining the position information of the originating protein.

Spatial resolution of peptide arrays is increasing and the field is expanding rapidly [47]. The potential for peptide array technology is very high, especially in terms of sensitivity and information density. By using bubble-jet or piezo printing technology, liquid fractions as small as 2 pL can be handled and deposited, leading to MALDI sample spots as small as 30 μm [13]. Arraying of more than 50 000 samples per cm^2 should become possible with this technique. Even higher density protein arrays can be manufactured in future using dip-pen nanolithography (DPN) technology. This method uses the cantilever of an atomic force microscope to print, for example, peptides, proteins or viruses onto a surface, with spot sizes as small as 80 nm [48]. As yet, however, there is no analytical readout technique besides fluorescence detection to characterize molecules captured by the printed bioactive structures with sufficient lateral resolution. SMALDI imaging is the method that most closely approaches the necessary lateral resolution for peptide/protein identification while being sufficiently sensitive when target molecules are affinity-enriched locally.

Today, one particularly important field of current research is to identify efficient ways of creating biologically meaningful protein arrays. Protein-DNA conjugates have been used as a new approach to form protein arrays based on well-established DNA array technology [49]. Capture proteins were conjugated with DNA to be bound non-covalently and reversibly to surface-bound DNA, and then used to extract biomarker molecules from complex solutions, with ultimate detection by MALDI-MS.

Another approach to optimize protein array formation uses bioactive nanoparticles as the structuring elements [50]. Here, capture proteins are immobilized onto the surface of nanoparticles which are then used to structure a MALDI target by ring-pin printing. Nanoparticles exhibit a high effective area per structural unit, and thus are able to increase analytical sensitivity at high-density arraying.

4.10
Identification and Characterization: Limitations Due to Mass Resolution and Accuracy

Information from published mass spectrometric images is limited to the visualization of semiquantitative intensity distributions of selected mass windows. The identification or characterization of underlying substances directly from image data, however, is not yet possible. Furthermore, analytical homogeneity of the selected mass windows is not assured, due to low mass resolution and low mass accuracy of the TOF mass spectrometers typically used for MALDI imaging. Images therefore might represent more than one component of the sample if mass resolution is insufficient, and consequently analytical validation (prior to any biomedical validation) is not possible. Most investigations published to date have used a parallel approach instead, in order to identify and characterize sample components by classical proteomics techniques. The identity of classically identified substances and imaged signals can then be assumed, but not proven.

In order to avoid these fundamental limitations, current efforts attempt to use mass spectrometric instrumentation with high mass resolution and high mass accuracy for MALDI imaging. FT-ICR mass spectrometry would allow the identification and characterization of substances directly from data of the imaged sample. Rapid *de-novo* sequencing approaches such as composition-based sequencing (CBS) [51] would then allow peptides to be rapidly sequenced and interpreted, and to be correlated to image information. Although these approaches are still at an early stage of development, the first results obtained have shown great promise [52,53].

4.11
Conclusions

MALDI imaging with high or low lateral resolution is rapidly becoming a technique of broad applicability and of ever-increasing potential. The areas of applicability are very broad, and the advantages of direct information transport to the human mind, and of the dense coding of information in multicolor images, should not be underestimated. Nonetheless, many problems remain to be solved, and many obstacles will inevitably emerge in the future. To combine MALDI imaging and SIMS imaging represents an important goal for the future, as both techniques can profit hugely from their combined mechanistic and technological research and development.

References

1 C.M. Henry. Drawing with mass spec. *Chem. Eng. News* **2004**, *82*(46), 33–35.

2 Special section "Biological Imaging". *Science* **2003**, *300*, 75–102.

3 F. Hillenkamp, E. Unsöld, R. Kaufmann, R. Nitsche. Laser microprobe mass analysis of organic materials. *Nature* **1975**, *256*, 119–120.

4 W.H. Schröder, G.L. Fain. Light dependent calcium release from photoreceptors measured by laser micro mass analysis. *Nature* **1984**, *309*, 268–270.

5 A.H. Verbueken, F.J. Bruynseels, R.E. Van Grieken. Laser microprobe mass analysis: a review of application in the life sciences. *Biomed. Mass Spectrom.* **1985**, *12*, 438–463.

6 S.G. Ostrowski, C.T. Van Bell, N. Winograd, A.G. Ewing. Mass spectrometric imaging of highly curved membranes during *Tetrahymena* mating. *Science* **2004**, *305*, 71–73.

7 M. Karas, F. Hillenkamp. Laser desorption ionization of proteins with molecular masses exceeding 10 000 daltons. *Anal. Chem.* **1988**, *60*, 2299–2301.

8 J.B. Fenn, M. Mann, C.K. Meng, S.F. Wong, C.M. Whitehouse. Electrospray ionization for mass spectrometry of large biomolecules. *Science* **1989**, *246*, 64–71.

9 B. Spengler, M. Hubert, R. Kaufmann. MALDI ion imaging and biological ion imaging with a new scanning UV-laser microprobe. Proceedings of the 42nd ASMS Conference on Mass Spectrometry and Allied Topics, Chicago, Illinois, May 29–June 3, **1994**, p. 1041.

10 J.C. Jurchen, S.S. Rubakhin, J.V. Sweedler. MALDI-MS imaging of features smaller than the size of the laser beam. *J. Am. Soc. Mass Spectrom.* **2005**, *16*, 1654–1659.

11 A. Hester, W. Bouschen, A. Leisner, K. Maass, C. Paschke, B. Spengler. MIRION: A data analysis software package for imaging MS. Proceedings of the 53rd ASMS Conference on Mass Spectrometry and Allied Topics, San Antonio, TX, June 5–9, **2005**.

12 D.C. Giancoli. *Physics for Scientists and Engineers.* 3rd edition. Prentice-Hall, London, **2000**.

13 B. Spengler, M. Hubert. Scanning Microprobe Matrix-Assisted Laser Desorption Ionization (SMALDI) mass spectrometry: instrumentation for sub-micrometer resolved LDI and MALDI surface analysis. *J. Am. Soc. Mass Spectrom.* **2002**, *13*, 735–748.

14 S.L. Luxembourg, T.H. Mize, L.A. McDonnell, R.M.A. Heeren. High-spatial resolution mass spectrometric imaging of peptide and protein distributions on a surface. *Anal. Chem.* **2004**, *76*, 5339–5344.

15 S.L. Luxembourg, L.A. McDonnell, T.H. Mize, R.M.A. Heeren. Infrared mass spectrometric imaging below the diffraction limit. *J. Proteome. Res.* **2005**, *4*, 671–673.

16 W. Bouschen, D. Kirsch, B. Spengler (in preparation).

17 R.M. Caprioli, T.B. Farmer, J. Gile. Molecular imaging of biological samples: Localization of peptides and proteins using MALDI-TOF MS. *Anal. Chem.* **1997**, *69*, 4751–4760.

18 M. Stoeckli, T.B. Farmer, R.M. Caprioli. Automated mass spectrometry imaging with a matrix-assisted laser desorption ionization time-of-flight instrument. *J. Am. Soc. Mass Spectrom.* **1999**, *10*, 67–71.

19 M. Stoeckli, P. Chaurand, D.E. Hallahan, R.M. Caprioli. Imaging mass spectrometry: A new technology for the analysis of protein expression in mammalian tissues. *Nature Med.* **2001**, *7*, 493–496.

20 T.C. Rohner, D. Staab, M. Stoeckli. MALDI mass spectrometric imaging of tissue sections. *Mech. Ageing Develop.* **2005**, *126*, 177–185.

21 D. Touboul, H. Piednoel, V. Voisin, S. De La Porte, A. Brunelle, F. Halgand, O. Laprevote. Changes in phospholipid composition within the dystrophic muscle by matrix-assisted laser desorption/ionization mass spectrometry and mass spectrometry imaging. *Eur. J. Mass Spectrom.* **2004**, *10*, 657–664.

22 D. Touboul, F. Kollmer, E. Niehuis, A. Brunelle, O. Laprevote. Improvement of biological time-of-flight-secondary ion mass spectrometry imaging with a bismuth cluster ion source. *J. Am. Soc. Mass Spectrom.* **2005**, *16*, 1608–1618.

23 J. Bunch, M.R. Clench, D.S. Richards. Determination of pharmaceutical compounds in skin by imaging matrix-assisted laser desorption/ionisation mass spectrometry. *Rapid Commun. Mass Spectrom.* **2004**, *18*, 3051–3060.

24 P.A. van Veelen, C.R. Jimenez, K.W. Li, W.C. Wildering, W.P.M. Geraerts, U.R. Tjaden, J. van der Greef. Direct peptide profiling of single neurons by matrix-assisted laser desorption-ionization mass spectrometry. *Org. Mass Spectrom.* **1993**, *28*, 1542–1546.

25 C.R. Jimenez, P.A. Van Veelen, K.W. Li, W.C. Wildering, W.P.M. Geraerts, U.R. Tjaden, J. van der Greef. Neuropeptide expression and processing as revealed by direct matrix-assisted laser desorption ionization mass spectrometry of single neurons. *J. Neurochem.* **1994**, *62*, 404–407.

26 S. Jespersen, P. Chaurand, F.J.C. van Strien, B. Spengler, J. van der Greef. Direct sequencing of neuropeptides in biological tissue by MALDI-PSD mass spectrometry. *Anal. Chem.* **1999**, *71*, 660–666.

27 L. Li, R.W. Garden, E.V. Romanova, J.V. Sweedler. In situ sequencing of peptides from biological tissues and single cells using MALDI-PSD/CID analysis. *Anal. Chem.* **1999**, *71*, 5451–5458.

28 L.F. Marvin-Guy, S. Parche, S. Wagniere, J. Moulin, R. Zink, M. Kussmann, L.B. Fay. Rapid identification of stress-related fingerprint from whole bacterial cells of *Bifidobacterium lactis* using matrix assisted laser desorption/ionization mass spectrometry. *J. Am. Soc. Mass Spectrom.* **2004**, *15*, 1222–1227.

29 U. Seydel, B. Lindner. Qualitative and quantitative investigation on mycobacteria with LAMMA. *Fresenius Z. Anal. Chem.* **1981**, *308*, 253–257.

30 R.D. Holland, J.G. Wilkes, F. Rafii, J.B. Sutherland, C.C. Persons, K.J. Voorhees, J.O. Lay. Rapid identification of intact whole bacteria based on spectral patterns using matrix-assisted laser desorption/ ionization with time-of-flight mass spectrometry. *Rapid Commun. Mass Spectrom.* **1996**, *10*, 1227–1232.

31 T. Krishnamurthy, P.L. Ross. Rapid identification of bacteria by direct matrix-assisted laser desorption/ionization mass spectrometric analysis of whole cells. *Rapid Commun. Mass Spectrom.* **1996**, *10*, 1992–1996.

32 R.J. Arnold, J.P. Reilly. Fingerprint matching of *E. coli* strains with matrix-assisted laser desorption/ionization time-of-flight mass spectrometry of whole cells using a modified correlation approach. *Rapid Commun. Mass Spectrom.* **1998**, *12*, 630–636.

33 C. Fenselau, P.A. Demirev. Characterization of intact microorganisms by MALDI mass spectrometry. *Mass Spectrom. Rev.* **2001**, *20*, 157–171.

34 R.A. Gieray, P.T.A. Reilly, M. Yang, W.B. Whitten, J.M. Ramsey. Real-time detection of individual airborne bacteria. *J. Microbiol. Methods* **1997**, *29*, 191–199.

35 A.L. van Wuijckhuijse, M.A. Stowers, W.A. Kleefsman, B.L.M. van Baar, C.E. Kientz, J.C.M. Marijnissen. Matrix-assisted laser desorption/ionisation aerosol time-of-flight mass spectrometry for the analysis of bioaerosols: development of a fast detector for airborne biological pathogens. *J. Aerosol Sci.* **2005**, *36*, 677–687.

36 S.S. Rubakhin, W.T. Greenough, J.V. Sweedler. Spatial profiling with MALDI-MS: Distribution of neuropeptides within single neurons. *Anal. Chem.* **2003**, *75*, 5374–5380.

37 L. Li, R.W. Garden, J.V. Sweedler. Single-cell MALDI: a new tool for direct peptide profiling. *Trends Biotechnol.* **2000**, *18*, 151–160.

38 W. Bouschen, D. Kirsch, K. Maass, B. Spengler. Automated scanning microprobe MALDI mass spectrometry for characterization of bioactive surfaces with 1 μm lateral resolution. 16th International Mass Spectrometry Conference, August 31–September 5, **2003**, Edinburgh, Scotland. Poster contribution and abstract.

39 B.J. Xu, R.M. Caprioli, M.E. Sanders, R.A. Jensen. Direct analysis of laser capture microdissected cells by MALDI mass spectrometry. *J. Am. Soc. Mass Spectrom.* **2002**, *13*, 1292–1297.

40 P. Chaurand, S. Fouchecourt, B.B. DaGue, B.G.J. Xu, M.L. Reyzer, M.C. Orgebin-Crist, R.M. Caprioli. Profiling and imaging proteins in the mouse epididymis by imaging mass spectrometry. *Proteomics* **2003**, *3*, 2221–2239.

41 D.E. Palmer-Toy, D.A. Sarracino, D. Sgroi, R. Levangie, P.E. Leopold. Direct acquisition of MALDI-ToF mass spectra from laser capture microdissected tissues. *Clin. Chem.* **2000**, *46*, 1513–1516.

42 H. Baker, V. Patel, A.A. Molinolo, E.J. Shillitoe, J.F. Ensley, G.H. Yoo, A. Meneses-Garcia, J.N. Myers, A.K. El-Naggar, J.S. Gutkind, W.S. Hancock. Proteome-wide analysis of head and neck squamous cell carcinomas using laser-capture microdissection and tandem mass spectrometry. *Oral Oncol.* **2005**, *41*, 183–199.

43 S.S. Rubakhin, J.V. Sweedler. Proceedings of the 53rd ASMS Conference on Mass Spectrometry and Allied Topics, San Antonio, TX, June 5–9, **2005**.

44 M. Mank, B. Stahl, G. Boehm. 2,5-Dihydroxybenzoic acid butylamine and other ionic liquid matrixes for enhanced MALDI-MS analysis of biomolecules. *Anal. Chem.* **2004**, *76*, 2938–2950.

45 S. Ekström, D. Ericsson, P. Önnerfjord, M. Bengtsson, J. Nilsson, G. Marko-Varga, T. Laurell. Signal amplification using spot-on-a-chip technology for the identification of proteins via MALDI-TOF MS. *Anal. Chem.* **2001**, *73*, 214–219.

46 M. Müller, R. Gras, R.D. Appel, W.V. Bienvenut, D.F. Hochstrasser. Visualization and analysis of molecular scanner peptide mass spectra. *J. Am. Soc. Mass Spectrom.* **2002**, *13*, 221–231.

47 M.F. Lopez, M.G. Pluskal. Protein micro- and macroarrays: digitizing the proteome. *J. Chromatogr.* **2003**, *787*, 19–27.

48 J.-M. Nam, S.W. Han, K.-B. Lee, X. Liu, M.A. Rattner, C.A. Mirkin. Bioactive protein nanoarrays on nickel oxide surfaces formed by dip-pen nanolithography. *Angew. Chem. Int. Ed.* **2004**, *43*, 1246–1249.

49 C.F.W. Becker, R. Wacker, W. Bouschen, R. Seidel, B. Kolaric, P. Lang, H. Schroeder, O. Müller, C.M. Niemeyer, B. Spengler, R.S. Goody, M. Engelhard. Direct readout of protein-protein interactions by mass spectrometry from protein-DNA microarrays. *Angew. Chem. Int. Ed.* **2005**, *44*, 7635–7639.

50 W. Bouschen, K. Borchers, H. Brunner, T. Flad, D. Kirsch, C.A. Mueller, G. Tovar, B. Spengler. Analysis of structured biological and bioactive surfaces with SMALDI MS. Justus Liebig Anniversary Symposium, "Bioanalytical Quantum Steps", Schloss Rauischholzhausen, May 12–15, **2003**, p. 71.

51 B. Spengler. De novo sequencing, peptide composition analysis, and composition-based sequencing: a new strategy employing accurate mass determination by Fourier transform ion cyclotron resonance mass spectrometry. *J. Am. Soc. Mass Spectrom.* **2004**, *15*, 703–714.

52 M. Koestler, B. Spengler, M. Köstler, D. Kirsch, A. Leisner, B. Spengler. Development of an imaging Atmospheric Pressure (AP) MALDI source for the Finnigan LTQ-FT mass spectrometer. 53rd ASMS Conference on Mass Spectrometry and Allied Topics, San Antonio, TX, June 5–9, **2005**.

53 V.A. Thieu, D. Kirsch, T. Flad, C. Mueller, B. Spengler. Direct protein and biomarker identification from non-specific peptide pools by high accuracy MS data filtering. *Angew. Chem. Int. Ed.*, **2006**, *45*, 3317–3319.

5
MALDI-MS of Nucleic Acids and Practical Implementations in Genomics and Genetics

Dirk van den Boom and Stefan Berkenkamp

5.1
Challenges in Nucleic Acid Analysis by MALDI-MS

The mass spectrometry of nucleic acids has a similar history compared to that of proteins. Although some investigations have been reported mainly using fast atom bombardment (FAB) for ionization, the field did not really take off until the soft ionization techniques of electrospray ionization (ESI) [1] and matrix-assisted laser desorption/ionization (MALDI) [2] became available. However, even with these new ionization techniques the analysis of nucleic acids turned out to be substantially more difficult compared to proteins. As a result, mass spectrometry (MS) is not nearly as much of an enabling technique in genetics and genomics as it is in proteomics. Many other powerful competing techniques for the analysis of nucleic acids, such as fluorescence detection after hybridization, are also reasons for this less-widespread use. MALDI-MS has found its place in these fields only recently, and it is important that we compare its performance with that of the other techniques for any given analytical task.

The successful use of ESI in the analysis of nucleic acids has been described in a number of publications, mostly in conjunction with ion trap or FT-ICR-MS, though by far the majority of applications uses MALDI as the technique of choice [3–9]. The main reason for this development is that MALDI is less dependent on highly purified samples and, most importantly, lends itself more easily to high-throughput applications.

Ion yield, fragmentation and signal heterogeneity due to multiple cationization of the phosphate backbone are the main problems that need to be solved for successful MALDI-MS of nucleic acids. At physiological pH, nucleic acids are polyanions in solution. Nonetheless they form singly charged ions as the base peak even in negative ion mode pointing to the dramatic difference of ions formed by dissociation in solution and gas phase ions of the same neutrals. It has been speculated that the ion yield of nucleic acids is substantially less than that for peptides and proteins, again independent of the polarity, though well-founded experimental proof of this assumption is still missing. Smaller signals can also result from a higher degree of fragmentation.

MALDI MS. A Practical Guide to Instrumentation, Methods and Applications.
Edited by Franz Hillenkamp and Jasna Peter-Katalinić
Copyright © 2007 Wiley-VCH Verlag GmbH & Co. KGaA, Weinheim
ISBN: 978-3-527-31440-9

During matrix crystallization in sample preparation for the MALDI analysis, either all (positive ion mode) or all but one (negative ion mode) of the phosphate backbone groups must be neutralized. Depending on the availability in the solution, the cations for this process can be protons, quaternary amines, or metal ions (Na$^+$ and K$^+$ in particular). If all are available in comparable concentrations, a heterogeneous ion mixture with masses spread out over a considerable mass range will result, thereby degrading the signal-to-noise (S/N) ratio. In the higher mass range the serious peaks will not be resolved, while and the single peak, shifted in mass to a higher value, will not allow for a determination of the mass of the free acid (see Fig. 5.1). Small quaternary amines such as ammonium ions are not a problem, because they quantitatively lose ammonia upon desorption, leaving the free acid group behind. Sodium and potassium, however, impose a rather strict limitation for MALDI sample preparation, particularly as they are essential ingredients for most molecular biological chemistry. Even if the reaction cocktails are modified to replace sodium or potassium salts by their ammonium moieties, careful purification of the samples before MALDI preparation is usually necessary [10], as will be discussed later for the different assays. Other reagents in the enzymatic reactions, such as detergents or proteins, can also seriously reduce sensitivity – either because of competition for the available charges, or because they interfere with correct matrix crystallization. For the same reason, the glycerol concentration should be kept at a minimum.

By far the biggest problem is fragmentation of the ions upon desorption. Indeed, this fragmentation constitutes the main difference between MALDI-MS of nucleic acids as opposed to proteins of comparable mass. It has been shown in a serious of experiments involving H/D exchange that for positive as well as negative (singly charged) MALDI ions the first step of fragmentation is in almost all cases the loss of a base, followed by a backbone fragmentation at the 3'-oxygen [11,12]. The model assumes that base loss is initiated by a protonation of the base, most probably involving the proton of the nearest 5'-phosphate group. This base protonation is known to destabilize the N-glycosidic bond, leading to the base loss. The model is supported among others by the observation that the high-proton affinity bases adenine (A), cytosine (C) and guanine (G) are lost much more facile than the thymine (T)-base. Poly-T, in fact, forms a rather stable ion. For the same reason, RNA forms substantially more stable ions than DNA [13], because the 2'-hydroxyl group stabilizes the N-glycosidic bond. The likelihood of a fragmentation increases with the number of bases – that is, the mass of the ion. It is due to this fragmentation that the mass range of DNA accessible to MALDI-MS with good sensitivity and mass resolution is limited to about 25 kDa or about an 80mer – a far cry from the >500-kDa value for proteins. A few spectra of DNA up to 150 kDa have been reported in the literature, but their quality does not suffice for a useful analysis [14,15]. Any assay design for the MALDI-MS of nucleic acids (which is described in detail below) must take the mass limitation into account.

Besides these difficulties, MALDI-TOF-MS of nucleic acids has one important advantage over that of peptides/proteins – namely that their structure is much more homogeneous, consisting of only four relatively similar building blocks.

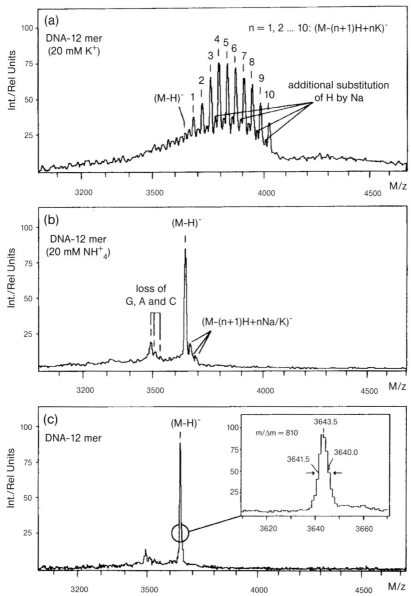

Fig. 5.1 Negative ion-mode mass spectra of a 12mer DNA oligonucleotide, demonstrating the importance of careful sample preparation and conditioning of the phosphate backbone of nucleic acids for efficient analysis by MALDI-TOF MS. (a) At concentrations of sodium or potassium similar to those used conventionally in molecular biological assays, the 12mer shows a distribution of up to 10 potassium ions attached to the phosphate backbone. Sodium and potassium can be efficiently replaced by employing ammonium-based buffers. (b) The 12mer oligonucleotide was dissolved in 20 mM ammonium acetate solution. The Na$^+$ and K$^+$ adduct formation is significantly reduced. (c) The use of cation-exchange polymer beads further leads to a significant reduction in adduct formation and improved signal-to-noise ratios.

Because of this structural simplicity, relative and even absolute (with an internal standard) quantification is possible and is used in a number of assays (see below).

Only two out of the dozen or so truly useful matrices for MALDI-MS have proven to be optimal for the analysis of nucleic acids. 3-Hydroxypicolinic acid (3HPA) [16,17] is the preferred matrix for the analysis of DNA, while a mixture of 2,3,4- and 2,4,6-trihydroxy-acetophenone (THAP) or 3-HPA is best-suited for RNA [18]. For short, charge-tagged segments of DNA up to about a 10mer – as is typically analyzed in the GOOD assay; see Section 5.4.1.4) – α-cyano-4-hydroxycinnamic acid methyl ester (CNME) can also be used. For 3HPA, the yield of positive and negative ions is comparable, but detection in the positive-ion mode is usually preferred because of a lower yield of the doubly charged ions and because of instrumental considerations.

Time-of-flight (TOF) mass analyzers with delayed ion extraction have been used almost exclusively for MALDI-MS of nucleic acids [19]. The facile fragmentation of nucleic acids also placed some constraints on the instrument configuration [17,20,21]. Up to about a 25mer or ca. 8000 Da, a reflectron TOF can be used to achieve a mass resolving power of up to 15000, with a mass accuracy as good as 10–50 ppm (i.e., 0.05–0.25 Da for a molecule with MW 5000 Da). Above this mass, fragmentation will degrade the resolving power in reflectron mode, and linear-TOF spectrometers become the instruments of choice (see Fig. 5.2). In the field-free drift tube, metastable post-source-decay fragments have the same arrival time at the detector of the linear tube as their parent ion, as described in Chapter 2. The mass-resolving power in linear TOFs is typically 500–1000 in the mass range above 10000 Da, but a resolving power of ca. 3000 can be achieved in the lower mass range. Even more of a problem is the achievable mass accuracy for linear TOF instruments. Mainly because of the heterogeneous surface topology, even of nanoliter 3HPA preparations, the mass accuracy for external calibration is only ±1–3 Da. However, with internal calibration a mass accuracy of ±1 Da can be achieved.

So far, essentially all applications are restricted to UV-MALDI at the wavelength of 337 nm of the N_2-laser. Infrared (IR) MALDI is known to be considerably softer than UV-MALDI, and the fragmentation of nucleic acids is indeed much reduced for a desorption at a wavelengths of 2.94 μm or 10.6 μm and glycerol or succinic acid as a matrix [22,23]. As shown in Figure 5.3, the spectra of DNA restriction enzyme fragments up to 1400mer single-stranded and 700mer double-stranded DNA have been obtained with this desorption mode [24,25]. Unfortunately, this combination of matrix and desorption wavelength also leads to an excessive adduct formation of glycerol and its desorption products, thus limiting the mass resolution to about 50. Worse, it even prevents the determination of the exact mass of the DNA free acid to any useful degree of accuracy.

The remainder of this chapter is mainly devoted to descriptions of the different assays that have been developed for the analysis of nucleic acids under the somewhat restrictive boundary conditions of MALDI-TOF-MS.

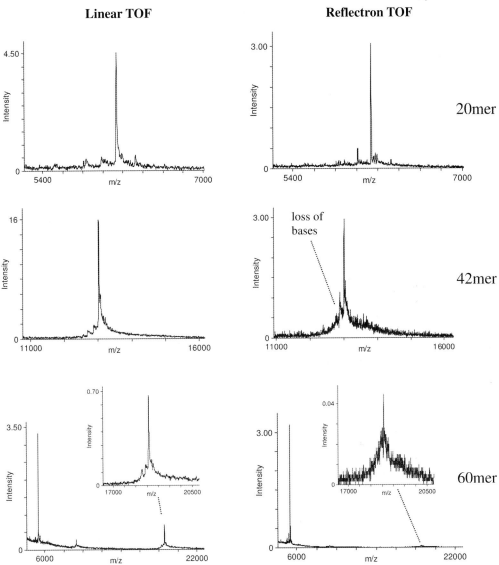

Fig. 5.2 Comparison of linear and reflectron time-of-flight (TOF) MALDI-MS for analysis of oligonucleotides of length up to 60 nucleotides. For shorter oligonucleotides up to 25mer length (ca. 8000 Da), higher resolution (mass resolving power of 15 000) and mass accuracy (10–50 ppm) can be achieved with reflectron TOF. At higher masses (>10 000 Da), the fragmentation degrades the resolving power in reflectron TOF MS, and linear TOF MS become the preferred instrument configuration. Linear TOFs can achieve a mass resolving power of 500–1000 for oligonucleotides above 10 000 Da, and of 3000 for smaller oligonucleotides.

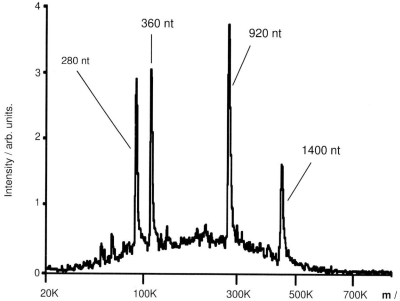

Fig. 5.3 Infra-red (IR) MALDI mass spectrum of large DNA analyzed from glycerol as a matrix. The spectrum depicts single-stranded DNA molecules of length up to 1.4 kB generated from restriction-digested plasmid DNA purified by ethanol precipitation. The achieved mass resolution is limited to about 50 for DNA molecules of this size caused by excessive adduct formation of glycerol and its desorption products. The demonstrated mass accuracy of 1% and sensitivity in low femtomole range is remarkable.

5.2
Genetic Markers

During the past decade, intensive worldwide efforts have been dedicated to establishing a reference sequence of the human genome. A draft sequence of the human genome was announced in 2000 [26,27], but by September 2002 many of the remaining gaps had been filled and the sequence was further refined such that over 90% of it could be presented in its final form, with an accuracy of greater than 99.99%.

Meanwhile, the genomes of a substantial number of other species have been published. As a result of this, emphasis has shifted from *de-novo* sequencing to the analysis of differences, for example between individuals, ethnic groups and diseased versus healthy populations. It is in this arena rather than *de-novo* sequencing that MALDI mass spectrometry has found multiple applications.

In this sense, although the completion of the Human Genome Project constitutes a major scientific milestone, it must also be seen as a starting point for the "real" exploration of our genome. We still have a rather "poor" understanding of the architecture, functionalities and regulation of the human genome. Although we have almost a complete DNA sequence, debate persists over the total number of genes, with recent estimates pointing to a number of approximately 30000 – much less than originally estimated.

The "variety" and complexity seen at the mRNA and protein levels is much higher than the number of genes suggest. One of the processes contributing to the estimated number of well over 100000 proteins arising from the rather low number of genes is alternative splicing. Among the processes not yet fully explored are epigenetic modifications such as methylation and histone acetylation, as well as RNA editing and certain forms of somatic recombination.

Aside from questions focusing on how genomic information translates into the mRNA and protein level, there is another important aspect: Is there something like "the reference sequence" for a species? How variable are our genomes?

It is currently estimated that the genetic diversity among humans is around 0.1%. Thus, two unrelated persons share about 99.9% of their DNA sequence, but given a genome size of three billion bases this means that they differ in millions of bases. The elucidation of this intra-species genetic diversity remains an important and still formidable task. It is important because the assessment of genetic variability will provide insights into genotype–phenotype interaction, and this will help us to understand the contribution of genes to diseases, disease predisposition, and for example drug response.

It is formidable because cataloguing these genotypic variations requires the analysis of large numbers of individuals either for a large number of known genetic markers or, preferably, for their complete individual genome sequences.

The genetic diversity within the human genome manifests itself mostly in single nucleotide polymorphisms (SNPs), restriction fragment length polymorphisms (RFLPs, forming a subset of all SNPs), short tandem repeats (STRs), and "random" base changes. The use of genome-scanning technologies has recently uncovered another type of genotypic variation that is mainly characterized by larger genomic deletions, duplications, and copy number variations [28,29]. In recent years, substantial efforts have focused on the development of suitable technologies to enable cost-efficient, large-scale assessment of these markers in multiple populations. In fact, MS has been among the technologies promising to sustain the high-throughput analysis of genetic markers. Naturally, most methods applying MS have focused on the development of assay formats suited to the analysis of these markers. To help understand the focus of most research groups developing tools for DNA analysis by MS, and the strategy upon which they embark, the nature and importance of the most common genetic markers and their assays are explained in more detail in the following section.

5.2.1
Restriction Fragment Length Polymorphisms (RFLPs)

Restriction endonucleases are enzymes, which recognize specific DNA sequences in a genome or given stretch of DNA and cleave the DNA in a predefined manner. Consequently, changes in a given DNA sequence may create new such cleavage sites, or abolish existing ones. Restriction enzymes were increasingly used for genome analysis during the late 1970s, when visualization of the restriction cleavage patterns was achieved by Southern blotting of the cleavage products and hybridization with labeled probes. When the restriction patterns of different individuals were compared, it became apparent that the nucleotide variation rate between individuals was higher than expected. DNA-based variations at restriction sites, which alter the restriction fragment length pattern, were accordingly termed RFLPs [30]. With the knowledge available today, RFLPs can be defined as a subset of SNPs located in the recognition sequence for the restriction site or the cleaved sequence.

RFLPs are of bi-allelic nature, which means that usually only two (of the four possible) nucleotides occur in a given population. The allele-status renders the DNA permissive to cleavage or not, and thus provides a "yes" or "no" answer, which can be used as a simple marker to follow a genetic trait. However, isolated RFLPs alone are not very informative and their abundance is too low to allow in-depth genetic studies. Although RFLPs have had their merits as a valuable tool in genome analysis and genetics, they have now been superseded by either more informative or more abundant markers. Furthermore, the process of Southern blotting is cumbersome and difficult to automate.

5.2.2
Microsatellites/Short Tandem Repeats (STRs)

Microsatellites are stretches of DNA, which consist of repeating units of two, three or four nucleotides (also referred to as di-, tri-, or tetra-nucleotide repeats). The length of such a microsatellite, which is based on the number of repeated units, can vary between the homologues of each chromosome of an individual and also between individuals, thus constituting a polymorphism [31]. The different repeat lengths of the microsatellite constitute the alleles of the polymorphism. Usually, multiple alleles are observed for any particular microsatellite in a given population, and thus the information content of this type of marker is much higher than for bi-allelic markers. This is one of the features which made them useful for genetic linkage studies. The probability of identifying heterozygous individuals for a selected marker is very important in linkage studies and in forensic applications. Tens of thousands of microsatellites were identified, and a portfolio with genome-wide coverage was developed for genetic linkage studies [32–34]. The analysis of microsatellites can be enabled by rather simple polymerase chain reaction (PCR) amplification techniques. Microsatellite marker analysis does not require Southern blotting, and thus offers a reasonable potential for automation. The separation and detection of the PCR products is usually performed by gel or capillary electrophoresis [35,36], and the screening of thousands of individuals for ten thousands of microsatellite alleles proved to be both challenging and cumbersome. Hence, in recent years the focus has shifted away from microsatellites such that they are now mainly employed in the field of identification (forensics), where less-informative markers are not easily applicable, or in the analysis of genomes of other organisms, where other types of markers are, as yet, scarce.

5.2.3
Single Nucleotide Polymorphisms (SNPs)

SNPs are usually defined as single base-pair (bp) positions in genomic DNA for which two sequence alternatives (called "alleles") exist, and the frequency of the least abundant allele is greater than 1%. Usually, SNPs are bi-allelic. About two-thirds of all SNPs are C/T sequence variations, while the other third is shared among the other three types (C/A, C/G, and T/A) with comparable frequencies [37–39]. Sequence changes occurring with a frequency of less than 1% are usually referred to as "mutations".

SNPs are the genetic marker with the highest abundance in the human genome, and occur with a frequency of about 1 in every 1000 bp. Given a human genome size of three billion base pairs, the genome of individuals will differ in millions of bases. Most of these SNPs will not reside in coding regions, and among the coding SNPs a fraction will be non-synonymous. Interestingly, about half of all coding SNPs identified so far are non-synonymous and thus lead to an amino acid exchange. The combinations of SNPs in regulatory regions as well as amino acid-coding SNPs have a substantial influence on the inter-individual difference in expression and in the proteome.

The high abundance of SNPs and the ease of developing assays for their analysis have recently attracted much attention from the scientific community. The importance of SNPs to our understanding of genetic diversity and the usefulness of SNPs for establishing genotype-to-phenotype correlations has been recognized by many academic and commercial research entities [40,41]. This has led in turn to the foundation and implementation of large-scale SNP discovery programs, among which the SNP Consortium (TSC) has received the most public attention. The TSC is a collaboration between 13 multinational companies and prominent academic institutions. The initial goal of the TSC was the delivery of 300 000 evenly spaced SNPs to the public domain by the end of 2001. The human diversity program of the Human Genome Project (HGP) is another program providing additional SNP markers. Currently about four million SNPs have been published in dbSNP (http://www.ncbi.nlm.nih.gov/SNP/), a public database of the National Center for Biotechnology information (NCBI), and thus the initial goals have been exceeded by far.

The tremendous effort associated with the generation of a SNP catalogue has furthered technology development on a broad scale, with efficient methods for the discovery of SNPs having been developed. Among these, indirect methods such as *in-silico* mining of existing data for polymorphisms through sequence overlays were employed. Direct methods have also been developed, which include for example reduced representative shotgun sequencing [42] (the method used primarily by the TSC), denaturing gradient HPLC [43], single strand conformation polymorphism analysis (SSCP) [44,45], re-sequencing hybridization chips [46,47], and direct DNA sequencing [48]. The discovery phase is usually followed by experimental verification of these SNPs and the assessment of their allelic frequencies in various populations. This step employs a different set of technologies, which were specifically developed for SNP analysis.

Mass spectrometry plays an important role in cataloguing SNPs and assessing their allele frequency, and the assay formats developed for this purpose are discussed in detail in the next sections.

5.2.3.1 The HapMap

With a catalogue of almost four million SNPs available, the scientific community has now entered into the next large-scale project: the HapMap. The focus of the HapMap project is to identify and characterize haplotype blocks in the human genome (haplotypes are defined as groups of closely linked genetic markers that tend to be inherited together and are rarely separated by recombination events) [49]. Having large amounts of SNPs available is a first step; the next step is to determine how these SNPs are grouped together, and how these groups (blocks) are inherited.

The initial question is how these SNPs are used in genome research, and how would the knowledge of haplotype group's advance genome research? The majority of SNPs do not reside in coding regions but, by virtue of their abundance, they are still the prominent percentage of SNP sets used in genome-wide association and linkage disequilibrium studies. Both are geared towards the elucidation of the polygenic origins of common diseases, the basis being a comparison of SNP allele frequencies among groups of affected and unaffected individuals. Statistically significant frequency changes are thought to be associated with phenotypic differences between groups, and the associated SNPs pinpoint to the genomic region within which the "causative" genomic polymorphism or mutation of interest resides.

SNP-based genome-wide studies have not only been suggested as methods to identify the genetic underpinning of common diseases, but they may also allow for the elucidation of individual responses to drug treatment, which would be another important piece in the puzzle of achieving individualized medicines.

Genome-wide linkage and association studies require the analysis of hundreds of thousands to millions of SNPs in thousands of individuals or DNA samples. This is a daunting task, and there are several technical ways to go about it. Another consideration is the HapMap, since if haplotype groups can be described for the human genome, and each haplotype group can be characterized by a subset of "tagSNPs", then the overall number of SNPs required for genome-wide analysis can be reduced drastically.

5.3
Assay Formats for Nucleic Acid Analysis by MALDI-MS

The development of assay formats for mass spectrometric analysis has followed the needs in genetic marker analysis and DNA diagnostics. In this section, a brief overview will be provided of the plethora of assay formats developed to analyze nucleic acids from biological samples in general, and various genetic markers in particular. A more detailed description follows in subsequent sections. Several recent reviews have summarized developments in the field of mass spectrometric nucleic acid analysis [50–53], and in order to avoid excessive redundancy, this chapter will only briefly summarize the early developments and devote special attention to recent developments, which make large-scale use of MALDI-TOF MS in genomic science.

As outlined earlier, sample preparation is a key aspect for the robust analysis of nucleic acids by MALDI-TOF MS. The challenge of sample preparation can be subdivided into three categories: (i) which purification format is suited to providing nucleic acids products sufficiently conditioned for MALDI-TOF MS analysis?; (ii) which enzymatic reaction generates nucleic acid products of a size (and molecular mass) suitable for robust analysis by MALDI-TOF MS?; and (iii) which preparation of the matrix/analyte mixture provides the sufficient performance for the various applications of nucleic acid analysis by MALDI-TOF MS?

Significant efforts were initially devoted to the purification of nucleic acid products from enzymatic reactions prior to MALDI-TOF MS. As will be indicated later (when the enzymatic assay formats are described in more detail), several components of the enzymatic reactions are detrimental to the MALDI-process. Purification formats must therefore be able to remove these components, such as detergents, surfactants, proteins and unincorporated nucleotides, and must also provide the means for efficient removal of high concentrations of salt commonly used in enzymatic reactions (sodium-, potassium-, magnesium-, chlorides and sulfates).

Among the purification formats evaluated for their potential in MALDI-TOF MS analysis, spin-columns with size-separation capabilities as well as solid-phase purification systems such as reversed-phase beads/columns and streptavidin-coated beads have proved very efficient [54–58]. The assay formats described in the following section very often rely on one of these purification formats.

While these methods have provided efficient means for purification of the enzymatic products, they have also proved to be a hurdle for high-throughput applications, with centrifugation steps (such as those required for spin columns) being difficult to automate. Solid-phase purification systems usually require several reagent addition and washing steps, and therefore require more elaborate pipetting systems.

It should also be considered that solid-supports – in particular streptavidin-coated paramagnetic beads – have a restricted binding capacity that varies with the length of the nucleic acid product bound to the surface. Hence, the overall analyte concentration will be limited by the binding capacity of the solid support, and if multiple gene regions are to be analyzed simultaneously (referred to as multiplexing) then differences in the lengths of the amplification products can lead to unwanted biases in their representation after purification. This cannot necessarily be offset by increasing the amounts of coated beads, mainly because of high costs but also because of leakage of the capture protein (usually streptavidin) into the analyte solution (leading to protein signals that dominate the mass spectrum).

It was noted more recently, however, that the effect of detergents/surfactants and proteins can be minimized in most assay concepts simply by dilution of the sample. Because of the exponential nature of the PCR commonly used to amplify gene regions prior to the analysis for mutations or SNPs, the nucleic acid analytes usually have a sufficiently high concentration that supports the concept of sample dilution prior to MALDI analysis. The main challenge therefore resides in the ion-exchange step to condition the nucleic acid phosphate backbone for MALDI. The following section will describe two concepts addressing the conditioning of the phosphate backbone. The first, introduced by Gut and Beck in the mid-1990s, uses DNA alkylation to generate non-ionic nucleic acids and therefore avoids the issue altogether [59]. The second (now commonly used) method employs the simple addition of porous ion-exchange beads to the diluted analyte-solution prior to mixing the analyte with matrix.

Lastly, the preparation of matrix and matrix/analyte mixture has received considerable attention. Here, the driving forces were the reproducibility and homogeneity of the crystallization process, as well as the potential for automation of the analyte transfer process. With respect to both, two developments of note helped significantly in allowing MALDI-TOF MS to be used routinely for nucleic acid analysis by non-experts, and in high-throughput settings. Both developments aimed to address the reproducibility of the final step of sample preparation by using prefabricated miniaturized arrays of matrix spots and transferring only sub-microliter amounts of sample onto these arrays [60–62]. This development was a cornerstone to enable the quantitative analysis of nucleic acids, as will be described later.

5.4
Applications in Genotyping

For each of the most common genetic markers described in the previous sections, attempts have been made to develop solutions using MALDI-TOF MS. Here, we will briefly outline those investigations involving the analysis of RFLPs [63,64] and microsatellites [58,65–69] by MALDI-TOF MS, and will focus on those applications, many of which have found widespread use in genomics/genetics research.

5.4.1
MALDI-TOF MS SNP and Mutation Analysis

Assay formats specifically suited to SNP analysis by MALDI-TOF MS have been a prime focus of recent method development. This is in part the case, because the analysis of large DNA molecules has not progressed sufficiently to allow high-throughput use of MS (see Section 5.1). However, with major genomics activities focusing on SNP analysis, new opportunities have arisen for MS, with multiple new assays having been developed, all of which had a common feature – namely that the size and nature of the generated products are well suited to mass spectrometric analysis in general, and to MALDI-TOF MS in particular.

Earlier reports on the use of MALDI-TOF MS to detect mutations/sequence changes employed more conventional techniques, such as restriction endonuclease digests [70] or the ligase chain reaction [71]. Initially, allele-specific hybridization was also combined with mass spectrometric detection. Smith and colleagues, for example, used peptide nucleic acid probes hybridized against an immobilized PCR template for genotyping [72]. Finally, the direct measurement of PCR products has been attempted as a means of nucleic acid analysis and genotyping [14,24,73–75].

The majority of recently published methods and applications use a primer extension concept in order to determine the genotype present in a particular SNP position and sample. One of the drivers for these assay concepts was the basic limitation in mass resolution and mass accuracy, which did not allow for "simple" mass determination of PCR products to accurately define genotypes.

A common scheme among these methods begins with the amplification of a target region by PCR. A detection primer is then annealed immediately adjacent to the mutation site/polymorphic site, and extended by a polymerase. The extension is terminated either on the polymorphic site or within a few bases thereafter through the incorporation of terminator nucleotides [usually dideoxynucleotides (ddNTPs); see also Fig. 5.4]. In the simplest case the products generated in these reactions are between 17 and 25 nucleotides in length (ca. 5000–8000 Da) and are thus readily amenable for MS analysis. The different concepts introduced deviate in the exact way that the primer extension reaction is performed, and in the purification formats applied.

5.4.1.1 The PinPoint Assay

In an approach termed the PinPoint assay, introduced by Haff and coworkers, four dideoxy terminators are used in the post-PCR primer extension reaction [76]. The primer is extended by only one nucleotide, and termination occurs directly at the polymorphic site. The genotype is determined from the molecular mass of the extension/termination product as this is directly correlated to the type of nucleotide incorporated at the SNP site. Discrimination of the alleles is based on the mass difference between the four terminators (9 Da for ddA/ddT; 40 Da for ddC/ddG; 16 Da for ddA/ddG; 25 Da for ddT/ddG; 15 Da for ddT/ddC; and 24 Da for ddC/ddA). A calculation of the molecular mass of the possible combinations of the four nucleotides for a string length between 17 and 30 shows that a single MALDI-TOF read could, in principle, resolve multiples of such primer extension products. In this way, Ross and coworkers were able to achieve a 12-plex analysis [77]. Intelligent assay design combined with "mass tuning" of primers, where nontemplated nucleotides were used to allow efficient mass-intercalation of multiple primer and primer extension products, increased the plexing level even further, though in most cases the molecular biology limited these efforts. Equal amplification of multiple target regions in one reaction is difficult to achieve.

Single base extension methods such as the PinPoint assay have an advantage for multiplexed SNP marker analysis as they can combine all assays, irrespective of the alleles to be genotyped, simply by using the four ddNTPs. However, the

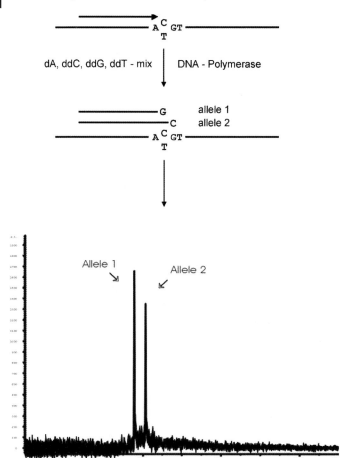

Fig. 5.4 Schematic representation of analysis of single nucleotide polymorphisms (SNPs) by primer extension and MALDI-TOF MS. The genomic target region is first amplified by the polymerase chain reaction (PCR). Then, an oligonucleotide primer is annealed immediately adjacent to the polymorphic site. This primer is extended allele-specifically using a DNA polymerase and a nucleotide mix that leads to a termination of the primer extension reaction either after one or two nucleotide additions. The length of the extension products is determined by the allele present in the analyzed sample. In the depicted case, the primer is extended by one nucleotide (incorporation of a ddG) for the G allele, and by two nucleotides (incorporation of dA and ddC) for the C allele. Analysis of the extension products by MALDI-TOF MS allows the unambiguous determination of the genotype present in the sample. Several variations of this basic concept have been developed where the primer is extended only by a single base and where mass-modified nucleotides are employed. All these variations aim to increase the ability to interrogate more then one genomic region at a time and therefore use the available mass window more efficiently.

concurrent use all four ddNTPs presents a technical challenge. Depending on the length of the primer, it is not trivial to resolve 9 Da for a heterozygous A/T mutation. For primers with MW larger than 5000 Da, mass accuracy might not even routinely be sufficient to distinguish between a ddA or ddT extension in homozygous samples. Furthermore, the mass differences between C-A (24 Da) and C-G (40 Da) fall very close to sodium and potassium adducts. This introduces a risk of false-positive heterozygote genotypes when salts are not completely removed from the analyte. A simple way to avoid ambiguities related to the "inconvenient" mass difference between pinpoint products is to use mass-modified terminator nucleotides, as introduced by Smith and coworkers [78]. A rationale design of terminator nucleotides with a "perfect" mass separation can be achieved by selecting appropriate mass-modifying chemical moieties.

5.4.1.2 The PROBE Assay

The primer oligo base extension assay (PROBE assay) introduced by Little and coworkers avoids mass resolution issues by employing a different terminator strategy [79,80]. The post-PCR primer extension reaction is performed in the presence of one ddNTP and three dNTPs. Mixes are chosen in a way that the two allele-specific extension products always differ in length and mass by at least one nucleotide. The mass difference of roughly 300 Da reduces the demand on the mass resolution and accuracy of the TOF-MS under high-throughput conditions, and adds a considerable safety margin. Assays can be multiplexed by intercalation in the same way as in the PinPoint. However, assays can only be multiplexed if their alleles can be differentiated with the same ddNTP/dNTP combination, as termination with a mix containing only one dideoxynucleotide bears another risk. Polymerase-pausing artifacts, where the extension reaction is not specifically terminated, can have the same mass as "real" termination products. For example, pausing on dATP (primer + 313.2 Da) leads to the same mass as termination on ddGTP (primer + 313.2 Da), and this could lead to a misinterpretation of results (again false heterozygous genotypes). These situations can, however, be circumvented through the design of tri-terminator mixes which exhibit clear mass differences between all possible products.

An interesting approach for the increase of multiplexing levels in MALDI-TOF MS genotyping has recently been introduced by Ju and coworkers [81,82]. The basis of the genotyping assay is a conventional single base primer extension reaction. Terminator nucleotides are used, which carry a biotin group, and the chemical linker between the biotin group and each of the four ddNTPs has been selected such that the mass difference between the terminator nucleotides is doubled. The reaction products are incubated with streptavidin-coated magnetic particles. Only extended primers carry a biotin moiety (at their 3′ end) and will be immobilized; the unextended primers and reaction components can then be efficiently separated with the supernatant. After release from the streptavidin beads, the extension products can be analyzed by MS. In this way, the benefits of solid-phase purification, which usually yields very pure analytes, was combined with the selection of only extended primers and with mass-modified terminator nucleotides.

The absence of unextended primers removes undesirable additional masses/signals and allows a more efficient use of the available mass window. As mentioned above, the use of terminators optimized for mass difference avoids ambiguities in spectra interpretation at limited mass resolution. Unfortunately, although this approach is technically quite attractive, it suffers from high processing costs and complications in automation.

Sample purification requirements of nucleic acids in MS have always been seen as a limiting factor. Adduct formation (mainly sodium and potassium as the predominant cations in enzyme reaction buffers) is probably the biggest concern in MALDI-MS. However, buffer detergents and large amounts of protein add significantly to the list of issues. Various methods to remove these ingredients from the analyte solution prior to MS analysis have been suggested, and solid-phase approaches such as coated magnetic beads have probably been the most successful. Their automation is feasible with reasonable efforts in instrumentation and robotics, and the samples are usually very "clean". However, when processing tens to hundreds of 384-well microtiter plates each day, the multitude of pipetting steps associated with the addition and removal of washing liquids present a bottleneck. In addition, coated particles are usually costly. The issue of the liquid handlers being the rate-limiting step (especially given the high acquisition speed of MS) is very often dealt with by integrating multiple pipettors/dispensers for parallel processing, but thus is expensive.

Several groups have made the development of less-demanding purification schemes for SNP analysis a focal point of their research. Meyer and coworkers employed the addition of NH_4^+-conditioned ion-exchange resin subsequent to completion of post-PCR primer extension reactions [83]. The overnight incubation with ion-exchange resin removes sodium and potassium sufficiently from primer extension products (usually these are only 17–25 nucleotides in length), and no further sample purification is required. The addition of ion-exchange resin to the solution containing analyte and matrix had been reported previously [10,84]. Upon matrix crystallization, the resin particles accumulate in the center of microliter preparations and can easily be removed manually before the sample is introduced into the mass spectrometer. For nanoliter preparations and automated analysis, as are commonly applied nowadays, commercially available resin particles of about 100 μm interfere with correct matrix crystallization, the result being unacceptably inhomogeneous samples.

5.4.1.3 The MassEXTEND, hME Assay

The PROBE assay has been further developed into a *homogeneous assay* – now called homogeneous MassEXTEND, hME – using the same purification principle: that all sequential enzymatic steps are performed by "simple" addition of the reagents to the reaction well. No washing steps are required, and ion-exchange resin is added to the crude solution in the last step [85]. To circumvent long incubation times, which are undesirable for high-throughput processes, the use of ion-exchange resin was coupled with a dilution of the sample with de-ionized water and optimization of resin types (mesh size and conditioning of the resin prior to use), which yields incubation times of only 10 min. Current processing volumes in the low microliter range allow such a step without the necessity of transferring the sample into new reaction vials. Since homogeneous assays do not allow for the removal of all components affecting the MALDI process, the dilution plays an important role. If sufficient amounts of analyte are produced, then dilution can be used as a vehicle to reduce the concentration of buffer and reaction components to the point that they do not interfere significantly with the sample crystallization and desorption process. Reduction of analyte volumes (as little as 10 nL are currently used in combination with micro-arrayed MALDI targets/chips) also aided the reduction of sample purity requirements [62,86].

Originally, homogeneous assays were not an obvious solution for the development in nucleic acid analysis by MS, as they represent a strong contrast to earlier reports of the impact of common ingredients of molecular biological reactions on MS performance. The final analyte solution contains almost everything (protein, salts, detergents, large nucleic acid PCR templates, genomic DNA, nucleotides) which was reported previously as degrading the spectra quality. As described, recent technological improvements in MALDI-TOF MS, as well as in the miniaturization of MALDI sample preparation, surely aided in this development.

The emergence of homogeneous assays must be attributed to the necessity for simple, cost-efficient processes in high-throughput operations. They represent a compromise between performance (analytical yield and accuracy) and cost for the assay, the process and the operation, which includes equipment time and resources. Figure 5.5 provides a representative mass spectrum of a 12-plex Mass-EXTEND reaction generated with the described homogeneous assay format. The extension products span the mass range from around 5000 to 8000 Da, and are sufficiently separated for clear assignment of genotypes based on the presence of mass signals. The successful use of MALDI-TOF MS as a high-throughput genotyping method at multiple sites worldwide shows that this compromise can be deemed successful.

The issue in high-throughput processing has been addressed in various ways. Gut and Beck for example, introduced the concept of charge-tagging and subsequent alkylation of the phosphate backbone as a means of increasing ion-yield and avoiding analyte purification [59]. Early experiments demonstrated improvements of the mass spectra when the nucleic acid analyte carries a preferred site for a single positive or negative charge [87]. In order to avoid heterogeneous cationization of the phosphate backbone by sodium or potassium, the phosphate backbone is alkylated. However, this approach relies on a quantitative alkylation which becomes increasingly difficult for species larger than 8mers. The reduction of the primer extension products of >18 nucleotides to small fragments can be achieved by the incorporation of phosphorthioate linkages near the 3′ end of the extension primer and the addition of an exonuclease digest step which stops at the

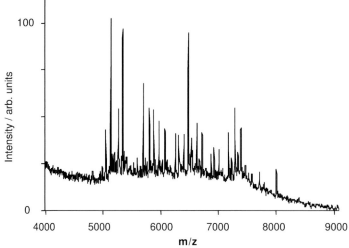

Fig. 5.5 MALDI-TOF mass spectrum of a 12-plexed SNP genotyping reaction. Multiplexed analysis of SNPs builds on the same scheme as depicted in Figure 5.4. However, to make more optimal use of the available mass window, multiple genomic regions are co-amplified in the same reaction and the amplified SNPs are each targeted with a specific extension primer. These primers are designed so that their molecular mass and the molecular mass of all their extension products do not overlap and still allow unambiguous identification of the SNP they were targeting and the allele that was present in the sample. Each assay occupies three mass signal positions representing the unextended primer, the first, and the second extension product. With careful consideration of the mass resolving power up to 40 SNPs can be interrogated in a single reaction and mass spectrometric readout.

phosphorthioate link. The concept was originally evaluated using an oligonucleotide model system. A streamlined version of this concept has been implemented as the "GOOD" assay for SNP analysis [88].

5.4.1.4 The GOOD Assay

The GOOD assay was adapted to cope with the peculiarities of oligonucleotide synthesis, enzymatic reactions, and alkylation efficiencies. The positive-ion MALDI-MS versions of the GOOD assays will be described here in more detail. The charge-tag concept for positive-ion MALDI-MS lends itself to a more facile assay development. Common to other approaches, the target region is first amplified by PCR, and the remaining dNTPs from the amplification are degraded by the addition of shrimp alkaline phosphatase (SAP). In a subsequent simple add-on step, an extension primer is annealed adjacent to the polymorphism/mutation. This primer carries a charge-tag near the 3′ end and a phosphorthioate bridge on the 5′ side of the charge tag. A DNA polymerase extends the primer allele-specifically using a suitable mixture of alpha-thio dNTPs/ddNTPs. A subsequent treatment with a 5′-specific phosphodiesterase digests the primer sequence down to the first phosphorthioate, leaving a small oligonucleotide with a charge tag and the SNP-specific nucleotide. The phosphorthioate bridges of these oligomers are then quantitatively alkylated using methyl iodide. The alkylated products are significantly less susceptible to cationization as the backbone is now neutralized.

In summary, the GOOD assay combines charge-tagging, backbone neutralization by alkylation and enzymatic treatments to allow for a homogeneous primer extension assay with subsequent detection of the products by MS. The addition of reagents during the subsequent steps "dilutes" those ingredients which hamper the MALDI process. This is similar to the dilution by addition of water as described above for the MassEXTEND assay. Furthermore, the allele-specific products are rather small after phosphodiesterase treatment (usually the length of tetramers/pentamers), and this helps the efficiency of the alkylation reaction.

One of the limitations of the original GOOD assay was the limited availability of modified nucleotides for subsequent coupling of positive charge-tags. However, this issue has been addressed with the introduction of new β-cyanoethyl phosphoramidites, which allow for easier charge tagging [89].

One further serious issue limiting more widespread use of the GOOD assay was the need for toxic reagents for alkylation of the phosphorthioate backbone. Two improvements have been introduced recently to avoid the necessity of alkylation [90]. Primers carrying methylphosphonate groups towards the 3′ end yield an "alkylated", charge-neutral 3′-end by synthesis. These provide the same inhibition of the phosphodiesterase digest as primers carrying phosphorthioate bridges. Incorporation of standard ddNTPs or α-thio-ddNTPs during the primer extension reaction generate a negative charge for MALDI-MS analysis. Methylphosphonate-containing primers are usually not the preferred substrate for DNA polymerases. The introduction of Tma31 FS, an enzyme capable of extending methylphosphonate primers and preferring the incorporation of ddNTPs over dNTPs, aided the development of an improved GOOD assay.

Some of the general issues in multiplexed MALDI-TOF MS genotyping related to mass accuracy, resolution and sensitivity could also be resolved if the products of primer extension reactions were to fall in a more benign mass window. Several approaches were conceived to generate short analytic fragments by introducing site-specific cleavage points into the primers. The recently introduced *GenoSNIP assay*, for example, uses photocleavable linkers in the primer to shorten the extension product prior to MALDI-TOF MS (http://www.bsax.de). Primers used in the GenoSNIP assay contain a photocleavable *o*-nitrobenzyl moiety, which replaces a nucleotide. This linker produces an abasic site within the primer, which does not inhibit its correct annealing to the target region. Exposure to UV irradiation cleaves the extended primer at the introduced moiety and allows for a separation of the informative extended 3′ end from the non-informative primer. The GenoSNIP assay uses streptavidin-coated particles for purification of the extended primer from the reaction prior to release of the 3′ end by photocleavage.

Sauer and coworkers merged the concepts of photocleavage for primer shortening and the GOOD assay to avoid the necessity of solid-phase purification and to overcome some shortcomings of the GOOD assay [91]. The combined approach uses primers containing phosphorthioates and a photocleavable linker towards the 3′ end of the extension primer. The allele-specific extension/termination is performed with α-thio-ddNTPs. Photocleavage and subsequent alkylation generate short assay-/allele-specific DNA fragments. The photocleavage leaves a 5′ phosphate group at the 3′ fragment and thus conveniently generates singly charged molecules without the introduction of real charge-tags or any need for phosphodiesterase treatment, as in previous versions of the GOOD assay.

Although shorter DNA fragments are easier to analyze, it must be noted here that this also reduces the compositional space available for multiplex design. A restriction to a length of 4mers with the last position fixed to a specific ddNTP leaves not more than 20 possible compositions with mass differences large enough to be resolved in the MALDI-TOF mass spectrometer. This presents a rather serious restriction if the assay panel to choose from is limited. The most efficient way for multiplexing might be to allow for cleavage sites to be distributed anywhere in the extension primer. In this way, compomer type and length can be used to spread assays over a defined mass range. However, to what extent the alkylation reaction can be efficiently employed on longer cleavage products remains to be explored.

5.4.1.5 The Invader Assay

The Invader assay, which is one of the few non-primer extension-based methods for MALDI-based genotyping [92], uses two sequence-specific oligonucleotides (an "invader" oligonucleotide and a probe oligonucleotide) which hybridize to a target sequence. The principle is of the assay is depicted in Figure 5.6. The probe oligonucleotide and target region form a duplex which includes the polymorphic site and a non-complementary 5′ overhang of the probe. The invader is designed such that its 3′ nucleotide invades into this duplex at the polymorphic site, forming a sequence overlap at this position. DNA repair enzymes called flap endonucleases (FENs) specifically recognize and cleave the unpaired region on the 5′ end of the probe oligonucleotide. This generates a reporter molecule, which can be used to signal the cleavage event. Each genotyping assay requires three oligonucleotides: the invader and two allele-specific probes. The non-complementary sequence of the 5′ end of these probes can be designed to allow discrimination between the probes that have been cleaved. If thermostable enzymes are used, this process can be run near the melting temperature of the duplex formed between probe and target sequence, so that cleaved products cycle off and are replaced by non-cleaved probes. This process is efficient enough that the Invader assay does not require prior amplification of the target region by PCR.

Smith and coworkers combined this genotyping method, which works directly on genomic DNA, with MALDI-MS analysis. In a reaction called the "Invader Squared Assay", where the primary cleavage product serves as an Invader molecule for a second Invader reaction, the "squared" cleavage product is purified via streptavidin–biotin. This allows separation and conditioning of the cleavage products for MALDI-TOF MS [93].

5.4.1.6 Incorporation and Complete Chemical Cleavage Assay

A process called the Incorporation and Complete Chemical Cleavage (ICCC) assay was introduced recently as an alternative genotyping principle to the most

Allele 1 (R)

Allele 2 (Y)

primary system

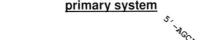

ARM 1

ARM 2

cleavage site

cleavage site

5′—AACGAGGCGCACTTR

5′—AGCAGGCACACGTTTY

Invader oligonucleotide   Primary **Allele 1** probe 3′

Invader oligonucleotide   Primary **Allele 2** probe 3′

5′

5′

3′   5′

3′   5′

Genomic DNA representing **Allele 1 (R)**

Genomic DNA representing **Allele 2 (Y)**

secondary system

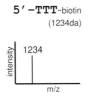

cleavage site

cleavage site

5′—TTTTTCATCTCAGAC—3′ Secondary probe for ARM 1

5′—TTTTCTCTTACACTGT—3′ Secondary probe for ARM 2

5′—AACGAGGCGCACTTR   ARM 1

5′—AGCAGGCACACGTTTY   ARM 2

3′ NH2—AAUUGCTCCGCGTGAAAAAGTAGAGTCTG—5′ Secondary target for ARM 1

3′ NH2—AAUCGTCCGTGTGCAAAAGAGAATGTGACA—5′ Secondary target for ARM 2

Identifier molecule

5′—**TTT**-biotin

5′—**TTTT**-biotin

(1234da)

(1538da)

intensity | 1234

intensity | 1234   1538

intensity | 1538

m/z

m/z

m/z

Sample Homozygote
for Allele 1 (R)

Sample Heterozygote

Sample Homozygote
for Allele 2 (Y)

Fig. 5.6 Schematic representation of the Invader assay for MALDI-TOF MS-based genotyping. The Invader assay uses two sequence-specific oligonucleotides (an "Invader" oligonucleotide and a probe oligonucleotide) that hybridize to the target sequence. The oligonucleotides form a duplex which includes the polymorphic site and a non-complementary 5′ overhang. The invader oligonucleotide creates a sequence overlap at the polymorphic site that is specifically recognized by a DNA repair enzyme (a flap endonuclease (FEN)) and cleaved. The cleavage releases the 5′ overhang of the probe oligonucleotide, and this released sequence can then be detected by MALDI-TOF MS. For each SNP, two allele-specific oligonucleotides are used that also differ in their 5′ overhang. MALDI-TOF MS analysis of the reaction products allows identification of the released overhangs and therefore identification of the alleles present in the sample. In most cases, a so-called "squared invader" assay is used. This variations uses the first released 5′ overhang as an invader oligonucleotide for a separate, generic reporter system that releases a second 5′ overhang specifically designed for detection by MALDI-TOF MS.

commonly used primer extension methods [94]. During PCR, one of the natural dNTPs is completely replaced by a chemically labile nucleotide. Following PCR, the generated amplicon can be cleaved specifically at the incorporation sites, and this generates a set of cleavage products that can be analyzed by MS. The genotype can be deduced from specific cleavage products and their corresponding mass signals. Usually, one of the PCR primers is designed adjacent to the polymorphic site such that the cleavage products containing the primer sequence (which do not contain any modified nucleotides) up to the first cleavage site generate the allele-determining mass signals. Assay design is a function of the alleles to be discriminated, the primer position and the choice of modified nucleotide. The combination of these three factors determines the signals pairs, which can be used for genotyping. 7-Deaza-7-nitro-dATP/dGTP and 5-hydroxy-dCTP/dUTP have been proposed as chemically labile nucleotides for base-specific cleavage. Both nucleotide sets form standard Watson–Crick base pairing, which is important to keep artifacts generated by modified nucleotides at the lowest minimum. Cleavage is mediated through an oxidant and incubation with an organic base. Two important factors determine the utility of this process: (i) the modified (chemically labile) nucleotide must be incorporated efficiently enough by a DNA polymerase for it to completely replace its natural counterpart in PCR without reducing the PCR yield; and (ii) the cleavage reaction must be complete for each incorporated modified nucleotide (to avoid potential misinterpretation of mass signals) and must not change the chemical nature of any other nucleotide in the amplicon.

There are several process-inherent advantages for the proposed method. The incorporation of modified nucleotides during PCR allows the amplification and genotyping process to be combined in one reaction and eliminates the need for subsequent enzymatic reactions. This simplifies the processing steps, which is important for automation. Since the double-stranded amplicon is used directly, the cleavage reaction generates products (i.e., mass signals) from both strands. This provides redundant information for determination of genotypes and, by the internal confirmation, increases accuracy of the result.

The underlying principle of this method – base-specific cleavage – is now used more widely for the sequence analysis of target regions (for further applications of this principle, see Section 5.5). The main concern surrounding ICCC as a general method for MS-based sequence analysis can be attributed to the slightly lower amplification yield with modified nucleotides, which might not prove sufficient for an analysis of 500+bp. Furthermore, oxidative agents used for the chemical modification of nucleotides are often hazardous and thus not desirable for high-throughput processing. Both issues can, however, be tackled by identifying DNA polymerases with increased incorporation rates for modified nucleotides, as well as alternative cleavage reagents. This is very similar to the evolution of the GOOD assay described earlier.

5.4.1.7 The Restriction Fragment Mass Polymorphism Assay

One of the most recent assay formats used for genotyping by MALDI-TOF MS is termed the restriction fragment mass polymorphism (RFMP) assay [95]. This was developed to allow the genotyping of polymorphic regions, where the close proximity of multiple polymorphisms renders the development of post-PCR primer extension readouts extremely difficult. In the RFMP assay, the region of interest is amplified with PCR primers carrying IIS restriction endonuclease recognition sites. Following the PCR amplification, the PCR product is digested with a IIS restriction endonuclease. This enzyme cleaves at a specified position up to 20 nucleotides distance from the recognition site. If both PCR primers carry a IIS recognition site, this process allows for enzymatic removal of the primer sequences from the PCR product and generates a cleavage product of a length suitable for direct analysis by MALDI-TOF MS. Kim et al. used this approach for genotyping polymorphic regions of the hepatitis C virus (HCV) genome [95].

5.4.2
MALDI-TOF MS for Haplotyping

The use of SNPs as polymorphic markers for association/linkage disequilibrium (LD) studies now supersedes conventional studies using microsatellite markers. The huge number of available SNPs, coupled to their relatively homogeneous distribution in the human genome, compensates their biallelic nature. In some instances, however, haplotype structures (the collection of genotypes found in a single allele or chromosome) rather than individual SNPs can be the principal determinant of a phenotypic consequence. Notably, they can provide additional statistical power in mapping disease genes, especially when the SNPs contributing to a disease are not directly observed, or when their interaction does not follow a simple additive effect [96–100]. Correspondingly, the determination of the allelic phase of SNPs (the haplotype) is an important component of genetic studies. Determining the haplotype for several SNP markers in a diploid cell is challenging. The availability of pedigree genotype information, for example, can be used to determine offspring haplotypes. In addition, computational algorithms such as the EM algorithms and PHASE have been developed to impute haplotypes based on available genotype information in a population [96,101]. Both methods, however, can fail to reconstruct the correct haplotypes.

An alternative to these methods is *direct molecular haplotyping*. All genotypes determined after the physical separation of the two homologous chromosome sets/genomes are in phase by default, and thus represent the haplotype. Physical separation requires rather elaborate methods such as cloning or somatic cell hybrid construction [102,103]. Other methods make use of allele-specific PCR to amplify only one homologue of a gene/gene region or to use single-molecule PCR, where each of the two homologues is represented on a statistical basis [104–107]. To date, however, reports on true high-throughput haplotyping methods are few in number.

Two recent reports have evaluated the use of MALDI-TOF MS for high-throughput haplotyping. Gut and coworkers combined allele-specific PCR for selective amplification of only one of the two homologues of a gene with multiplex genotyping of further SNPs contained in the PCR product by the GOOD assay [107]. Allele-specific PCR uses primers that hybridize with the 3′ end on the polymorphic site and, by choosing the nucleotide, the amplification can be directed towards the desired allele. DNA polymerases will recognize mismatches in the hybridization of the primer to the undesired allele and will not extend these primer template structures.

If the allele frequency of SNPs in the genomic region of interest is unknown, however, multiple SNPs have first to be typed to find the informative heterozygous positions. These can then be used for allele-specific amplification, followed by re-genotyping of SNPs contained in the allele-specific amplicon, which then yields the haplotype information. If a sufficient number of SNPs with known heterozygosity is already available from databases, a brute-force approach can be chosen. Each high-polymorphic SNP can be used as an anchor position to perform allele-specific PCR on all samples. Samples homozygous for the selective SNP will be easily identified by producing heterozygous genotypes in the mass spectrometric analysis and can be excluded from haplotype analysis.

The combination of allele-specific PCR with primer extension and MALDI-TOF MS for haplotyping may help to overcome some of the throughput hurdles of current technologies. MALDI-TOF MS offers multiplex SNP haplotyping capabilities while maintaining a high analysis speed and accuracy.

Although allele-specific PCR can be established as a robust method, it has often been observed that reactions "leak": the undesired allele sometimes co-amplifies through primer mismatch and this results in ambivalent genotypes. Thus, careful adjustment of primer design (GC-tails, mismatch design) and amplification conditions must be performed in order to render the reactions more specific. These optimizations are not feasible in brute-force approaches. To avoid issues associated with allele-specific PCR, Ding and Cantor used single-molecule dilutions of genomic DNA to separate two homologous genomic DNAs, and combined this with the MassEXTEND assay [108]. Their scheme proposes the dilution of genomic DNA to about 1 genome copy per PCR aliquot. Estimation of copy numbers of very dilute DNA concentrations follows the Poisson distribution [109]. Accordingly, multiple replicates of each individual must be analyzed until the haplotype for multiple SNPs can be constructed. Ding and Cantor estimated from their results that the PCR efficiency from single molecules was about 90–95% for amplicons of about 100 bp length. This high PCR efficiency obtained with current amplification systems is one of the reasons why the approach, which was proposed originally a decade ago but soon abandoned due to its impracticability, now seems feasible. When multiplexing PCR and MassEXTEND, Ding and Cantor achieved a haplotyping efficiency of 40–45% per reaction. Thus, four replicates should increase the haplotyping efficiency to about 90%.

An additional advantage of single-molecule amplification for haplotyping is the independence of the method from the distance between SNPs. Allele-specific PCR-based methods require SNPs to be adjacent enough that they can be amplified on the same PCR product. Single-molecule methods combined with intelligent design of overlapping multiplex assay allow haplotyping of more than 20 kB length, provided that the genomic DNA does not have substantial physical breaks.

As highlighted earlier, the described methods would in principle also be applicable to other detection platforms, but the accuracy and speed of mass spectrometric analysis combined with its multiplexing capabilities simplifies the implementation in high-throughput settings.

5.5
Applications of Comparative Sequence Analysis

MALDI-TOF MS was proposed as a separator and detector for Sanger sequencing ladders during the early 1990s, when new technologies to sequence the human genome were in demand. The appeal of TOF MS was the extremely high duty cycle, which suggested acquisition and analysis times of seconds, as well as a high potential for automation. For comparison, the separation and detection speed of state-of-the-art sequencing equipment using the Sanger concept is still in the range of hours. In addition to the replacement of laborious electrophoresis steps, MS is interesting for the field of nucleic acid sequencing, as it does not require labeling (either radioactive or fluorescent). Moreover, MS measures an inherent physical property – the molecular mass – and this should increase accuracy and minimize false interpretation of artifacts.

The basic concept of sequence determination using mass spectrometric analysis of Sanger sequencing ladders relies on the superposition and alignment of mass signals obtained in the four base-specific termination reactions originally described by Sanger and coworkers [110]. The molecular mass of termination products, as well as their mass difference, can be used to calculate the sequence. In essence, all mass signals arising from Sanger sequencing should be composed of a combination of the four nucleotides A, C, G and T, plus one terminator nucleotide (ddA, ddC, ddG, or ddT). Discrimination of termination events from polymerase artifacts is feasible, if a sufficient mass resolution and mass accuracy is provided by the instrumentation. An early report by Lloyd Smith exemplified the concept of sequence analysis with MALDI-TOF MS using a mock DNA sequencing ladder produced by mixing oligonucleotides [111].

Since then, several groups have elaborated on the approach with various schemes and assay formats [54–56,94,112–122]. One of the longest sequencing read lengths obtained by MALDI-TOF MS analysis of Sanger sequencing ladders was reported by Taranenko and coworkers [123,124], who were able to identify sequencing products of up to 120 bases. With the primer length subtracted, this represents an effective read length of 100 bases. At the same time, the report by Taranenko et al. illustrates some of the basic and often-ignored technological limitations. First, in most reports a model of known sequence was used to evaluate the method, but this bears the risk of supervised data analysis geared towards the known end result. Second, current mass accuracy and mass resolution present a challenge to accurate sequence determination. The biggest challenge is to discriminate between "real" termination products and polymerase-pausing events. A mass difference of 16 Da must be resolved in a mass window ranging from 5 to 25 kDa. Even more challenging situations can occur in case of, for example, sporadic false termination signals interfering with "true" termination, in which case mass differences less than 16 Da have to be resolved.

The two most recent approaches attempted to address these issues. In the first of these, Kang and coworkers employed transcriptional synthesis of chain-terminated RNA ladders for sequencing [124]. The approach has two potential advantages: (i) the isothermal transcription process increases the yield of analytes several fold more than a cycle sequencing reaction; and (ii) the analyte is now comprised of RNA, which is less prone to ion fragmentation. As expected, the mass separation power reported by Kang and coworkers was less limiting as compared to that for standard DNA sequencing ladders, though the read length could not be significantly extended. One reason or this might be related to the processivity and fidelity of the enzyme when non-canonical nucleotides, such as the 3'-deoxyribonucleotide terminators, are employed. Abortive cycling (the unspecific premature termination of the transcription process within the first 10 nucleotides) is another factor, which surely influences spectra interpretation and analyte yield.

The most recent report by Ju and coworkers introduced the use of biotinylated dideoxy nucleotides for MALDI-based DNA sequencing [125,126]. The use of biotinylated chain termination nucleotides addresses the issue of polymerase pausing. This biochemical artifact is not a specific issue for MALDI-based sequencing; in fact, it became eminent very early in Sanger sequencing with fluorescence-labeled sequencing primers. All extension products generated from the sequencing primer were carrying the same fluorescent label, and distinction between specifically terminated fragments and unspecific byproducts was impossible. In order to increase the accuracy and ease of sequence analysis, fluorescent-labeled chain terminators (ddNTPs) were developed. Biotinylated dideoxynucleotides follow the same principle by allowing the separation of specifically terminated sequencing products from unspecific/unwanted byproducts through solid-phase purification systems such as streptavidin-coated magnetic beads. In this way the mass spectra appear "clearer", and the risk of misinterpretation of mass signals is greatly reduced. A representative result of a Sanger sequencing mass spectrum generated with the approach of biotinylated dideoxynucleotides is shown in Figure 5.7.

Even today the read length is still far lower than can be achieved with current state-of-the-art sequencing equipment. Despite the recent developments MALDI-based sequencing, the use of the Sanger concept never found its way into the production units of genome centers.

Nonetheless, a new set of methods for MALDI-TOF MS-based sequencing has recently been implemented. These new methods rely on the generation of rather short oligonucleotide fragments by complete endonucleolytic cleavage, such that each cleavage product features at least one defined terminal base. This concept of base-specific cleavage resembles a sequencing method previously developed by Maxam and Gilbert. From a MS point of view, base-specific cleavage also relates to peptide mapping used for protein identification, because mass signals are matched to compositional explanations and sequence strings are reconstructed from the explanations. The degradation of a DNA analyte into short oligonucleotide fragments for nucleic acid sequence analysis avoids some of the mass accuracy, resolution and sensitivity issues, which are inherent to primer extension methods.

All methods described in this section require the a priori knowledge of a validated reference sequence, and cannot be used for de-novo sequencing. Nevertheless, they can be applied to experimental questions for DNA-based identification

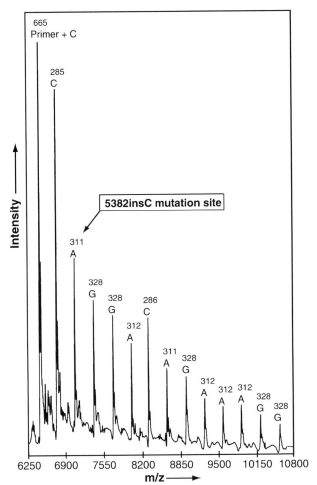

Fig. 5.7 MALDI-TOF mass spectrum of a Sanger sequencing ladder generated by primer extension from PCR products using a mixture of normal elongators (dNTPs) and biotinylated terminator nucleotides (ddNTPs). The termination products are purified on a streptavidin-coated solid support. The sequence of the target region is derived by calculating the mass difference between the termination products. Each mass signal in the spectrum is marked with the mass difference to the preceding mass signal and the corresponding nucleotide that this mass difference represents (C = 289.2 Da; A = 313.2 Da; G = 329.2 Da; T = 304.2 Da).

or re-sequencing, because in these applications an experimentally determined sequence is cross-compared to a known reference sequence.

The race for the human genome sequence is over. Thus, the question must be asked if there remains a need for high-throughput sequencing methods based on MS.

As eluded to earlier, genome sequencing projects have made tremendous progress during the past few years with conventional sequencing technology [127], and the sequence data accumulated to date has led to the discovery of the most abundant type of genetic marker, SNPs. Today, over four million SNPs have been deposited in databases. This number of markers may prove sufficient for current efforts in genome-wide association studies and the elucidation of linkage dise-quilibrium in the human genome (i.e., to build a map of the most common haplotypes). However, once particular genomic regions are identified as being associated or in-linkage with a specific phenotypic trait, there remains a need to identify the exact gene leading to the explored phenotype difference. Usually, higher-density marker panels are required within the particular genomic region. However, public databases very often do not contain sufficient SNPs for higher-density panels, and consequently new SNPs must be discovered in the target region and study population. This is usually accomplished by sequencing the target region in a sufficient number of individuals (the term "sufficient" being dependent on the desired allele-frequency of SNP markers in the panel).

Once higher-density panels have confirmed linkage or association and identi-fied the disease gene(s), the disease-causing sequence variants (mutations) must be discovered. This usually involves sequencing the respective genes in multiple affected individuals and leads to project efforts of considerable size. It seems fair to conclude that there is still a tremendous demand for large-scale sequencing in the human genome.

Given our genetic variability, the results of genetic studies will most likely tran-scend into multiple diagnostic sequencing applications. The continuing demand for high-throughput sequencing may become even more obvious, if we include other organisms, especially those with much faster generation times (as would be necessary in the field of infection and healthcare).

5.5.1
Base-Specific Cleavage of Amplification Products

The core of the new methods for MALDI-TOF MS-based sequencing is the base-specific cleavage of amplification products. Hillenkamp and coworkers first intro-duced such a concept in 1997 [128], though newer approaches have slightly modified the concept to avoid some of the initial problems. Today, a single-stranded copy of the target sequence is usually generated and cleaved to comple-tion in up to four separate base-specific reactions. Each reaction reduces the original sequence into a set of oligonucleotides, which is readily separated and analyzed by MALDI-TOF MS. The analysis utilizes the concept that a set of com-pomers (combinations of the nucleotides A, C, G and T) can be assigned to each mass signal of a base-specific cleavage reaction. The sequence can be reconstructed from the set of compomers by combining the information of all four base-specif-ic cleavage reactions and comparing it to a predicted set of mass signals derived from *in-silico* cleavage of the supposedly known reference sequence.

Several methods have been developed to obtain base-specific cleavage of ampli-fication products, the main difference between them being related to the type of analyte: base-specific cleavage occurs either on DNA or RNA.

5.5.1.1 **DNA-Based Methods**

DNA-based methods obtain base-specific cleavage directly from a DNA amplification product (such as a PCR product). As described earlier in Section 5.4, this can be achieved for example by the incorporation of chemically modified nucleotides during PCR, which generates site-specific cleavage sites in further post-PCR reactions [129]. This concept has some disadvantages when applied to longer target regions, such as those normally used in SNP discovery or mutation detection. Chemically modified (non-natural) nucleotides are often incorporated with a lower efficiency during PCR, and this normally has a more profound impact on amplification yield when products over 500 bp length need to be generated. Another issue is related to the fact that base-specific cleavage products are generated from both strands. Although this is desirable for genotyping, because it provides redundancy in the analysis and increases accuracy, it can also lead to issues in the analysis of longer amplicons. Without strand separation prior to cleavage, the mass signals of sense and anti-sense cleavage products may overlap and may compromise the reconstruction of sequence changes. Both issues were addressed in an approach published by Shchepinov and coworkers [130]. The group introduced the use of acid-labile P–N bond nucleotides to obtain base-specific cleavage. For this, 5'- phosphoramidate analogues were chosen, because they exist as stable triphosphates, they are incorporated by DNA polymerases, and their cleavage proceeds under acidic conditions such that the addition of MALDI matrix generates the site-specific cleavage products.

Due to the instability of the nucleotides under PCR cycling conditions and issues with incorporation efficiencies, the process features immobilization of the PCR product to a solid support, thereby generating single-stranded templates followed by an isothermal primer extension reaction for incorporation of the modified nucleotides. In this way a specific strand of the PCR product can be selected for base-specific cleavage and the yield of products is not compromised.

An alternative approach introduced by two different groups [131] makes use of the natural enzyme repair mechanisms in prokaryotic and eukaryotic cells. The base uracil is exclusive to RNA, and is not a part of the DNA alphabet. Deoxyuracil can, however, be generated when mutagenic chemicals modify DNA, as for example in the deamination of cytosine. DNA glycosylases, such as uracil DNA glycosylase (UDG), recognize uracil residues and remove the base from the DNA strand. This triggers further enzymatic reactions to repair the affected basepair. Consequently, the incorporation of dUTP (as an analogue of dTTP) during an amplification reaction (PCR) will tag all T positions in the PCR product for attack by UDG. The UDG treatment generates abasic sites (an exposed phosphate backbone), which can be cleaved under alkaline conditions. The concept has been used on a wider scale for prevention of PCR contamination in diagnostic settings. In the context of this section, the method can be used to generate T-specific cleavage patterns of PCR amplification products. The target region of interest can be amplified with a nucleotide mix containing dUTP instead of dTTP (full replacement). In the methods introduced by von Wintzingerode and co-workers, and also by Elso and coworkers, the dUTP-containing PCR product is immobilized to a solid-support (e.g., using the streptavidin–biotin system), denatured, and the generated single-strand is subjected to UDG treatment as well as backbone cleavage. The resulting T-specific cleavage products are analyzed by MALDI-TOF MS. In principle, the reaction can also be performed with the double-stranded PCR product [130]. This generates cleavage products from both strands simultaneously. As mentioned earlier, the number of potentially coinciding products could lead to issues in the discovery of sequence changes. For applications such as pathogen identification, however, the simultaneous generation of sense and anti-sense cleavage patterns could increase the number of discriminatory signals.

5.5.1.2 RNA-Based Methods

More recently developed methods have focused on RNA transcription and RNase cleavage as a means of generating base-specific cleavage. In two reports, which appeared almost simultaneously, G-specific cleavage patterns were generated with RNase T1 cleavage and used for the identification of sequence polymorphisms and for the generation of bacteria-specific mass signal patterns. The concept uses amplification of genomic target regions with primer pairs containing promoter tags. The PCR product is transcribed in separate reactions from forward and reverse directions, and the transcripts are cleaved at every G upon addition of RNase T1 [123,125,132]. Although at first glance the concept seems complicated, there are some inherent advantages over DNA-based approaches. As discussed in Section 5.2.2, the RNA transcription step amplifies the target molecule at least 50- to 100-fold, which means that about 50- to 100-fold more analyte is available for the mass spectrometric analysis. Correspondingly, the S/N ratios are significantly higher. Since RNA is less prone to base loss, the spectra are also virtually devoid of base-loss signals. The stringency for adjusting appropriate laser fluence is reduced, and this simplifies fully automated data acquisition. The use of RNA transcription enabled the development of a homogeneous assay format for base-specific cleavage. All reagents for subsequent steps are simply added to the PCR product (provided that the starting volume for the PCR reaction is sufficiently small that the total volume does not exceed the maximum volume of the MTP wells). The high concentration of analyte allows dilution of the sample prior to sample transfer on a MALDI target with precrystallized matrix. This eliminates the detrimental effect that some of the buffer components of enzymatic reactions exhibit for the MALDI process.

The use of RNase T1 provides considerable power in the detection of single nucleotide polymorphisms. G-specific cleavage reactions performed from both, forward and reverse strands, allow in aggregate the detection of 80–90% of all possible SNPs within target regions up to 500 basepair length. Exact numbers vary with sequence context, and also depend largely on the allelic nature of the SNP (heterozygous or homozygous). To increase the sensitivity of base-specific cleavage for the detection of sequence changes, a scheme featuring cleavage all four bases is preferred. Stanssens and coworkers recently introduced an RNase A-based concept to achieve base-specific cleavage at all four bases [133]. RNase A naturally cleaves RNA at every pyrimidine residue (C and U); hence, cleavage at all pyrimidines would reduce the target sequence to an uninformative mixture of extremely short oligonucleotides. Stanssens and coworkers avoided this issue by rendering RNase A base-specific through the incorporation of non-cleavable nucleotides during RNA transcription. A mutant of T7 and SP6 RNA polymerase has the ability to in-corporate non-canonical nucleotides, such as dNTPs [134]. Correspondingly, the incorporation of dCTP during transcription can be used to obtain U (T) -specific cleavage. The incorporation of dUTP (or dTTP) leads to a C-specific cleavage. Virtually all four bases can be covered, if C- and T-specific cleavages are performed on the forward and reverse transcript. Typical base-specific cleavage spectra gener-ated through the process described above are shown in Figure 5.8.

5.5.1.3 Analysis of Base-Specific Cleavage Patterns

The ability to generate cleavage patterns of virtually all four bases increases the sensitivity of base-specific cleavage for the detection of sequence changes. An ini-tial simulation using a 4-Mb sequence region surrounding the *ApoE* gene as a model system revealed that about 99% of all possible single nucleotide changes (substitutions, insertions, deletions) could be detected at amplicon length of 500 base pairs. Exact numbers vary with amplicon length (decrease with increasing length of the amplicon) and are sequence context-dependent [73,135–137].

The analysis of base-specific cleavage patterns is significantly more complex than that of primer extension products, and in order to interpret spectral changes the results of the four complementary cleavage spectra must be integrated. In genotyping, one can simply define the molecular mass of potential extension products and evaluate the presence of mass signals at either defined mass (or at both, for a heterozygous individual). In base-specific cleavage, different and math-ematically more challenging concepts must be applied, as the simple comparison of the sample spectrum with *in-silico* spectra from all possible sequence changes would, computationally, not be time-efficient. In principle, a single base substitu-tion can lead to the following observations: it can remove a cleavage site generat-ing a new, larger fragment; it can introduce a cleavage site generating two new shorter fragments; it does not alter a cleavage site, but leads to a mass shift in one of the fragments leading to a new mass signal with either lower or higher mass. In aggregate, the combination of four cleavages can result in a maximum of five such observations for a heterozygous sequence change, and in up to 10 observa-tions for a homozygous sequence change (since the information of a missing pre-dicted signal is also an observation). Single base insertions and deletions generate a maximum of nine observations in a homozygous sequence change.

A concept for automated analysis of cleavage spectra has been published recently [138]. Given the availability of a reference sequence and a defined method of cleav-ing the nucleic acid amplicon, one can simulate the expected mass signals for each cleavage reaction in an *in-silico* experiment. Spectra are then evaluated by com-paring the *in-silico* mass signal pattern with the experimental mass signal pattern. For reasons explained later, additional mass signals (defined as signals present in the sample spectrum but not in the reference spectrum) are selected as indicators for a deviation of the experimental sequence from the reference sequence. As a

T specific cleavage forward strand

C specific cleavage forward strand

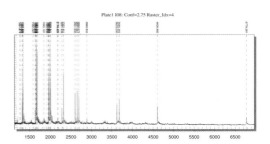

T specific cleavage reverse strand

C specific cleavage reverse strand

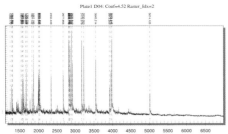
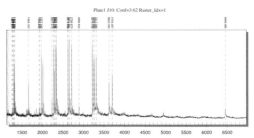

Fig. 5.8 Mass spectra derived from comparative sequence analysis by complete base-specific cleavage of amplification products and analysis of the cleavage products by MALDI-TOF MS. In this concept the target region is first amplified and then cleaved base-specifically into small oligonucleotides. The oligonucleotides are analyzed by MS. Sequence changes are identified by comparing the experimental mass signal pattern to a simulated mass signal pattern derived from a reference sequence. Sequence changes can introduce or remove cleavage sites from the target region and therefore change the mass signal pattern. The nucleobase composition of new mass signals can be derived from their molecular mass and allows identification of the nature and location of the sequence change. Usually, up to four cleavage reaction are employed for highly confident identification of sequence changes. Compared to sequencing by MALDI-TOF MS using the Sanger method (usually restricted to a read length of ca. 25 bases), this approach provides analysis of target regions up to 500 bp in length, or more. In the depicted case, cleavage has been performed at every cytosine (C) and thymine (T) of the forward and reverse strand (in four separate reactions). Dotted lines represent the expected mass signals (as calculated from the reference sequence). The T-specific forward cleavage shows a new, non-expected mass signal and one expected mass signal is not detected (missing). This information is computationally deconvoluted to assign a sequence change.

next step, it is determined which nucleic acid fragments can account for the additional mass signals in a spectrum. Although it is not possible to assign an exact nucleotide sequence to a mass signal in a spectrum, one can calculate the potential compositions of the four nucleotides A/C/G/T (multiplicity of nucleotides, but with unknown order), which could correspond to the mass signal. The compositional analysis of mass signals is challenging if either the number of building blocks with similar mass is high (as is the case for amino acids, the building blocks of proteins), or if the mass accuracy is limiting. However, the simplicity of DNA/RNA keeps the complexity manageable. DNA (and RNA) is comprised of "only" a four-letter alphabet, and base-specific cleavage even reduces this four-letter alphabet essentially to three. Complete base-specific cleavage provided, each cleavage product contains at most three out of the four possible bases (the cut base being fixed in terms of compositional analysis). According to an early calculation by Pomerantz and coworkers, there exist only a limited number of base-compositions with a mass difference of up to 2 Da [139–143]. Thus, even with current linear MALDI-TOF MS, we can assign a composition to a mass signal and generate a list of base compositions with a mass sufficiently close to the observed additional signal. The combined list of additional signals and the corresponding compositions can now be mapped back to the reference sequence and used to search the space of sequence variations matching the observed compositional changes. Despite the challenge to perform this search in a time-efficient manner for all possible single base substitutions, insertions and deletions, there are further layers of complication. First, the generation of a useful list of compomers for each additional mass signal is greatly simplified if the mass accuracy is at least ±1 Da. Despite advances in sample preparation, this mass accuracy can only be obtained with internal recalibration. However, the *in-silico* mass signal pattern predicted from the reference sequence can be used for efficient internal recalibration over the full mass range, since sequence changes will only change a minority of the mass signals. Additional challenges related to MALDI-TOF MS are the reduced detectivity in the higher mass range (cleavage products with masses exceeding 8000 Da tend not to be detected) and the varying ionization/desorption behavior of cleavage products of the same length, but different composition or sequence. T-rich fragments, for example, tend to show extremely low S/N ratios compared to signals of other composition of comparable mass. Both aspects compromise the ability of algorithms to use missing signals (those predicted from the reference sequence but not observed in the sample spectrum) as a reliable indicator for sequence changes. According to simulations, however, the majority of sequence changes not detectable are related to mass resolution/accuracy, which means that the corresponding base-compositions cannot be assigned uniquely with linear MALDI-TOF MS. Additional challenges are posed by process biochemistry. Any biochemical artifact might contribute to the mass signal pattern in an unpredictable manner, and thus could lead to false interpretation (false positive indication of a sequence change). Finally, biology and genetics each contribute their share of complication. For example, the longer the target region scanned, the more likely one will observe multiple sequence changes co-occurring in an amplicon. The detailed sequence analysis of several gene regions revealed SNP densities as high as one SNP per 200 base pairs. Mathematical algorithms must be able to differentiate between multiple events per amplicon and sample [144].

The ability to scan larger target regions for sequence changes with MALDI-TOF MS represents a major milestone. The question must be asked, however, as to how this method compares with state-of-the art competing technology, and what the incentives are to use base-specific cleavage and MALDI-TOF MS versus any other non-MS technology. At a laser repetition rate of 20 Hz, a sample spot can be measured in about 5 s if the sample is rastered up to five times at different positions. If we assume that the average target length scanned is 500 base pairs and that we need four cleavage reactions to identify a maximum of sequence changes, a single mass spectrometer would be able to read over 2 Megabases in 24 h. From a throughput standpoint, this compares favorably with state-of-the art capillary sequencers. If we ignore the rate at which we can process the biochemical reactions, the limiting factor in current MALDI-TOF instruments is the laser repetition rate. The use of 200-Hz lasers would shift the bottleneck to other parts of the instrument, but it can be estimated that the throughput would increase to well over 4 Megabases.

The real advantage, however, is not necessarily the throughput. More interesting is the combination of both, speed of signal acquisition/analysis and availability of the molecular mass information (interpreted here as high accuracy). Fast acquisition/analysis times are very important in diagnostic applications (e.g., for pathogen identification), where the sample-in, result-out timeframe must be short, and where massive parallelization cannot substitute for inherently long process times of a technology. With signal acquisition speed in the microsecond range, MALDI-TOF MS seems well suited to these applications.

Additionally – and perhaps even more importantly – base-specific cleavage offers collateral security: in a majority of cases, the detection and identification of sequence changes is based on multiple "observations". This aspect of the analysis has been described above. In aggregate, the use of four base-specific cleavage reactions can provide an inherent redundancy of up to 10 observations (five additional mass signals supporting the sequence change and five "missing" signals predicted from the reference sequence), which is an important aspect in diagnostic applications.

The principle of base-specific cleavage significantly expands the application portfolio of MALDI-TOF MS because larger target regions can now be analyzed. Essentially, a majority of applications currently dominated by capillary electrophoresis separation/detection now become amenable to mass spectrometric analysis. In addition to SNP discovery/re-sequencing, which has been discussed above, the use of base-specific cleavage has also been described for genotypic identification of pathogens [145]. Further applications may include the qualitative and quantitative analysis of cytosine methylation in genomic DNA [141–143,146–148], the identification and characterization of cDNAs and their splice variants, for example to identify potential targets for antibody development against cancer cells, and finally also mutation screening [149,150].

Recent years have also witnessed approaches to increase the apparent mass resolution [139–143,146]. Most genotyping assay formats were designed such that the instrumental mass resolution of ca. ±1 Da, available for linear MALDI-TOF instruments, does not compromise the accuracy. More challenging assays and higher degrees of multiplexing, however, would benefit from increased mass resolution and accuracy.

On one hand, the use of isotopically depleted nucleotides for MALDI-TOF MS represents one option to increase the apparent mass resolution. In combination with linear TOFs, however, the instrumental limitation in mass accuracy is still the dominant factor. A gain in mass resolution of factor 2 does not merit the costs associated with the production of isotopically depleted nucleotides. Newer instruments such as orthogonal TOFs, on the other hand, will alone not overcome all of the current limitations, because the isotopic envelope of the analytes will be limiting when natural nucleotides are used. Hence, the success of more challenging applications in nucleic acid analysis is linked to both, instrumental as well as biochemical improvements – that is, to the large-scale use of isotopically depleted nucleotides.

5.6
Applications in Quantitation of Nucleic Acids for Analysis of Gene Expression and Gene Amplification

5.6.1
Analysis of DNA Mixtures and Allele Frequency Determinations in DNA Pools

Ross and coworkers were the first to describe the determination of ratios of primer extension products by MALDI-TOF MS for the analysis of DNA mixtures [151]. This approach opened the door to new ways of applying MALDI-TOF MS in DNA analysis, including quantitative aspects. The main application for the approach is the relative quantitation of allele frequencies, and this is achieved by calculating the ratio of the peak area associated with allele-specific extension products. The combined allele frequency of a biallelic SNP in a sample pool (and equimolar mixture of genomic DNA from multiple individuals) is 1 (100%). To keep the influence of the sample heterogeneity low, even relative quantification requires that multiple spectra are averaged from several locations of a given sample. Correspondingly, various statistical models can be used to integrate multiple mass spectra from different raster positions on a sample spot or even from multiple dispenses of the same analyte. The allele frequency is then expressed as the ratio of the peak areas of any given allele to the total/combined peak area of all alleles.

Several groups have assessed the quantitative capabilities of primer extension assays combined with MALDI-TOF MS [152]. It was found that allele frequencies can routinely be measured down to frequencies of 5%. A comparison with other technologies using fluorescence or other detection methods showed that mass spectrometric results were of comparable accuracy and reproducibility. Depending on the assay quality, even frequencies below 5% are detected, though the accuracy of these values is lower as the peak areas would exceed a 50:1 ratio. On a routine basis and using automated processing, frequencies between 10 and 90%

are detected and analyzed, with a standard deviation of 2–3%. A detailed analysis of the major contributors to this standard deviation showed that DNA sample generation and amplification have a higher impact on accuracy and reproducibility than the heterogeneity of the crystallization and the MALDI-process. This might be surprising to specialists in MALDI-TOF MS, but it must be noted that most reports used miniaturized sample preparation on silicon chips, which has been shown to minimize these effects. Unstable PCR amplification can have a dramatic impact on the reproducibility of semi-quantitative analysis. However, such impact is independent of the detection technology, and usually these assays are identified after multiple reactions and excluded from further consideration.

Amexis and coworkers reported an interesting application of semi-quantitative analysis [153] by using a MALDI-TOF MS analysis of primer extension products in their vaccine quality control process. Ratios of viral quasi-species of the mumps virus were determined between Jeryl Lynn substrains in live, attenuated mumps/measles vaccine. Determination of the ratio of two substrains was performed at five discriminative nucleotide positions within the viral genome. Methods such as this are important for maintaining vaccine safety.

Mueller and coworkers used semi-quantitative analysis of primer extension products by MALDI-TOF MS to compare the allelic expression between healthy tissue and tumor tissue [154]. The approach first screened for informative cases (heterozygous individuals) on genomic DNA, and then compared the expression of the corresponding alleles on the mRNA level. Using this approach, these authors could determine that changes in allelic expression of their gene of interest correlated with tumor development and progression. In particular, the study evaluated the impact of parental imprinting and so-called "loss-of-imprinting" in tumor development. Interestingly, the study also revealed that previously used technologies mainly had insufficient sensitivity in assessing the effects accurately. Tumor samples normally represent DNA mixtures, which also contain the remainders of healthy cells; this makes their analysis a challenging application for technologies with insufficient sensitivity.

The assessment of allele frequencies in DNA pools has been extensively used as a convenient way to validate SNPs and to characterize their allele frequency in different ethnic groups. With the rapid assay development capacity of primer extension-based assays, the approach allows for a large-scale initial discrimination not only between "real" and "false" SNPs derived from databases, but also between common and rare SNPs.

To exploit the approach in large-scale association studies, the accuracy of allele frequency estimation in DNA pools by MALDI-TOF MS had to be verified by comparison with the true allele frequency derived from individual genotyping. Figure 5.9(a) shows a scatter plot with allele frequencies determined for 24 assays in a DNA pool of 96 individuals versus the "true" allele frequency of the same assays observed in individual genotyping. These data show that the correlation between allele frequencies estimated from a DNA pool and the frequencies determined by individual genotyping is not perfect. There are numerous factors contributing to this effect. Most of them are technology-independent, such as the accuracy with

(a)   Genotype vs. Pooled (Uncorrected) Allele Frequencies

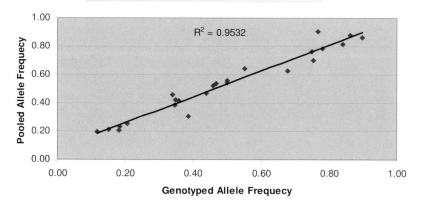

◆ Genotype vs. Pooled Allele Frequencies
— Linear (Genotype vs. Pooled Allele Frequencies)

$R^2 = 0.9532$

Pooled Allele Frequecy (y-axis)

Genotyped Allele Frequecy (x-axis)

(b)   Genotype vs. Pooled (Corrected) Allele Frequencies

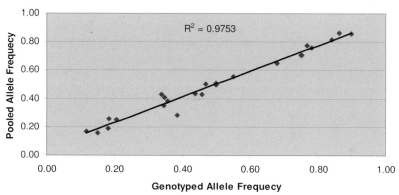

◆ Genotype vs. Pooled Allele Frequencies
— Linear (Genotype vs. Pooled Allele Frequencies)

$R^2 = 0.9753$

Pooled Allele Frequecy (y-axis)

Genotyped Allele Frequecy (x-axis)

Fig. 5.9 Exemplification of the use of semi-quantitative analysis of mass signal peak area ratios for the measurement of the relative abundance genotypes in a population. The allele-frequencies derived from the relative peak area ratios of SNP-specific extension products from a pool of individual samples are compared to those determined by individual genotyping of each sample to estimate the precision of the measurement. (a) Depicted is a scatter plot of genotyped population allele frequencies (x-axis) versus allele frequencies calculated using pooled population DNAs (y-axis). Results for 48 unique assays are shown. The DNA population pool consisted of 96 individual DNAs at equimolar concentrations (260 ng per individual DNA $\mu L^{-1} = 25$ ng μL^{-1}). The calculated allele frequency for each assay represents the average of four replicate reactions. The best-fit line and the coefficient of determination (R^2) were calculated using Excel 2000 (Microsoft). The allele-frequencies derived from pooled DNAs were also corrected

for amplification bias of individual alleles, which can contribute to differences between the allele-frequencies derived from pools versus the individual genotypes. For this purpose, the allele ratio of individual heterozygote reaction can be used as a "correction" factor for the allele frequencies determined in the pool reaction. The individual heterozygote should have a 0.50:0.50 (1:1) allele ratio. Any deviation from this expected ratio represents a "skewing" factor in that reaction. After correction with the heterozygote allele ratios the allele frequencies from the pool match the genotyped population frequency exactly. (b) The second scatter plot shows compares the genotyped population allele frequencies (x-axis) versus allele frequencies calculated using pooled population DNAs (y-axis) after correction. The coefficient of determination (R^2) improves from $R^2 = 0.9532$ to $R^2 = 0.9753$, and hence leads to an increase of the precision of allele-frequency estimates.

which the DNA pool was prepared, or the preferential amplification of one of the alleles over the other during PCR. Due to the importance of allele frequency estimation in genetic studies, recent reviews have provided a detailed account of strategies and issues for analysis of DNA pools.

It was frequently also found that individual DNAs, heterozygous for a particular SNP, showed "skewed" distributions of the two alleles. Genetically, they have a 1:1 distribution of the two genetic informations (one chromosome carrying one allele, the second chromosome carrying the other allele), so one would expect a 1:1 ratio of the two alleles in the SNP assay (or 50% frequency for both alleles in terms of peak areas). This skewing, which is most likely introduced either during PCR (preferential amplification of one allele) or during the post-PCR primer extension, leads to a deviation of the estimated allele frequency from the true frequency when the DNA pools are analyzed. A further MALDI-specific factor might be a slightly lower desorption/ionization efficiency of the higher mass allele versus the lower mass allele. However, a "correction" factor can be applied for each individual assay based on the peak area distributions observed in individual heterozygous DNAs. Deviations from the expected 1:1 ratio can be computed into a correction factor, which should be applied to allele frequency estimations from pools. Figure 5.9(b) shows a scatter plot similar to that in Figure 5.9(a), but here the allele frequency estimations have been corrected by assay-specific correction factors. With the correction applied, the coefficient of correlation is seen to improve.

Although some assay-specific deviations between true and estimated frequencies remain, the approach has been successfully applied to semi-quantitative SNP allele frequency analysis in samples pools, and also to differential protein binding to mRNA associated with allelic variants of a gene [154].

As mentioned in the introduction to this section, allele frequency information derived from DNA pools not only allows the rapid collection of validated SNP sets. Rather, the comparison of the abundance of alleles between different populations represented as DNA pools is also a suitable way to identify genotype–phenotype correlations. In this respect, SNPs are used as genetic markers, which allow the identification of causative genetic loci in complex disease through the linkage between SNP marker and genetic locus (the SNP being in linkage disequilibrium with the causative genetic locus).

Several research groups have identified the use of DNA pools as a potential short-cut to identify associations between genetic loci and phenotypes [155]. Instead of costly and cumbersome individual genotyping of hundreds of thousands of SNPs, the allele frequency of SNPs in various DNA pools stratified by phenotype is used to find such as association. For this approach, the correctness of the allele frequency estimation compared to the "true" frequency is less of an issue, as long as the relative abundance between the two pools can be assessed accurately and with high reproducibility.

Buetow and coworkers were the first to apply allele-frequency estimations by MALDI-TOF MS in DNA pools on a genome-wide basis. Recently, several other groups have dissected the major process variables from amplification to

MALDI-TOF MS analysis and provided proof-of-principle that the approach allows for identification of major genetic contributors to complex diseases [156].

Recently, Ding et al. took these applications one step further. By combining the sensitivity and specificity of the combination of PCR and analysis by MALDI-TOF MS (that they had demonstrated with the molecular haplotyping approach) with the quantitative features established in the experiments described above, these authors were able to demonstrate an analysis of mutations in circulating nucleic acids. This discovery proved to be an important milestone in the quest for non-invasive prenatal diagnoses.

5.6.2
Analysis of Gene Expression

The field of gene expression analysis has only recently become of broader interest to users of MS. The reason for this lag in interest was simply the lack of methods allowing absolute quantification of nucleic acids in general, and mRNA in particular. MS allows only an endpoint analysis, and even then a standard for normalization (such as a second allele) is required. Thus, approaches have focused on the semi-quantitative analysis of allelic expression, as described above, which is a rather narrow – but expanding – field of research.

The first method to enable gene expression analysis by MS was reported only recently by the group of Smith et al., who applied the Invader assay (see Section 5.4.1.5; see also Fig. 5.5) for the detection and relative quantification of RNA. In order to allow inter-spectrum (and thus inter-sample) correlation, Invader assays of multiple target genes were multiplexed and a reference gene was included for normalization of the data.

Using a different approach, Ding and Cantor recently described the use of an internal standard added to cDNA samples as a means of quantifying nucleic acids with technologies lacking real-time monitoring capabilities (as are used for real-time PCR). The approach uses a synthetic oligonucleotide designed to match the target sequence (i.e., a sequence stretch of the cDNA to be investigated) in all positions except one single nucleotide. The internal standard is added to the sample of interest (i.e., a cDNA preparation) in a known concentration prior to amplification (hence, the process is called "competitive PCR"). The introduction of a single nucleotide change creates a way of differentiating between target cDNA and internal standard by means of a post-PCR primer extension assay. The internal standard and cDNA are co-amplified during PCR, but the efficiency of amplification should be equivalent as both share very nearly the same sequence and, in particular, the same primer binding sites.

The post-PCR primer extension reaction targets the nucleotide difference between the internal standard and cDNA, and generates two specific products resembling the same reaction performed for allele frequency determination. An analysis of the peak areas allows not only for a relative comparison of cDNA amount versus internal standard, but also for absolute quantification, as the concentration of the internal standard is known. The use of an internal standard with

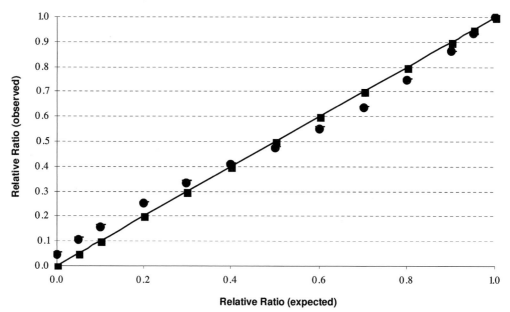

Fig. 5.10 Quantitation of gene expression by PCR/primer extension and MALDI-TOF MS in a mixture model system. Depicted is a scatter plot of calculated allele ratios (based on mass signal peak area rations of primer extension products) for expression analysis of the CETP gene. Data points represent the average of triplicate reactions. Expected ratios are depicted as a solid line, and observed values as dots. In this model system, two artificial templates (90 bp) were designed based on the sequence of the CETP gene mRNA (Accession# AC023825). One of the templates matched this region of the CETP gene exactly. The second had a 1-bp mismatch introduced to mimic a mutation and to serve as a second allele in a primer extension reaction. Each template and allele is co-amplified at equal rates as shown in the graph. Deviation from an exact fit to expected allele frequencies represent a "skew" as detailed in Figure 5.9, and can be corrected in the same manner using a heterozygote (in this case artificial). The concentration of each template added to the reaction is known and therefore the amount of wild-type mRNA (or cDNA) can be determined when the two alleles are at a 1:1 ratio (0.5:0.5 allele frequency).

the same sequence as the target sequence alleviates common issues related to quantification. Due to the same PCR amplification efficiency, the process is PCR-cycle-independent, and real-time monitoring of the amplification process becomes obsolete, as it will impact on both (standard and cDNA) in the same way.

The results of a model system used to establish and validate this concept are shown graphically in Figure 5.10. A 90-bp region of the CETP gene region was selected to demonstrate the principle. Two sequences were designed: one exactly matched the gene sequences serving as the unknown; and one carried a single base change (C to G substitution) in the central region of the oligonucleotide (this was the internal standard). Both molecules were used as templates for competitive PCR in various template ratios. Post-PCR primer extension and MALDI-TOF MS was used to track the relative ratio of the two standards after PCR amplification. In Figure 5.10 the observed allele ratios derived from the standards are plotted against the expected allele ratios, as calculated from the mixture used for PCR. The uncorrected observed allele ratio tracks the expected ratio closely, which supports the validity of the approach. Of further note is the low standard deviation. All data points were acquired in triplicate (on the PCR level), and the standard deviations resembled those values obtained for allele frequency determination in DNA pools (~3%). If these standard deviations were to be obtained also for multiple studies carried out in different laboratories, the method would allow for the accurate analysis of rather subtle changes in expression analysis.

The technique described by Ding and Cantor was reported very recently, and further use of the proposed method in various laboratories will surely demonstrate the utility and accuracy of the approach. A more recent report validated the competitor PCR approach described by Ding and Cantor against real-time PCR. The study, conducted by Elvidge et al., showed a good correlation between real-time PCR data and MALDI-TOF MS data. However, there was a significant difference in sensitivity, and low-abundance mRNAs were much more readily detected and quantified by the combination of rtPCR and mass spectrometry. It can be envisioned that this approach will find application in studies with a focused interest in the expression of a set of genes, because the combination of the internal standard with PCR and primer extension can be easily multiplexed when MALDI-TOF MS is used for the analysis. Thus, massive amounts of data comparing the relative and absolute expression of multiple genes can be generated for large sample sets. One drawback of conventional microarray-based gene expression analysis is that it delivers a large number of data points (or even genome-wide data points) for individual samples. However, once genes or gene sets of interest are selected, the microarrays do not permit the scanning of large numbers of samples in a cost-efficient manner. Elvidge et al. also noted that the development of real-time PCR assays for expression analysis may more frequently require time-consuming and potentially costly assay optimization rounds.

The basic concept described by Ding et al. has recently been extended to exciting new aspects of gene expression, including the quantitation of allele-specific expression and the quantitative analysis of splice variants.

The use of internal standards and competitive PCR is not limited to gene expression analysis. Rather, there are many more scientific questions that will require the absolute quantification of nucleic acids, and the concept of competitive PCR will help to open these applications to mass spectrometric analysis.

Acknowledgment

The authors are grateful to Jingue Ju for providing Figure 5.7. They also thank Christian Jurinke, Christiane Honisch, Mathias Ehrich, Paul Oeth and April Kinsler for their help in the preparation of this chapter, and their valuable comments.

References

1 J.B. Fenn, M. Mann, C.K. Meng, S.F. Wong, C.M. Whitehouse. Electrospray ionization for mass spectrometry of large biomolecules. *Science* **1989**, *246*, 246.

2 M. Karas, F. Hillenkamp. Laser desorption ionization of proteins with molecular masses exceeding 10000 Daltons. *Anal. Chem.* **1988**, *60*, 2299–2301.

3 D.C. Muddiman, A.P. Null, J.C. Hannis. Precise mass measurement of a double-stranded 500 base-pair (309 kDa) polymerase chain reaction product by negative ion electrospray ionization Fourier transform ion cyclotron resonance mass spectrometry. *Rapid Commun. Mass Spectrom.* **1999**, *13*, 1201–1204.

4 J.C. Hannis, D.C. Muddiman. Detection of double-stranded PCR amplicons at the attomol level electrosprayed from low nanomolar solutions using FTICR mass spectrometry. *Fresenius J. Anal. Chem.* **2001**, *369*, 246–251.

5 S. Hahner, A. Schneider, A. Ingendoh, J. Mosner. Analysis of short tandem repeat polymorphisms by electrospray ion trap mass spectrometry. *Nucleic Acids Res.* **2000**, *28*, 18.

6 S.J. Laken, et al. Genotyping by mass spectrometric analysis of short DNA fragments. *Nat. Biotechnol.* **1998**, *16*, 1352–1356.

7 Y. Jiang, S.A. Hofstadler. A highly efficient and automated method of purifying and desalting PCR products for analysis by electrospray ionization mass spectrometry. *Anal. Biochem.* **2003**, *316*, 50–57.

8 M.J. Doktycz, et al. Analysis of polymerase chain reaction-amplified DNA products by mass spectrometry using matrix-assisted laser desorption and electrospray: current status. *Anal. Biochem.* **1995**, *230*, 205–214.

9 S.A. McLuckey, J. Wu, J.L. Bundy, J.L., Jr. Stephenson, G.B. Hurst. Oligonucleotide mixture analysis via electrospray and ion/ion reactions in a quadrupole ion trap. *Anal. Chem.* **2002**, *74*, 976–984.

10 E. Nordhoff, et al. Ion stability of nucleic acids in infrared matrix-assisted laser desorption/ionization mass spectrometry. *Nucleic Acids Res.* **1993**, *21*, 3347–3357.

11 J. Gross, F. Hillenkamp, K.X. Wan, M.L. Gross. Metastable decay of negatively charged oligodeoxynucleotides analyzed with ultraviolet matrix-assisted laser desorption/ionization post-source decay and deuterium exchange. *J. Am. Soc. Mass Spectrom.* **2001**, *12*, 180–192.

12 J. Gross, et al. Investigations of the metastable decay of DNA under ultraviolet matrix-assisted laser desorption/ionization conditions with post-source-decay analysis and hydrogen/deuterium exchange. *J. Am. Soc. Mass Spectrom.* **1998**, *9*, 866–878.

13 F. Kirpekar, et al. Matrix assisted laser desorption/ionization mass spectrometry of enzymatically synthesized RNA up to 150 kDa. *Nucleic Acids Res.* **1994**, *22*, 3866–3870.

14 K. Tang, N.I. Taranenko, S.L. Allman, L.Y. Chang, C.H. Chen. Detection of 500-nucleotide DNA by laser desorption mass spectrometry. *Rapid Commun. Mass Spectrom.* **1994**, *8*, 727–730.

15 H. Lin, J.M. Hunter, C.H. Becker. Laser desorption of DNA oligomers larger than one kilobase from cooled 4-nitrophenol. *Rapid Commun. Mass Spectrom.* **1999**, *13*, 2335–2340.

16 K.J. Wu, A. Steding, C.H. Becker. Matrix-assisted laser desorption time-of-flight mass spectrometry of oligonucleotides using 3-hydroxypicolinic acid as an ultraviolet-sensitive matrix. *Rapid Commun. Mass Spectrom.* **1993**, *7*, 142–146.

17 E. Nordhoff, et al. Comparison of IR- and UV-matrix-assisted laser desorption/ionization mass spectrometry of oligodeoxynucleotides. *Nucleic Acids Res.* **1994**, *22*, 2460–2465.

18 Y.F. Zhu, et al. The study of 2,3,4-trihydroxyacetophenone and 2,4,6-trihydroxyacetophenone as matrices for DNA detection in matrix-assisted laser desorption/ionization time-of-flight mass

spectrometry. *Rapid Commun. Mass Spectrom.* **1996**, *10*, 383–388.

19 R.J. Cotter. *Time-of-Flight Mass Spectrometry.* ACS Professional Reference Books. Am. Chem. Soc. Washington, DC, USA **1997**.

20 K. Schneider, B.T. Chait. Matrix-assisted laser desorption mass spectrometry of homopolymer oligodeoxynucleotides. Influence of base composition on the mass spectrometric response. *Org. Mass Spectrom.* **1993**, *28*, 1353–1361.

21 K. Schneider, B.T. Chait. Increased stability of nucleic acids containing 7-deaza-guanosine and 7-deaza-adenosine may enable rapid DNA sequencing by matrix-assisted laser desorption mass spectrometry. *Nucleic Acids Res.* **1995**, *23*, 1570–1575.

22 S. Berkenkamp, C. Menzel, M. Karas, F. Hillenkamp. Performance of infrared matrix-assisted laser desorption/ionization mass spectrometry with lasers emitting in the 3 µm wavelength range. *Rapid Commun. Mass Spectrom.* **1997**, *11*, 1399–1406.

23 C. Menzel, S. Berkenkamp, F. Hillenkamp. Infrared matrix-assisted laser desorption/ionization mass spectrometry with a transversely excited atmospheric pressure carbon dioxide laser at 10.6 um wavelength with static and delayed ion extraction. *Rapid Commun Mass Spectrom.* **1999**, *13*, 26–32.

24 S. Berkenkamp, F. Kirpekar, F. Hillenkamp. Infrared MALDI mass spectrometry of large nucleic acids. *Science* **1998**, *281*, 260–262.

25 F. Kirpekar, S. Berkenkamp, F. Hillenkamp. Detection of double-stranded DNA by IR- and UV-MALDI mass spectrometry. *Anal. Chem.* **1999**, *71*, 2334–2339.

26 J.C. Venter, et al. The sequence of the human genome. *Science* **2001**, *291*, 1304–1351.

27 E.S. Lander, et al. Initial sequencing and analysis of the human genome. *Nature* **2001**, *409*, 860–921.

28 D.A. Hinds, A.P. Kloek, M. Jen, X. Chen, K.A. Frazer. Common deletions and SNPs are in linkage disequilibrium in the human genome. *Nat. Genet.* **2006**, *38*, 82–85.

29 L. Feuk, A.R. Carson, S.W. Scherer. Structural variation in the human genome. *Nat. Rev. Genet.* **2006**, *7*, 85–97.

30 B. de Martinville, A.R. Wyman, R. White, U. Francke. Assignment of first random restriction fragment length polymorphism (RFLP) locus (D14S1) to a region of human chromosome 14. *Am. J. Hum. Genet.* **1982**, *34*, 216–226.

31 G.R. Taylor, J.S. Noble, J.L. Hall, A.D. Stewart, R.F. Mueller. Hypervariable microsatellite for genetic diagnosis. *Lancet* **1989**, *2*, 454.

32 C. Epplen, J. Buitkamp, H. Rumpf, M. D'Souza, J.T. Epplen. Immunoprinting reveals different genetic bases for (auto)immuno diseases. *Electrophoresis* **1995**, *16*, 1693–1697.

33 S.L. Sunden, et al. Chromosomal assignment of 2900 tri- and tetranucleotide repeat markers using NIGMS somatic cell hybrid panel 2. *Genomics* **1996**, *32*, 15–20.

34 C. Dib, et al. A comprehensive genetic map of the human genome based on 5,264 microsatellites. *Nature* **1996**, *380*, 152–154.

35 J.S. Ziegle, et al. Application of automated DNA sizing technology for genotyping microsatellite loci. *Genomics* **1992**, *14*, 1026–1031.

36 P.W. Reed, et al. Chromosome-specific microsatellite sets for fluorescence-based, semi-automated genome mapping. *Nat. Genet.* **1994**, *7*, 390–395.

37 L. Kruglyak, D.A. Nickerson. Variation is the spice of life. *Nat. Genet.* **2001**, *27*, 234–236.

38 J.C. Stephens, et al. Haplotype variation and linkage disequilibrium in 313 human genes. *Science* **2001**, *293*, 489–493.

39 D.E. Reich, S.B. Gabriel, D. Altshuler. Quality and completeness of SNP databases. *Nat. Genet.* **2003**, *33*, 457–458.

40 N. Risch, K. Merikangas. The future of genetic studies of complex human diseases. *Science* **1996**, *273*, 1516–1517.

41 E.S. Lander. The new genomics: global views of biology. *Science* **1996**, *274*, 536–539.

42 D. Altshuler, et al. An SNP map of the human genome generated by reduced representation shotgun sequencing. *Nature* **2000**, *407*, 513–516.

43 M.C. O'Donovan, et al. Blind analysis of denaturing high-performance liquid chromatography as a tool for mutation detection. *Genomics* **1998**, *52*, 44–49.

44 D. Glavac, M. Dean. Applications of heteroduplex analysis for mutation detection in disease genes. *Hum. Mutat.* **1995**, *6*, 281–286.

45 M. Orita, H. Iwahana, H. Kanazawa, K. Hayashi, T. Sekiya. Detection of polymorphisms of human DNA by gel electrophoresis as single-strand conformation polymorphisms. *Proc. Natl. Acad. Sci. USA* **1989**, *86*, 2766–2770.

46 J.G. Hacia, L.C. Brody, M.S. Chee, S.P. Fodor, F.S. Collins. Detection of heterozygous mutations in BRCA1 using high-density oligonucleotide arrays and two-color fluorescence analysis. *Nat. Genet.* **1996**, *14*, 441–447.

47 J.G. Hacia. Resequencing and mutational analysis using oligonucleotide arrays. *Nat. Genet.* **1999**, *21*, 42–47.

48 P.Y. Kwok, C. Carlson, T.D. Yager, W. Ankener, D.A. Nickerson. Comparative analysis of human DNA variations by fluorescence-based sequencing of PCR products. *Genomics* **1994**, *23*, 138–144.

49 J. Couzin. Human genome. HapMap launched with pledges of $100 million. *Science* **2002**, *298*, 941–942.

50 P.F. Crain, J.A. McCloskey. Applications of mass spectrometry to the characterization of oligonucleotides and nucleic acids. *Curr. Opin. Biotechnol.* **1998**, *9*, 25–34.

51 K.K. Murray. DNA sequencing by mass spectrometry. *J. Mass Spectrom.* **1996**, *31*, 1203–1215.

52 J. Tost, I.G. Gut. Genotyping single nucleotide polymorphisms by mass spectrometry. *Mass Spectrom. Rev.* **2002**, *21*, 388–418.

53 B. Guo. Mass spectrometry in DNA analysis. *Anal. Chem.* **1999**, *71*, 333R–337R.

54 T.A. Shaler, Y. Tan, J.N. Wickham, K.J. Wu, C.H. Becker. Analysis of enzymatic DNA sequencing reactions by matrix-assisted laser desorption/ionization time-of-flight mass spectrometry. *Rapid*

Commun. Mass Spectrom. **1995**, *9*, 942–947.

55 T.A. Shaler, J.N. Wickham, K.A. Sannes, K.J. Wu, C.H. Becker. Effect of impurities on the matrix-assisted laser desorption mass spectra of single-stranded oligodeoxynucleotides. *Anal. Chem.* **1996**, *68*, 576–579.

56 E. Nordhoff, C. Luebbert, G. Thiele, V. Heiser, H. Lehrach. Rapid determination of short DNA sequences by the use of MALDI-MS. *Nucleic Acids Res.* **2000**, *28*, E86.

57 C. Jurinke, et al. Recovery of nucleic acids from immobilized biotin-streptavidin complexes using ammonium hydroxide and applications in MALDI-TOF mass spectrometry. *Anal. Chem.* **1997**, *69*, 904–910.

58 P.L. Ross, P. Belgrader. Analysis of short tandem repeat polymorphisms in human DNA by matrix-assisted laser desorption/ionization mass spectrometry. *Anal. Chem.* **1997**, *69*, 3966–3972.

59 I.G. Gut, S. Beck. A procedure for selective DNA alkylation and detection by mass spectrometry. *Nucleic Acids Res.* **1995**, *23*, 1367–1373.

60 M. Schuerenberg, et al. Prestructured MALDI-MS sample supports. *Anal. Chem.* **2000**, *72*, 3436–3442.

61 D.P. Little, et al. MALDI on a chip: analysis of arrays of low-femtomole to subfemtomole quantities of synthetic oligonucleotides and DNA diagnostic products dispensed by a piezoelectric pipet. *Anal. Chem.* **1997**, *69*, 4540–4546.

62 D.P. Little, A. Braun, M.J. O'Donnell, H. Koster. Mass spectrometry from miniaturized arrays for full comparative DNA analysis. *Nat. Med.* **1997**, *3*, 1413–1416.

63 D.P. Little, et al. Direct detection of synthetic and biologically generated double-stranded DNA by MALDI-TOF MS. *Int. J. Mass Spectrom. Ion Processes* **1997**, *169/170*, 323–330.

64 Y. Wada. Separate analysis of complementary strands of restriction enzyme-digested DNA. An application of restriction fragment mass mapping by matrix-assisted laser desorption/ionization mass spectrometry. *J. Mass Spectrom.* **1998**, *33*, 187–192.

65 J.M. Butler, J. Li, T.A. Shaler, J.A. Monforte, C.H. Becker. Reliable genotyping of short tandem repeat loci without an allelic ladder using time-of-flight mass spectrometry. *Int. J. Legal Med.* **1999**, *112*, 45–49.

66 D. van den Boom, C. Jurinke, M.J. McGinniss, S. Berkenkamp. Microsatellites: perspectives and potentials of mass spectrometric analysis. *Expert Rev. Mol. Diagn.* **2001**, *1*, 383–393.

67 S. Krebs, D. Seichter, M. Forster. Genotyping of dinucleotide tandem repeats by MALDI mass spectrometry of ribozyme-cleaved RNA transcripts. *Nat. Biotechnol.* **2001**, *19*, 877–880.

68 A. Braun, D.P. Little, D. Reuter, B. Muller-Mysok, H. Koster. Improved analysis of microsatellites using mass spectrometry. *Genomics* **1997**, *46*, 18–23.

69 Y. Wada, K. Mitsumori, T. Terachi, O. Ogawa. Measurement of polymorphic trinucleotide repeats in the androgen receptor gene by matrix-assisted laser desorption/ionization time-of-flight mass spectrometry. *J. Mass Spectrom.* **1999**, *34*, 885–888.

70 Y.H. Liu, et al. Rapid screening of genetic polymorphisms using buccal cell DNA with detection by matrix-assisted laser desorption/ionization mass spectrometry. *Rapid Commun. Mass Spectrom.* **1995**, *9*, 735–743.

71 C. Jurinke, et al. Analysis of ligase chain reaction products via matrix-assisted laser desorption/ionization time-of-flight-mass spectrometry. *Anal. Biochem.* **1996**, *237*, 174–181.

72 T.J. Griffin, W. Tang, L.M. Smith. Genetic analysis by peptide nucleic acid affinity MALDI-TOF mass spectrometry. *Nat. Biotechnol.* **1997**, *15*, 1368–1372.

73 X. Chen, Z. Fei, L.M. Smith, E.M. Bradbury, V. Majidi. Stable-isotope-assisted MALDI-TOF mass spectrometry for accurate determination of nucleotide compositions of PCR products. *Anal. Chem.* **1999**, *71*, 3118–3125.

74 C. Jurinke, et al. Detection of hepatitis B virus DNA in serum samples via nested PCR and MALDI-TOF mass spectrometry. *Genet. Anal.* **1996**, *13*, 67–71.

75 C. Jurinke, et al. Application of nested PCR and mass spectrometry for DNA-based virus detection: HBV-DNA detected in the majority of isolated anti-HBc positive sera. *Genet. Anal.* **1998**, *14*, 97–102.

76 L.A. Haff, I.P. Smirnov. Single-nucleotide polymorphism identification assays using a thermostable DNA polymerase and delayed extraction MALDI-TOF mass spectrometry. *Genome Res.* **1997**, *7*, 378–388.

77 L.A. Haff, I.P. Smirnov. Multiplex genotyping of PCR products with MassTag-labeled primers. *Nucleic Acids Res.* **1997**, *25*, 3749–3750.

78 Z. Fei, T. Ono, L.M. Smith. MALDI-TOF mass spectrometric typing of single nucleotide polymorphisms with mass-tagged ddNTPs. *Nucleic Acids Res.* **1998**, *26*, 2827–2828.

79 A. Braun, D.P. Little, H. Koster. Detecting CFTR gene mutations by using primer oligo base extension and mass spectrometry. *Clin. Chem.* **1997**, *43*, 1151–1158.

80 D.P. Little, A. Braun, B. Darnhofer-Demar, H. Koster. Identification of apolipoprotein E polymorphisms using temperature cycled primer oligo base extension and mass spectrometry. *Eur. J. Clin. Chem. Clin. Biochem.* **1997**, *35*, 545–548.

81 S. Kim, J.R. Edwards, L. Deng, W. Chung, J. Ju. Solid phase capturable dideoxynucleotides for multiplex genotyping using mass spectrometry. *Nucleic Acids Res.* **2002**, *30*, e85.

82 S. Kim, et al. Multiplex genotyping of the human beta2-adrenergic receptor gene using solid-phase capturable dideoxynucleotides and mass spectrometry. *Anal. Biochem.* **2003**, *316*, 251–258.

83 A. Harksen, P.M. Ueland, H. Refsum, K. Meyer. Four common mutations of the cystathionine beta-synthase gene detected by multiplex PCR and matrix-assisted laser desorption/ionization time-of-flight mass spectrometry. *Clin. Chem.* **1999**, *45*, 1157–1161.

84 E. Nordhoff, et al. Matrix-assisted laser desorption/ionization mass spectrometry of nucleic acids with wavelengths in the

ultraviolet and infrared. *Rapid Commun. Mass Spectrom.* 1992, *6*, 771–776.

85 N. Storm, B. Darnhofer-Patel, D. van den Boom, C.P. Rodi. MALDI-TOF mass spectrometry-based SNP genotyping. *Methods Mol. Biol.* 2003, *212*, 241–262.

86 M.J. O'Donnell-Maloney, D.P. Little. Microfabrication and array technologies for DNA sequencing and diagnostics. *Genet. Anal.* 1996, *13*, 151–157.

87 K. Berlin, I.G. Gut. Analysis of negatively "charge tagged" DNA by matrix-assisted laser desorption/ionization time-of-flight mass spectrometry. *Rapid Commun. Mass Spectrom.* 1999, *13*, 1739–1743.

88 S. Sauer, et al. A novel procedure for efficient genotyping of single nucleotide polymorphisms. *Nucleic Acids Res.* 2000, *28*, E13.

89 S. Sauer, et al. Full flexibility genotyping of single nucleotide polymorphisms by the GOOD assay. *Nucleic Acids Res.* 2000, *28*, E100.

90 S. Sauer, et al. Facile method for automated genotyping of single nucleotide polymorphisms by mass spectrometry. *Nucleic Acids Res.* 2002, *30*, e22.

91 S. Sauer, H. Lehrach, R. Reinhardt. MALDI mass spectrometry analysis of single nucleotide polymorphisms by photocleavage and charge-tagging. *Nucleic Acids Res.* 2003, *31*, e63.

92 R.W. Kwiatkowski, V. Lyamichev, M. de Arruda, B. Neri. Clinical, genetic, and pharmacogenetic applications of the Invader assay. *Mol. Diagn.* 1999, *4*, 353–364.

93 T.J. Griffin, J.G. Hall, J.R. Prudent, L.M. Smith. Direct genetic analysis by matrix-assisted laser desorption/ionization mass spectrometry. *Proc. Natl. Acad. Sci. USA* 1999, *96*, 6301–6306.

94 J.L. Wolfe, et al. A genotyping strategy based on incorporation and cleavage of chemically modified nucleotides. *Proc. Natl. Acad. Sci. USA* 2002, *99*, 11073–11078.

95 Y.J. Kim, et al. Population genotyping of hepatitis C virus by matrix-assisted laser desorption/ionization time-of-flight mass spectrometry analysis of short DNA fragments. *Clin. Chem.* 2005, *51*, 1123–1131.

96 A.R. Templeton, C.F. Sing, A. Kessling, S. Humphries. A cladistic analysis of phenotype associations with haplotypes inferred from restriction endonuclease mapping. II. The analysis of natural populations. *Genetics* 1988, *120*, 1145–1154.

97 R. Judson, J.C. Stephens, A. Windemuth. The predictive power of haplotypes in clinical response. *Pharmacogenomics* 2000, *1*, 15–26.

98 E.R. Martin, et al. Analysis of association at single nucleotide polymorphisms in the APOE region. *Genomics* 2000, *63*, 7–12.

99 L.R. Cardon, G.R. Abecasis. Using haplotype blocks to map human complex trait loci. *Trends Genet.* 2003, *19*, 135–140.

100 L. Kruglyak. Prospects for whole-genome linkage disequilibrium mapping of common disease genes. *Nat. Genet.* 1999, *22*, 139–144.

101 M. Stephens, N.J. Smith, P. Donnelly. A new statistical method for haplotype reconstruction from population data. *Am. J. Hum. Genet.* 2001, *68*, 978–989.

102 H. Yan, et al. Conversion of diploidy to haploidy. *Nature* 2000, *403*, 723–724.

103 K.L. Bentley, et al. Detailed analysis of a 17q21 microdissection library by sequence bioinformatics and isolation of region-specific clones. *Somat. Cell. Mol. Genet.* 1997, *23*, 353–365.

104 G. Ruano, K.K. Kidd, J.C. Stephens. Haplotype of multiple polymorphisms resolved by enzymatic amplification of single DNA molecules. *Proc. Natl. Acad. Sci. USA* 1990, *87*, 6296–6300.

105 G. Ruano, K.K. Kidd. Direct haplotyping of chromosomal segments from multiple heterozygotes via allele-specific PCR amplification. *Nucleic Acids Res.* 1989, *17*, 8392.

106 G. Ruano, W. Fenton, K.K. Kidd. Biphasic amplification of very dilute DNA samples via "booster" PCR. *Nucleic Acids Res.* 1989, *17*, 5407.

107 J. Tost, et al. Molecular haplotyping at high throughput. *Nucleic Acids Res.* 2002, *30*, e96.

108 C. Ding, C.R. Cantor. Direct molecular haplotyping of long-range genomic DNA

with M1-PCR. *Proc. Natl. Acad. Sci. USA* **2003**, *100*, 7449–7453.

109 J.C. Stephens, J. Rogers, G. Ruano. Theoretical underpinning of the single-molecule-dilution (SMD) method of direct haplotype resolution. *Am. J. Hum. Genet.* **1990**, *46*, 1149–1155.

110 F. Sanger, S. Nicklen, A.R. Coulson. DNA sequencing with chain-terminating inhibitors. *Proc. Natl. Acad. Sci. USA* **1977**, *74*, 5463–5467.

111 L.M. Smith. The future of DNA sequencing. *Science* **1993**, *262*, 530–532.

112 H. Koster, et al. A strategy for rapid and efficient DNA sequencing by mass spectrometry. *Nat. Biotechnol.* **1996**, *14*, 1123–1128.

113 M.T. Roskey, et al. DNA sequencing by delayed extraction-matrix-assisted laser desorption/ionization time of flight mass spectrometry. *Proc. Natl. Acad. Sci. USA* **1996**, *93*, 4724–4729.

114 F. Kirpekar, et al. DNA sequence analysis by MALDI mass spectrometry. *Nucleic Acids Res.* **1998**, *26*, 2554–2559.

115 F. Kirpekar, et al. 7-Deaza purine bases offer a higher ion stability in the analysis of DNA by matrix-assisted laser desorption/ionization mass spectrometry. *Rapid Commun. Mass Spectrom.* **1995**, *9*, 525–531.

116 D.J. Fu, et al. Sequencing exons 5 to 8 of the p53 gene by MALDI-TOF mass spectrometry. *Nat. Biotechnol.* **1998**, *16*, 381–384.

117 N.I. Taranenko, et al. Sequencing DNA using mass spectrometry for ladder detection. *Nucleic Acids Res.* **1998**, *26*, 2488–2490.

118 Y. Kwon, K. Tang, C. Cantor, H. Koster, C. Kang. DNA sequencing and genotyping by transcriptional synthesis of chain-terminated RNA ladders and MALDI-TOF mass spectrometry. *Nucleic Acids Res.* **2001**, *29*, E11.

119 J.R. Edwards, Y. Itagaki, J. Ju. DNA sequencing using biotinylated dideoxynucleotides and mass spectrometry. *Nucleic Acids Res.* **2001**, *29*, E104.

120 S. Broder, J.C. Venter. Whole genomes: the foundation of new biology and medicine. *Curr. Opin. Biotechnol.* **2000**, *11*, 581–585.

121 S. Hahner, et al. Matrix-assisted laser desorption/ionization mass spectrometry (MALDI) of endonuclease digests of RNA. *Nucleic Acids Res.* **1997**, *25*, 1957–1964.

122 M.S. Shchepinov, et al. Matrix-induced fragmentation of P3'-N5' phosphoramidate-containing DNA: high-throughput MALDI-TOF analysis of genomic sequence polymorphisms. *Nucleic Acids Res.* **2001**, *29*, 3864–3872.

123 F. von Wintzingerode, et al. Base-specific fragmentation of amplified 16S rRNA genes analyzed by mass spectrometry: a tool for rapid bacterial identification. *Proc. Natl. Acad. Sci. USA* **2002**, *99*, 7039–7044.

124 C. Elso, et al. Mutation detection using mass spectrometric separation of tiny oligonucleotide fragments. *Genome Res.* **2002**, *12*, 1428–1433.

125 R. Hartmer, et al. RNase T1 mediated base-specific cleavage and MALDI-TOF MS for high-throughput comparative sequence analysis. *Nucleic Acids Res.* **2003**, *31*, e47.

126 S. Krebs, I. Medugorac, D. Seichter, M. Forster. RNaseCut: a MALDI mass spectrometry-based method for SNP discovery. *Nucleic Acids Res.* **2003**, *31*, e37.

127 P. Stanssens, et al. High-throughput MALDI-TOF discovery of genomic sequence polymorphisms. *Genome Res.* **2004**, *14*, 126–133.

128 R. Sousa, R. Padilla. A mutant T7 RNA polymerase as a DNA polymerase. *EMBO J.* **1995**, *14*, 4609–4621.

129 P. Stanssens, et al. High-throughput discovery of genomic sequence polymorphisms. *Genome Res.* **2004**, *14*, 126–133.

130 S. Bocker. SNP and mutation discovery using base-specific cleavage and MALDI-TOF mass spectrometry. *Bioinformatics* **2003**, *19* (Suppl. 1), I44–I53.

131 S.C. Pomerantz, J.A. Kowalak, J.A. McCloskey. Determination of oligonucleotide composition from mass spectrometrically measured molecular weight. *J. Am. Soc. Mass Spectrom.* **1993**, *4*, 204–209.

132 M. Lefmann, et al. Novel mass spectrometry-based tool for genotypic identification of mycobacteria. *J. Clin. Microbiol.* **2004**, *42*, 339–346.

133 M. Ehrich, et al. Quantitative high-throughput analysis of DNA methylation patterns by base-specific cleavage and mass spectrometry. *Proc. Natl. Acad. Sci. USA* **2005**, *102*, 15785–15790.

134 C. Honisch, A. Raghunathan, C.R. Cantor, B.O. Palsson, D. van den Boom. High-throughput mutation detection underlying adaptive evolution of *Escherichia coli*-K12. *Genome Res.* **2004**, *14*, 2495–2502.

135 K. Tang, et al. Improvement in the apparent mass resolution of oligonucleotides by using 12C/14N-enriched samples. *Anal. Chem.* **2002**, *74*, 226–231.

136 F.A. Abdi, M. Mundt, N. Doggett, E.M. Bradbury, X. Chen. Validation of DNA sequences using mass spectrometry coupled with nucleoside mass tagging. *Genome Res.* **2002**, *12*, 1135–1141.

137 F. Abdi, E.M. Bradbury, N. Doggett, X. Chen. Rapid characterization of DNA oligomers and genotyping of single nucleotide polymorphism using nucleotide-specific mass tags. *Nucleic Acids Res.* **2001**, *29*, E61-1.

138 P. Ross, L. Hall, L.A. Haff. Quantitative approach to single-nucleotide polymorphism analysis using MALDI-TOF mass spectrometry. *Biotechniques* **2000**, *29*, 620–626, 628–629.

139 M. Werner, et al. Large-scale determination of SNP allele frequencies in DNA pools using MALDI-TOF mass spectrometry. *Hum. Mutat.* **2002**, *20*, 57–64.

140 S. Le Hellard, et al. SNP genotyping on pooled DNAs: comparison of genotyping technologies and a semi automated method for data storage and analysis. *Nucleic Acids Res.* **2002**, *30*, e74.

141 K.L. Mohlke, et al. High-throughput screening for evidence of association by using mass spectrometry genotyping on DNA pools. *Proc. Natl. Acad. Sci. USA* **2002**, *99*, 16928–16933.

142 A. Bansal, et al. Association testing by DNA pooling: an effective initial screen. *Proc. Natl. Acad. Sci. USA* **2002**, *99*, 16871–16874.

143 K.H. Buetow, et al. High-throughput development and characterization of a genomewide collection of gene-based single nucleotide polymorphism markers by chip-based matrix-assisted laser desorption/ionization time-of-flight mass spectrometry. *Proc. Natl. Acad. Sci. USA* **2001**, *98*, 581–584.

144 G. Amexis, et al. Quantitative mutant analysis of viral quasispecies by chip-based matrix-assisted laser desorption/ionization time-of-flight mass spectrometry. *Proc. Natl. Acad. Sci. USA* **2001**, *98*, 12097–12102.

145 S. Muller, et al. Retention of imprinting of the human apoptosis-related gene TSSC3 in human brain tumors. *Hum. Mol. Genet.* **2000**, *9*, 757–763.

146 N. Herbon, et al. High-resolution SNP scan of chromosome 6p21 in pooled samples from patients with complex diseases. *Genomics* **2003**, *81*, 510–518.

147 C.P. Rodi, B. Darnhofer-Patel, P. Stanssens, M. Zabeau, D. van den Boom. A strategy for the rapid discovery of disease markers using the MassARRAY system. *Biotechniques* **2002**, *Suppl.*, 62–66, 68–69.

148 J.C. Knight, B.J. Keating, K.A. Rockett, D.P. Kwiatkowski. In vivo characterization of regulatory polymorphisms by allele-specific quantification of RNA polymerase loading. *Nat. Genet.* **2003**, *33*, 469–475.

149 P. Sham, J.S. Bader, I. Craig, M. O'Donovan, M. Owen. DNA pooling: a tool for large-scale association studies. *Nat. Rev. Genet.* **2002**, *3*, 862–871.

150 B.J. Barratt, et al. Identification of the sources of error in allele frequency estimations from pooled DNA indicates an optimal experimental design. *Ann. Hum. Genet.* **2002**, *66*, 393–405.

151 C. Ding, et al. MS analysis of single-nucleotide differences in circulating nucleic acids: Application to noninvasive prenatal diagnosis. *Proc. Natl. Acad. Sci. USA* **2004**, *101*, 10762–10767.

152 W.T. Berggren, et al. Multiplexed gene expression analysis using the invader RNA assay with MALDI-TOF mass spectrometry detection. *Anal. Chem.* **2002**, *74*, 1745–1750.

153 C. Ding, C.R. Cantor. A high-throughput gene expression analysis technique using competitive PCR and matrix-assisted laser desorption ionization time-of-flight MS. *Proc. Natl. Acad. Sci. USA* **2003**, *100*, 3059–3064.

154 G.P. Elvidge, T.S. Price, L. Glenny, J. Ragoussis. Development and evaluation of real competitive PCR for high-throughput quantitative applications. *Anal. Biochem.* **2005**, *339*, 231–241.

155 C. Ding, E. Maier, A.A. Roscher, A. Braun, C.R. Cantor. Simultaneous quantitative and allele-specific expression analysis with real competitive PCR. *BMC Genet.* **2004**, *5*, 8.

156 R.M. McCullough, C.R. Cantor, C. Ding. High-throughput alternative splicing quantification by primer extension and matrix-assisted laser desorption/ionization time-of-flight mass spectrometry. *Nucleic Acids Res.* **2005**, *33*, e99.

6
MALDI-MS of Glycans

Dijana Šagi and Jasna Peter-Katalinić

6.1
Introduction

Protein glycosylation is a highly abundant and the most complex form of covalent protein modification, and is characteristic of eukaryotic cells. The cellular machinery devoted to the synthesis and modulation of oligosaccharide chains attached to proteins, called glycans, involves an estimated 1% of mammalian genes [1]. The biological role of glycosylation ranges from conformational stability and protection against degradation to molecular and cellular recognition in development, growth and cellular communication [2,3]. Changes in glycosylation are often a hallmark of disease states such as cancer or inflammation [4]. Moreover, naturally occurring mutations of the proteins involved in glycan biosynthesis in humans can lead to severe diseases with different syndromes, known under the collective name of "congenital disorders of glycosylation" (CDG) [5]. Both secreted and membrane-bound proteins can be glycosylated. Many of the recombinant proteins used as therapeutics are in fact glycoproteins, including erythropoietin, cytokines, and antibodies. In order for glycans to possess physiological and therapeutic activities, to have optimal biological function, and/or to avoid triggering an immune response, they must demonstrate the correct structure in terms of glycosylation site occupancy and the structure of glycoforms present. Likewise, in order to verify the structure of recombinant glycoproteins and to elucidate the biological function of glycans, a range of analytical methods that are both rapid and sensitive is required. Among these methods is included MALDI mass spectrometry (MS).

Synthesis of the polypeptide chain of a glycoprotein is under the direct control of genes, whereas oligosaccharides are attached to the protein and processed by a series of enzymatic reactions, without any direct genetic control. Consequently, a single glycoprotein normally emerges from the biosynthetic pathway as a mixture of glycosylation variants, known as glycoforms. At any given glycosylation site on a given protein, a range of variations can be found in the structure of N-glycans; this glycoprotein characteristic is called "microheterogeneity on the single glycosylation site". Furthermore, each potential glycosylation site in the glycoprotein

MALDI MS. A Practical Guide to Instrumentation, Methods and Applications.
Edited by Franz Hillenkamp and Jasna Peter-Katalinić
Copyright © 2007 Wiley-VCH Verlag GmbH & Co. KGaA, Weinheim
ISBN: 978-3-527-31440-9

may or may not be occupied, or it may only be partially occupied; this phenomenon is called "microheterogeneity of glycosylation", and is a common feature of protein glycosylation. The structure of glycans can vary among glycoproteins, as well as among cell types, tissues and species [6,7]. Consequently, the same protein, when derived from different cell types or tissues, can be differently glycosylated [8].

Mammalian and plant glycoproteins commonly contain three types of constituent oligosaccharide:

- N-linked oligosaccharides (*N*-glycans) which are attached to asparagine via an *N*-acetylglucosamine (GlcNAc) residue in an Asn-Xxx-(Ser, Thr) motif, where Xxx can be any amino acid except proline.
- O-linked oligosaccharides (*O*-glycans) that are attached to serine or threonine.
- Carbohydrate components of glycosyl-phosphatidylinositol (GPI) anchors attached near the C-terminal of a protein chain; these anchor the protein to the cell membrane via a glycan chain attached at the reducing end to a diacyl glycerol.

6.2
N-Glycosylation

N-linked oligosaccharides are formed in the endoplasmic reticulum (ER) of the cell, and further processed in the Golgi apparatus. Interestingly, within the ER all glycoproteins contain the same oligomannose sugars, and it is later – during passage through the Golgi cisternae – that the heterogeneity develops.

N-linked oligosaccharides contain a common trimannosyl-chitobiose core (Glc-NAc$_2$Man$_3$) with one or more antennae attached to each of the two outer mannose residues (3- or 6-linked). There are three general types of *N*-glycan (Fig. 6.1):

- high-mannose *N*-glycans, which contain only mannose residues appended to the core oligosaccharide;
- hybrid *N*-glycans in which the 6-antenna contains only mannose residues and the 3-antenna contains one or more disaccharides Gal (β1-4/3) GlcNAc (β1-), called *N*-acetyllactosamine (LacNAc) units; and
- complex *N*-glycans, which contain a variable number of LacNAc units attached to the outer mannose residues of the core oligosaccharide.

N-Acetyllactosamine units present in hybrid and complex *N*-glycans can sometimes be repeated, giving rise to polylactosamines. The chains may be terminated by a sialic acid moiety (NeuAc or NeuGc) linked (α2-6)- or (α2-3)- to the galactose

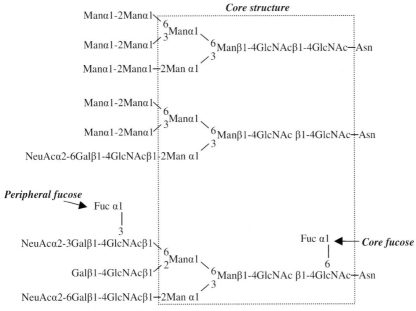

Fig. 6.1 The three subgroups of N-glycans: (a) high-mannose;
(b) hybrid; and (c) complex type. The structure within the box is
the trimannosyl-chitobiose core common to all N-glycans.

in LacNAc units. The fucose moiety, which is (α1-6)- or (α1-3)-attached to the as-
paragine-linked GlcNAc residue, is called the core fucose. If the fucose residue is
linked to the GlcNAc residues in the antennae by (α1-3/4) linkage it is termed "pe-
ripheral fucose". Another common modification is a GlcNAc residue attached to
the 4-position of the core branching mannose residue, referred to as a "bisecting"
GlcNAc residue.

Although the sites for potential N-linked glycosylation may be predicted from
the consensus sequence, it is not possible to predict whether the site will actually
be glycosylated. For O-linked glycosylation sites, there is no known general consensus sequence, and therefore the sites and structures must be identified and
characterized experimentally for both types of glycosylation, the molecular size of
the oligosaccharides being determined by MALDI-MS. Because most oligosaccharides derived from mammalian glycoproteins are comprised of relatively few
different monosaccharides with unique incremental masses (e.g., fucose, 146 Da;
hexose, 162 Da; N-acetylhexosamine, 203 Da; N-acetylneuraminic acid, 291 Da), a
molecular mass can be used to deduce the composition. Under particular conditions, the sequence (and in some cases also the branching) of the oligosaccharide
can be determined by using MALDI-MS, if the fragment ions are analyzed. Furthermore, a combined approach of degradation by exoglycosidase(s) and MALDI-
MS analysis can be used for carbohydrate "ladder" sequencing, the determination

of branching patterns, and for obtaining information on the configuration and anomericity of sugar residues.

6.2.1
Release of *N*-Glycans

The detailed overall structural analysis of *N*-glycans is usually carried out after release from the protein core and recovery of the carbohydrate moieties (mixtures) from protein. The linkage between the glycan moiety and the asparagines in the protein can be cleaved either chemically or enzymatically. The release of *O*-glycans from proteins is usually performed chemically, by β-elimination.

6.2.1.1   Chemical Release
The most common chemical procedure used to liberate *N*-glycans from glycoproteins is that of hydrazinolysis [9]. Hydrazine cleaves the amide bonds between *N*-glycans and asparagines, as well as amide bonds in the polypeptide backbone, thus causing a loss of protein integrity. *O*-glycans may also be released by hydrazinolysis, though the reaction requires a lower temperature than that *N*-glycan release. Although, depending on the reaction conditions, both *N*- and *O*-glycans can be released [10,11], the reaction is not absolutely specific, and may also lead to a modification of the reducing end sugar moieties, which enhances heterogeneity. Another drawback of this procedure is the cleavage of acyl groups from the *N*-acetylamino sugars and sialic acids. Normally, these acyl groups are replaced chemically in the acetylation step, on the assumption that they were originally acetyl groups, though this is not always true for the sialic acids which frequently contain *N*-glycolyl groups.

Although *O*-glycosidic linkages can be cleaved by hydrazinolysis, β-elimination is the most commonly used chemical cleavage method used to liberate *O*-glycans [12], which can be released in either reduced or non-reduced forms [13–16]. The latter form enables subsequent derivatization of the *O*-glycans by reductive amination. However, β-elimination may lead to a partial destruction of the peptide bond whereby the information relating to the protein – for example, the site of glycan attachment – is lost.

6.2.1.2   Enzymatic Release
The enzymatic treatment of glycoproteins with endoglucosidases or glycoamidases causes *N*-glycans to be released, while the glycan and protein/peptide chain is left intact [17–19].

The most popular enzyme for *N*-glycan release is peptide-*N*4-(*N*-acetyl-β-glucosaminyl)-asparagine amidase (PNGase F), which cleaves the linkage between GlcNAc and asparagines, thereby releasing the intact glycan and leaving aspartic acid in place of asparagine at the N-linked site of the protein. A mass difference of one unit between Asn and Asp in the peptide mass can be detected by MS on

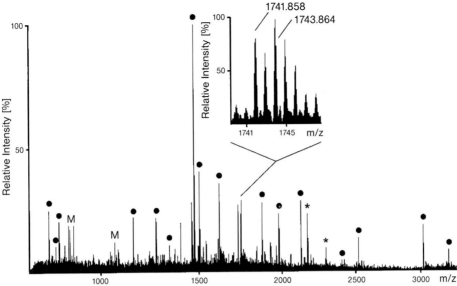

Fig. 6.2 Delayed-extraction MALDI peptide mass map obtained from bovine fetuin after in-gel deglycosylation using PNGase F in the presence of 50% ^{18}O-labeled water and followed by in-gel trypsin digestion. The insert shows an expanded view of the molecular ion region for one of the formerly glycosylated peptides. The partial incorporation of the ^{18}O label can be recognized by the 2 Da spacing between the monoisotopic peaks of the unlabeled and labeled peptide. Sequence-specific peptide ion signals are marked with bullets, matrix-related peaks are labelled "M", and trypsin autolysis products are denoted with an asterisk. From Ref. [20].

peptide sequencing, and provides information on the extent of occupancy at that site. By performing the reaction in H$_2$^{18}O, an ^{18}O label will be incorporated into Asp to aid detection (Fig. 6.2) [20,21]. When using PNGase F, most glycans will be released, except those with the fucose attached to the 3-position to the reducing-terminal GlcNAc, which can be released by PNGase A [22,23]. The corresponding glycoproteins are commonly found in plants and nematodes [24]. N-glycan release by PNGase F treatment of glycoproteins can be performed either in solution or in one- or two-dimensional gels, where single glycoprotein bands obtained by SDS-PAGE separation can be in-gel digested (Fig. 6.3) [25–27].

Endo-β-N-acetylglucosaminidase (endoglycosidase H or endo H) is another popular enzyme used for N-glycan release. Endo H cleaves the glycosidic bond between the two GlcNAc residues of the core, thus leaving the reducing-terminal GlcNAc residue, together with any substituents, attached to the protein/peptide. This enzyme is more specific than PNGase F as it cleaves only high-mannose and most hybrid-type chains [28,29]. Information on the presence of core fucosylation is, thus, not available from the spectra of the resulting glycans. Two other useful endoglycosidases are endo F$_1$ and endo F$_2$; these are produced by *Flavobacterium meningosepticum*, and have different substrate specificities [30,31].

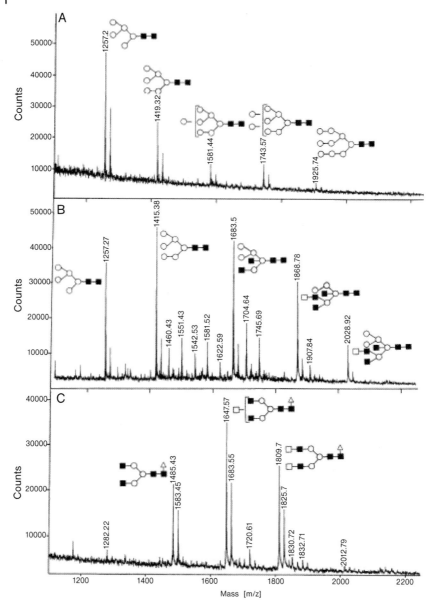

Fig. 6.3 MALDI mass spectra of neutral oligosaccharides obtained by in-gel PNGase F digestion from 100 pmol each of (A) RNase B, (B) ovalbumin, and (C) IgG. All molecular weights represent monoisotopic masses of the respective [M + Na]⁺ ions of a particular glycan species. [M + Na]⁺ ions are accompanied by a smaller signal of the respective [M + K]⁺ ion (16u above the [M + Na]⁺ ion). Only the major structures are shown in structural detail. □, galactose; ■, N-acetylglucosamine; ○, mannose; △, fucose. From Ref. [26].

6.2.2
Sample Preparation for MALDI-MS

6.2.2.1 Contaminant Removal

High purity of the sample is critical in order to obtain high-quality MALDI spectra. In this respect, carbohydrates appear to be more susceptible than do peptides to the negative effects of salts, buffers and other compounds. Although MALDI-MS is more tolerant to the salt contaminants than electrospray ionization (ESI), extensive purification prior to MS analysis is highly relevant to the quality of the spectra. Several methods can be used to remove the contamination, including ion-exchange resin microcolumns [26], membrane dialysis, fast protein liquid chromatography (FPLC) with gel filtration if larger amounts of an N-glycan sample are available [32], and graphitized carbon microcolumns [33].

For membrane dialysis, a molecular mass cut-off of 500 Da should be generally chosen. When microdevices or drop dialysis membranes are used [34–36], this method can be used to remove small contaminating quantities of N-glycan, such as those obtained by in-gel deglycosylation.

Ion-exchange resins are very useful when packed into disposable pipette tips as microcolumns [26], and can be used to clean-up N-glycans obtained via in-gel deglycosylation. For ion exchange prior to MALDI analysis, resin beads can be mixed with the MALDI matrix and removed thereafter from the MALDI target by a jet of air, directly before sample analysis [37].

The use of graphitized carbon for desalting carbohydrates, as introduced recently by Packer et al. [33], appears to be the method of choice for an efficient clean-up of carbohydrates. The oligosaccharides are adsorbed onto the graphitized carbon, and this allows the removal of salts and other contaminants from the sample. Self-prepared microcolumns of many sizes have been used successfully to desalt small amounts of N-glycans obtained by in-gel deglycosylation, as well as for large sample quantities. Alternatively, a high-pressure liquid chromatography (HPLC) set-up fitted with a graphitized carbon column can be used not only for desalting but also for the separation of glycans [38].

6.2.2.2 Derivatization

Derivatization, though not essential for MALDI analysis of carbohydrates can, nevertheless, be advantageous under certain circumstances, particularly for improving the ionization efficiency relative to those of the native molecules. Derivatization also improves the stability of the substrate (particularly sialylated glycans) in the ion source for the neutralization of acidic residues, thereby allowing all released glycans to be analyzed in a single mass spectrum. Fragmentation analysis is also simplified.

The most widely used derivatization approach is that of permethylation, which can improve the sensitivity by about an order of magnitude. Two permethylation techniques have been developed: in the first method, permethylation by methyl iodide is catalyzed by the methylsulfenyl carbanion [39]; in the second, the reaction

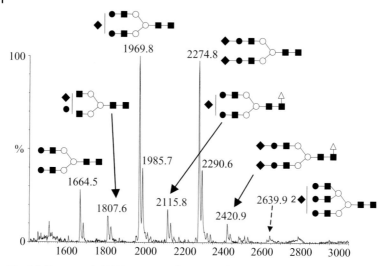

Fig. 6.4 Linear mode MALDI spectra of native *N*-glycans obtained by in-gel PNGase F digestion from the 2-D gel of α1-antitrypsin spots of a CDG-type-IIx patient in the positive ion mode after methylation of sialic acids. Mass spectra were recorded using THAP as a matrix. From Ref. [27]. Key to the symbols: ■ = GlcNAc, ● = galactose, ○ = mannose, ◆ = sialic acid, △ = fucose.

is catalyzed by sodium hydroxide [40]. Peracetylation of glycans is also effective, and can be used to determine the linkage of sialic acid [41].

The advantages of these derivatization procedures must be balanced against the disadvantages of the derivatization step itself. The derivatization chemistry places a lower limit on the quantity of sample that can be analyzed, and consequently does not appear to be appropriate for the analysis of samples available only in small quantities (e.g., from gel spots). Both permethylation methods cause a considerable increase in the molecular weight of glycans. Moreover, permethylation reactions are often accompanied by undesirable undermethylation, which can lead to wrong assignments. Alternatively, a selective methylation method of sialic acid residues to methyl esters has been described [42]. This allows the detection of sialylated oligosaccharides and ganglioside ions in the positive ion mode, without metastable decay [43]. Selective carboxylate esterification by a methyl group has also been used to facilitate MALDI-TOF analysis of N-linked glycans released from one- or two-dimensional gel spots with glycosidases (Fig. 6.4) [26,27].

Reductive amination Selective derivatization of reactive carbonyl groups on the reducing end of an *N*-glycan is frequently used to add a fluorescent/UV-absorbing tag for the detection of *N*-glycan by HPLC and capillary electrophoresis (CE). The most common derivatization technique is a reductive amination by condensation of the ring-opened (carbonyl) form of the carbohydrate with an amine, usually aromatic. Several amines have been used in this context, and a number of these have also been investigated as derivatives to extend mass spectrometric detection limits and minimize fragmentation over that produced by the native carbohydrates. Many of these derivatives have been discussed in recent reviews [44,45]. The choice of the amine influences the ionization efficiency, and the effects are different with MALDI versus ESI [46]. For example, the 2-(diethylamino)ethyl ester of 4-aminobenzoic acid was shown to increase the sensitivity 1000-fold relative to underivatized glycans under MALDI-MS conditions [47]. Also, 2-aminoacridine (2-AMAC) gave the highly intense MALDI signals of oligosaccharides by MALDI-MS [46,48,49]. Other frequently used derivatization agents include 2-aminopyridine (AP) [50,51], 2-aminobenzamide (AB) [52], and 2-aminobenzoic acid (ABA).

Several derivatives capable of forming negative charges, such as those from ABA [53], 8-aminonaphthalene-1,3,6-trisulfonic acid (ANTS) [54] and 1-aminopyrene-3,6,8-trisulfonate (APTS) [55], allow carbohydrates to be examined as negative ions, with good sensitivity.

A combination of HPLC separation of derivatized N-glycans and their consequent MALDI-MS analysis has often been used for biological studies. A recent example is a study of changes in N-glycan expression during *Caenorhabditis elegans* development (Fig. 6.5) [56].

6.2.3
Matrices

6.2.3.1   Matrices for Neutral N-Glycans

Neutral carbohydrates ionize easily in positive ion mode under the formation of $[M + Na]^+$ as major ionic species. This ion is frequently accompanied by a less-abundant $[M + K]^+$ ion. Other cationized species, such as $[M + Li]^+$ can also be generated by the addition of an appropriate inorganic salt to the matrix [57,58]. One of the first – and still the most common – matrices used for the examination of neutral carbohydrates is 2,5-dihydroxybenzoic acid (DHB), which was introduced for carbohydrate analysis by Stahl et al. [58]. When crystallizing on the target, DHB tends to form large crystals at the periphery of the spot and the microcrystalline central region. In contrast, proteins and peptides are more easily desorbed from large crystals, with carbohydrates being located within the inner microcrystalline region. In order to produce a more homogeneous sample/matrix spot, the crystals may be recrystallized from ethanol to form a microcrystalline surface which improves the sensitivity [59]. Alternatively, DHB can be mixed with various additives [60,61]. For example, 10% 2-hydroxy-5-methoxybenzoic acid improves sensitivity and resolution, most likely due to an altered crystallization behavior of the matrix [60].

Several other matrices have been used during the course of investigations into oligosaccharides. For example, osazones have been reported to produce superior spectra to those produced with DHB for neutral [62] as well as sialylated [63] and sulfated [64] glycans, due mainly to improved resolution and signal-to-noise (S/N) ratio.

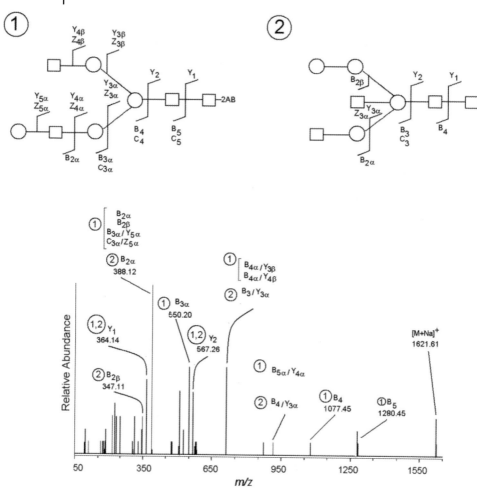

Fig. 6.5 MALDI QoTOF MS/MS analysis of *Caenorhabditis elegans* Hex₄HexNAc₄-2AB, [M+Na]⁺ m/z 1621.61. According to the fragmentation pattern, it can be postulated that two isomers originated from this precursor ion. The fragmentation pattern is consistent with the presence of LacNAc and core bisecting GlcNAc in structures ① and ②, respectively. From Ref. [56].

3-Aminoquinoline (3-AQ) [65] has been used for the ionization of plant inulins with masses of up to 6 kDa. The matrix provides a comparatively low background, and appears more efficient than DHB for sialylated glycans [66]. However, it sublimes too rapidly to be of general use [67], and appears to be more sensitive to the presence of contaminants than DHB [68]. 5-Chloro-2-mercaptobenzothiazole (CMBT) has proved useful for high-mannose N-linked glycans [69], the combination of CMBT and DHB having been used for the analysis of neutral oligosaccharides from human milk [70].

6.2.3.2 Matrices for Acidic *N*-Glycans

In contrast to neutral glycans, DHB is generally not recommended for the analysis of acidic glycans such as the sialylated N-linked sugars. One problem in the analysis of sialylated glycans is a frequent loss of sialic acid or CO_2 from the sialic acid moiety. In addition, sialylated glycans, particularly when analyzed in the positive ion mode, appear in the spectrum as a mixture of cation adducts due to a tendency for the carboxylic group of the sialic acid to form salts with one or more alkali metal cations present in the mixture. In such cases multiple peaks originating from the same molecular species are present in the spectra, thus reducing the sensitivity of detection to 1–10 pmol of acidic oligosaccharides compared to that of neutral oligosaccharides (10–100 fmol). Moreover, as the number of sialic acid residues incorporated into the oligosaccharide increases, these problems are further aggravated.

Many of these problems associated with loss of sialic acids by fragmentation and cation adducts can be overcome by a suitable choice of matrix and ionization mode. The sialic acid-containing glycans are generally better detected in the negative ion mode as [M-H]⁻ ions due to the lower abundance of cation attachment compared to the positive ion mode. In order to minimize the loss of sialic acid in the spectra, the sialylated glycans can be submitted to the linear ion mode analysis, avoiding the separation of metastable products. Two different matrices – 6-aza-2-thiothymine (ATT) [67,70] and 2,4,6-trihydroxyacetophenone (THAP) [67] – have been found to provide a significant increase in sensitivity for acidic glycans over that produced by other, more common, matrices (Fig. 6.6). One advantage of ATT over THAP was its greater tolerance to contaminants. However, some degree of fragmentation by loss of HCOOH from the [M-H]⁻ ion and (on occasion) a small amount of sialic acid, were observed in ATT spectra recorded in the linear mode. On the other hand, the THAP matrix, with additive ammonium citrate, provided equal sensitivity compared to ATT, with no evidence of fragmentation in the linear mode. The conditions used for sample preparation were found to be crucial to obtain maximum sensitivity. Vacuum drying of the sample spot was used to prevent the formation of large crystals, after which the sample was allowed to absorb water to promote the formation of small crystals. Alternatively, the recrystallization could be used for the formation of small crystals of THAP/DAC matrix [27]. Furthermore, THAP/DAC was found to be the matrix of choice for analysis of mixtures of neutral and acidic glycans in positive and negative ion MALDI analysis from a single sample preparation spot on the target [27].

6.2.4
Structural Analysis

6.2.4.1 Fragmentation of *N*-Glycans

The fragmentation of glycans can reveal information on monosaccharide sequences, branching and, eventually, on linkages. Fragment ion spectra of complex glycans are considerably more complicated than are those of peptides because of their branching structure. Carbohydrates ionized by different ionization

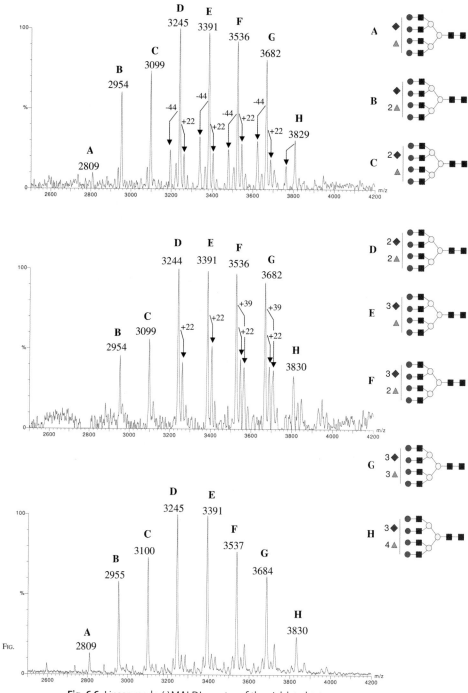

Fig. 6.6 Linear mode (-)MALDI spectra of the sialylated tetra-antennary *N*-glycan mixture obtained with: (a) ATT/DAC (15 gL⁻¹/20 mM) and (b) THAP/DAC (20 gL⁻¹/ 20 mM) before and (c) after purification with graphitized carbon (from Ref. [27]). Key to the symbols: ■ = GlcNAc, ● = galactose, ○ = mannose, ◆ = sialic acid, ▲ = fucose.

techniques, for example MALDI, ESI and FAB, tend to show similar fragmentation patterns. Mass spectrometry techniques for the ionization and fragmentation have been discussed extensively in a recent review [71]. The observed fragment ions depend on a number of factors such as the nature of the cation or anion (e.g., [M + H]$^+$ or [M + Na]$^+$ [72–74]), and the energy involved in the fragmentation process [75]. Glycans show two types of fragmentation: (i) glycosidic cleavages that result from the breaking of the bonds between the sugar rings; and (ii) cross-ring cleavages that involve the breaking of two bonds on the same sugar residue. The glycosidic cleavages are thought to be mainly charge-induced [76] and to provide information on the sequence and branching of constituent monosaccharides. The formation of cross-ring cleavages involves charge-remote processes. Those cleavages are usually weaker and may yield information on the linkage position of one residue to the next. Thus, the presence of the cross-ring cleavages ^{0,4}A and ^{3,5}A produced by cleavage of the core-branching mannose residue contains only the antenna attached to the 6-position and provides information on the composition of each antenna. If sugars are ionized by metal attachment, the fragment ions will also contain a metal cation. Additionally, there is also a tendency for simultaneous fragmentation processes in different regions of the molecule. The formation of the resulting "internal" fragments often involves more than one cleavage pathway. One of the very useful internal fragments is the so-called D ion introduced by Harvey [77]. The D ion refers to the ion formed by loss of the 3-antenna together with the chitobiose core. Preferential losses of groups attached to the 3-position of hexoses have frequently been observed in the fragmentation of glycans [78–82]. Because ion D contains the complete 6-antenna, the composition of this antenna can easily be determined, as can that of the 3-antenna, by difference and its detection in the spectrum.

The generally accepted nomenclature for description of carbohydrate fragmentation is that introduced by Domon and Costello [83]. Briefly, ions retaining the charge at the reducing terminus are called X (cross-ring), Y and Z glycosidic cleavages on both sides of the glycosidic bond (Fig. 6.7). Ions with the charge at the non-reducing terminus are labeled with A (cross-ring), B, and C (glycosidic).

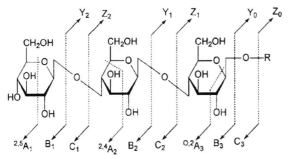

Fig. 6.7 The Domon And Costello nomenclature system for fragmentations of carbohydrate chains [83].

Sugar rings are numbered from the reducing end for X, Y and Z ions, and from the non-reducing end for A, B and C, and the subscript numerals denote the linkage cleaved. Where fragmentation is specific to an antenna, the given antenna is denoted by a Greek letter, starting with α for the largest antenna. Bonds broken in cross-ring fragmentations are denoted by superscript numerals placed before the letter, for example $^{2,4}A_2$. Where two glycosidic bonds are broken, the cleavages are separated by a slash, for example, B_2/Y_2.

The type of the parent ion produced from the carbohydrate has a significant effect on the fragmentation [73,76,84]. Protonated species decompose much more readily than metal-cationized species. Decay rates of the cationized species decrease in the order $Li^+ > Na^+ > K^+ > Cs^+$. The type of fragmentation also varies, depending on the ionization adduct. Whereas the protonated species fragment predominantly by glycosidic cleavages, the metal-containing ions give additional cross-ring cleavages [72,73,76,85]. It has been shown that the most informative fragment ions for sequence and linkage information are obtained from $[M + Na]^+$ or $[M + Li]^+$ ions [84].

Furthermore, the branched glycans produce less fragment ions than linear ones. Cancilla et al. [76] attributed this observation to the coordination of the metal ion to the sugar.

Fragmentation spectra in MALDI-MS can be produced under different conditions, namely in-source decay fragmentation (ISD), post-source decay fragmentation (PSD), and collision-induced decomposition (CID). ISD fragment ions are formed by a rapid decomposition in the ion source and, in contrast to PSD fragment ions, they are detected as focused fragments in the spectra. They can be observed with TOF instruments, particularly those fitted with delayed extraction that are operated with relatively long delay times [86] or with magnetic sector instruments [87]. Loss of the sialic acid by ISD fragmentation is frequently observed and can represent a major problem in analysis of glycan mixtures because they are indistinguishable from molecular ions. In order to overcome this problem, this derivatization or optimization of ionization conditions is used.

6.2.4.2 Post-Source Decay

Post-source decay (PSD) fragment ions correspond to the fragment ions formed between the ion source and reflectron in a reflectron TOF mass spectrometer. Since PSD ions are unfocused or metastable, they appear as broad peaks at higher apparent mass than their focused counterparts. PSD ions can be focused by stepping the reflectron voltage and combining the resulting partial spectra. These methods were used extensively for the structural determination of carbohydrates. Generally, PSD spectra are dominated by glycosidic and internal cleavages with very weak cross-ring fragments. Despite these problems, linkage information could be obtained from the relative abundance of glycosidic cleavage ions [88–91]. For example, Yamagaki and Nakanishi have used PSD fragment ions to distinguish between Lewis x and Lewis a type carbohydrates that have, respectively, fucose or galactose substituted to C-3 of the GlcNAc residue. Cross-ring cleavages, although weak, could also be detected in the PSD spectra. Spengler et al. have found ^{0,4}A ions in the PSD spectra of bi- and tri-antennary N-linked glycans, while Zal et al. have used ^{3,5}A and ^{0,4}A cross-ring cleavage ions produced from the branching mannose and internal fragments of type D to determine the antenna configuration in high-mannose N-linked glycans. Rouse et al. [92] proposed a "knowledge-based" strategy to characterize unknown isomeric N-glycan structures. The strategy is based on the comparison of the PSD spectra of the unknown glycan with the PSD spectra of the standards possessing similar structural features. This strategy was applied to distinguish between the respective isomers of several high-mannose and asialo complex N-glycans (Fig. 6.8).

PSD fragmentation is influenced by the matrix employed. Matrices such as 4-HCCA usually catalyze considerable fragmentation such as sialic acid loss, whereas "cooler" matrices such as DHB, ATT or 1,3,5-THAP do not. Furthermore, Pfenninger et al. have shown that using a composition of two-layer matrices, 5-chloro-2-mercaptobenzothiazole as the first layer and 3-aminoquinoline as a second, extensive fragmentation could be achieved [70].

Fragmentation can also be influenced by derivatization. Although permethylation simplifies the interpretation of fragment spectra, the fragmentation of permethylated glycans yields less PSD ions compared to non-derivatized glycans [93]. The cleavage of glycosidic bonds to sialic acid and GlcNAc are the only significant fragments. Nevertheless, Yamagaki and Nakanishi [94] observed that the cleavage of $\alpha(2-3)$-linked sialic acids in sialyllactoses occurs more readily than cleavage of the $\alpha(2-6)$-linkage, allowing the isomers to be distinguished by PSD due to B_1 ion abundance. Peracetylated glycans fragmented slightly better than permethylated ones giving rise to B- and Y-ions, but the fragments were usually accompanied by additional ions produced by loss of acetic acid.

Reductively aminated oligosaccharides modified with benzylamine or 2-aminopyridine produce a full series of fragment ions that contain the original reducing terminus, providing sequence and branching information [50,93]. The 4-aminobenzoic acid 2-(diethylamino)ethyl ester and 2-aminobenzamide derivatization have also been shown to be suitable for the structural analysis of oligosaccharides using PSD.

Generally, PSD suffers from relatively low sensitivity and only moderate resolution; moreover, the need to stitch together spectra derived from several reflector voltages makes calibration difficult. The recent availability of MALDI sources for analyzers such as the Q-TOF, TOF/TOF, FT and QIT, enable far more efficient production and detection of product ions than is possible with PSD.

6.2.4.3 Collision-Induced Decomposition

The CID technique enables good precursor-ion selection, controlled fragmentation and higher fragment ion resolution than PSD. CID of carbohydrates in combination with MALDI ionisation source was mostly performed on Q-TOF, ion trap, FT, TOF/TOF and magnetic sector mass spectrometers.

The first study on the ionization and fragmentation of N-glycans with MALDI Q-TOF was reported by Harvey et al. [95], who analyzed both non-derivatized and

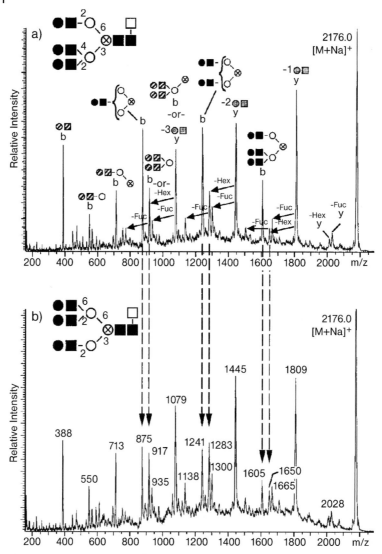

Fig. 6.8 MALDI-PSD spectra of: (a) a complex 3-branched triantennary N-linked glycan; and (b) a complex 6-linked glycan. From Ref. [92].

derivatized N-glycans. The MS/MS spectra were almost identical to those obtained by electrospray Q-TOF, and easier to interpret due to the presence of singly charged ions only in the spectra. The generation of fragment ions depended on the type of precursor ion, for example [M + Na]⁺ or [M + H]⁺. The fragmentation of [M + Na]⁺ was dominated by B- and Y-type glycosidic cleavages, and cross-ring cleavages were also present though less abundant. Wheeler and Harvey [64] also analyzed sulfated *N*-glycans using MALDI Q-TOF instrument, while Hanrahan et al. [96] analyzed neutral and acidic glycans released from glycoproteins separated by gel electrophoresis after reductive amination with 3-acetylamino-6-acetylaminoacridine. Fragmentation observed with MALDI QoTOF MS/MS in the positive mode showed that fragment ions containing the derivatized reducing terminus were predominant, thereby facilitating interpretation of the spectra (Fig. 6.9).

The detailed characterization of carbohydrate structures often requires multistage decomposition (MS^n), making the application of trapped ion-type instruments such as the QIT and FTMS advantageous. The FTMS has the considerable advantage over the QIT that ions are detected with high resolution and mass accuracy. FTMS can be used with several different dissociation techniques, including CID, sustained off-resonance irradiation (SORI) CID [97], infrared multiphoton dissociation (IRMPD) [98], and electron capture dissociation (ECD) [99].

When Penn et al. [100] analyzed milk sugars and N-linked carbohydrates by MALDI CID FT mass spectrometry, the loss of fucose was observed as the most favorable glycosidic cleavage, and was found to have the lowest relative dissociation threshold. Dissociation thresholds for the appearance of cross-ring cleavage ions were found to be higher than those of glycosidic cleavages. The ability to measure dissociation thresholds with the FT instrument suggested that such measurements might be used to determine the composition of a carbohydrate by distinguishing between isomers, for example, galactose and mannose. Another useful feature of this instrument was the ability to store ions for a considerable period of time, making possible the selection and further fragmentation of primary fragments. This feature was demonstrated for the B3 fragment ion from the trimannosyl-chitobiose core oligosaccharide of N-linked glycans.

The utility of IRMPD for the analysis of carbohydrates was examined by Xie and Lebrilla [101]. These authors suggested that the dissociation threshold might be useful for differentiating among isomeric oligosaccharides in milk. Furthermore, comparison of the CID and IRMPD spectra of oligosaccharide alditols revealed that IRMPD could be used as a complementary method to obtain structural information.

In order to obtain MS^n spectra of carbohydrates, Creaser et al. used atmospheric pressure (AP) UV-MALDI coupled with an ion trap (IT) mass spectrometer [102]. Moyer et al. used UV AP-MALDI IT for the negative mode analysis of sialylated carbohydrates. More recently, the analysis of sialylated carbohydrates using infrared (IR) AP-MALDI IT mass spectrometry has been reported [103,104].

High-energy CID could be performed on magnetic sector and recently developed TOF/TOF instruments; the benefits of high-energy CID of carbohydrates are cross-ring cleavages. Mechref et al. analyzed neutral and permethylated sialylated glycans by MALDI-TOF/TOF, and observed cross-ring and internal type of ions that enabled linkage informations [105,106].

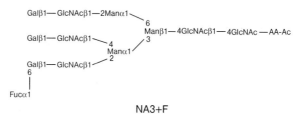

NA3F

NA3+F

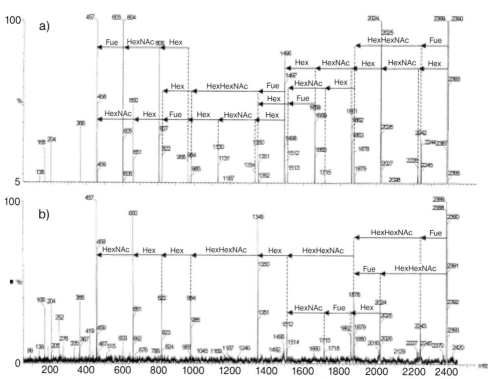

Fig. 6.9 MS/MS analysis of 3-acetylamino-6-acetylaminoacridine derivatives of the two isomers (A) NA3F and (B) NA3 + F, showing differences in fragmentation due to differences in the position of the fucose residue. From Ref. [96].

6.2.4.4 Use of Exoglycosidase Sequencing

The enzymatic sequencing of oligosaccharides is based on the ability of specific exoglycosidases to cleave terminal monosaccharides from the non-reducing end, providing information on the identity and linkage position of the monosaccharides. Furthermore, cleavage by α- or β-specific exoglycosidases indicates the anomericity of specific terminal sugar residues. The sequential application of a single exoglycosidase, or as arrays, can be used to obtain the complete sequence of an N-glycan [107,108], with the enzyme reaction products being monitored by MALDI-MS [109–111] and leading to the complete sequence of N-glycans. Important prerequisites for enzymatic sequencing are high specificity and purity of the exoglycosidase, since contaminating glycosidase activities can give rise to misleading interpretations.

Küster et al. modified the method so that all enzyme digests could be performed on the MALDI target [112]. Before each exoglycosidase incubation step the matrix was removed by drop dialysis. Furthermore, Geyer et al. [113] used a procedure in which 6-aza-2-thiothymine rather than the acidic DHB was used as the matrix, and the enzyme digests were performed in the presence of the matrix, which did not affect the enzyme activities (Fig. 6.10). Mechref and Novotny [114] have developed similar methods for the simultaneous release of glycans from the glycoprotein with PNGase F; these are then sequenced using exoglycosidase arrays on the MALDI target. For each glycoprotein the procedure was carried out on four different target spots with PNGase F and a different exoglycosidases array in order to avoid removal of the matrix, as required by the original sequential protocol. Arabinosazone was used as a matrix. This method was successfully applied for the characterization of major urinary protein N-glycans in the house mouse [114].

Küster et al. performed sequencing by the exoglycosidase treatment of N-glycans obtained after in-gel PNGase F digestion of SDS-PAGE-separated glycoproteins [26].

6.2.5
MALDI-MS of Intact Glycoproteins

To date, the analysis of intact glycoproteins by UV MALDI-MS has been only partially successful, mainly because the macro- and microheterogeneity of the attached carbohydrates makes the analysis complicated. MALDI-TOF mass spectrometers are generally able to resolve individual glycoforms only for small proteins containing a limited number of glycans, preferably attached only to one site. Examples are ribonuclease B (15 kDa) with five high-mannose glycans at a single glycosylation site [115,116] and Sf9-derived IL-4 receptor (30 kDa) [117]. Unresolved mass peaks are usually generated from larger glycoproteins and those with several glycosylation sites. However, UV MALDI-MS can be used with large molecules to determine the presence or absence of glycosylation [118–120]. Wada et al. [118] studied the glycosylation of human serum transferrin (79.6 kDa) from patients suffering with CDG syndrome (Fig. 6.11). Compared to controls, the

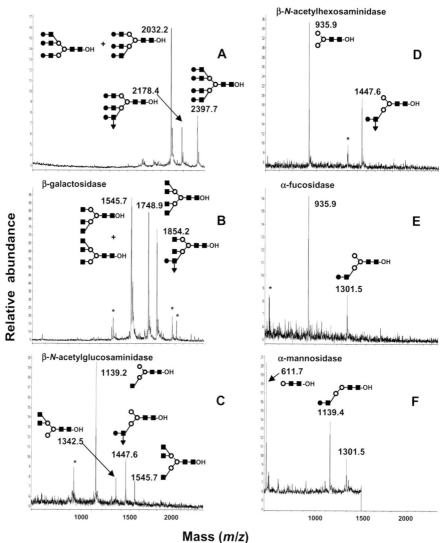

Fig. 6.10 Mass spectra of a mixture of oligosaccharide alditols (80 pmol) before and after sequential enzymatic treatment: native glycans (A); incubation with β-galactosidase from *Diplococcus pneumoniae* (B); β-*N*-acetylglucosaminidase from *D. pneumoniae*; (C) β-*N*-acetylhexosaminidase from jack beans (D); 0.08 mU α-fucosidase (5 h) (E); and α-mannosidase from jack beans (F). Key to the symbols: ■ = GlcNAc, ● = galactose, ○ = mannose, ◆ = sialic acid, ▲ = fucose. Peaks marked in parentheses reflect unknown impurities or unspecific degradation products. From Ref. [113].

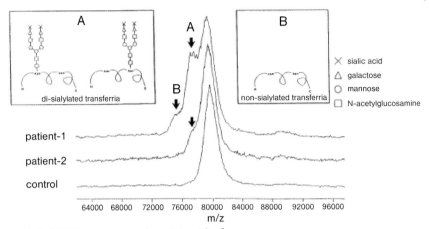

Fig. 6.11 MALDI mass spectra (sinapinic acid) of serum transferrins from a healthy control and two patients with CDG syndrome. From Ref. [118].

transferrin from CDG patients produced additional peaks at the lower m/z values, corresponding to species that lack one and two sialylated biantennary glycans.

The appropriate choice of parameters such as matrix, the sample-matrix preparation technique, the pH and instrumental condition can improve the quality of the spectra [121,122]. Bahr et al. have shown that delayed ion extraction significantly improved the resolution. The matrix DHBs gave by far the best results in glycoprotein analysis.

MALDI-TOF MS can also be used to determine protein molecular mass (MM) before and after removal of the attached glycans [19,123,124]. The difference in MM provides information on the state of glycosylation, even though the individual glycoforms may not be fully resolved. Glycan compositions, in terms of the constituent isobaric monosaccharides, can then be deduced from these masses as N-glycans contain only a limited number of monosaccharide types (hexose, HexNAc, etc.).

More recently, the analysis of intact glycoproteins using IR MALDI-TOF mass spectrometry with orthogonal extraction has been reported [125]. A significantly improved resolution – particularly in the high mass range – enables the resolution of individual glycoforms (Fig. 6.12).

6.2.6
Determination of N-Glycosylation Site Occupancy

The determination not only of which sites in the glycoprotein are glycosylated, but also of the extent of occupation at each site, are generally accomplished by performing enzymatic or other degradative reactions on the glycoprotein and isolating the individual glycopeptides. Mass spectrometric analysis of glycopeptides in a complex protein digest is complicated by the fact that the signals for glycosylated

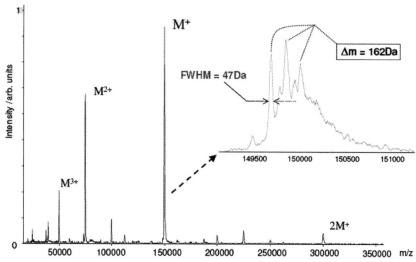

Fig. 6.12 Glycosylation pattern of an intact monoclonal antibody obtained on the IR-MALDI-TOF instrument with orthogonal extraction. From Ref. [125].

peptides tend to be suppressed in the presence of non-glycosylated peptides. The characterization of glycoproteins by MS is typically performed on glycopeptides isolated by liquid chromatography, by sequential treatment with specific endo-/exoglycosidases, and by monitoring the mass changes by MS [110]. This strategy requires detection of the glycopeptide in the peptide mass map, an availability of suitable glycosidases, and sufficient material to perform multiple digestion steps. The advantage of this method is that the specificity of the glycosidases provides information on the identity of the attached monosaccharides, as well as the linkage type. This approach has also been applied to small quantities of glyco- proteins available from electrophoretic gels [126,127]. Mills et al. [128] proposed a strategy employing several digestion protocols including a panel of proteases and deglycosylating enzymes to monitor the state of glycosylation of all puta- tive N-glycosylation sites on serum transferrin, 1-antitrypsin and recombinant β-glucosylceramidase.

Another strategy for the characterization of glycosylated proteins is based on an analysis of the blotted glycosylated proteins after on-blot release of the glycan struc- tures and tryptic digestion [129]. Küster et al. proposed an enzymatic release of the glycan in buffer containing 50% $H_2^{18}O$ [20]. The glycosylated peptide and the glycosylation site can be identified based on its isotope pattern (see Fig. 6.2); this method can also be combined with lectin affinity capturing and liquid chromatog- raphy-tandem MS [130]. A similar strategy involves enrichment of the glycopeptide by using hydrophilic interaction liquid chromatography columns, followed by iden- tification of the glycosylation site by treatment with a mixture of glycosidases that leaves a single GlcNAc residue on N-linked glycosylation sites. This technique was successfully applied to the identification of glycosylation sites in serum proteins

[131]. A recent quantitative technique utilizes glycoproteins that are linked covalently to a solid support by hydrazine chemistry, followed by proteolysis and stable isotope labeling of the bound glycopeptides [132]. The labeled glycopeptides are released by peptide-*N*-glycosidase F digestion and subsequently identified and quantified using a LC-tandem MS-based strategy. Larsen et al. [133] have described a method that allows the characterization of a small amount of gel-separated N-linked glycoprotein. This utilizes a sequential specific and non-specific enzymatic treatment followed by selective purification and characterization of glycopeptides using graphite powder microcolumns in combination with MS.

6.3
O-Glycosylation

It has been known for many years that most cell-surface and extracellular proteins are *O*-glycosylated. To analyze the *O*-glycosylation of proteins by MS is a complex task, due to the high structural diversity not only of the glycan chains but also of the protein factors. Determination of the full structure of glycoform(s) on single glycosylation sites is considered to be crucial for functional proteomics- and glycomics-type studies.

The parameters in structural analysis of *O*-glycans include the determination of [134]: (i) the occupancy of potential *O*-glycosylation sites; (ii) the type of monosaccharide moiety attached to a protein via *O*-glycosylation; (iii) core typing; (iv) the type and size of the oligosaccharide chains attached to the core monosaccharide; (v) branching patterns; (vi) monosaccharide glycosidic linkages; (vii) the anomericity; and (viii) the covalent modifications of the sugar backbone chains by carbohydrate- and non-carbohydrate-type substituents.

According to the monosaccharide attached to the protein, the following classes of *O*-glycans exist: *O*-GalNAc; *O*-GlcNAc; *O*-Fuc; *O*-Man; *O*-Glc; *O*-glycosaminoglycan-(GAG); and collagen-type.

The most abundant type of *O*-glycosylation in proteins is the GalNAc-type, where the GalNAc is linked to serine (Ser) or threonine (Thr) in the protein chain by an α-glycosidic linkage. Each of these groups is generated by a different, but specific regulated biosynthesis. A full identification by MALDI-MS by *de-novo* sequencing can be achieved on analyzers containing a collisional cell, such as hybrid QTOF instruments, the MALDI-ion traps, on FT-ICR analyzers by SORI-CID, ECD or IRMPD, and on MALDI-TOF/TOF instruments for high-energy sequencing. In most biological projects MALDI-TOF experiments will be introduced at an early stage of the protocol in order to obtain immediate information on the type, degree, and site of glycosylation. From the data obtained, the number of monosaccharide units involved and their stoichiometry can be calculated according to the different mass increments and their combinations. Such data can be combined with direct information on structural parameters obtained prior to, during, or after the MALDI mapping by analytical use of glycosidases, lectin or NMR analysis, or by combination with ESI experiments conducted in the tandem MS mode.

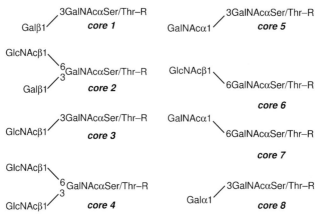

Fig. 6.13 Eight types of O-glycan GalNAc core structures. (Reprinted with permission from Ref. [151].)

A useful protocol including the incubation of glycans with exoglycosidases can provide a differential picture for determining the terminal units and their linkage, and can be applied in an array mode during the MALDI analysis [113].

Mucin-linked O-glycans exhibit a high degree of structural variation already at the level of the core types, and of chain elongations arising from the core extension (Fig. 6.13). The O-mannose-linked glycan chain be extended with up to eight monosaccharide units and modified by sialylation, glucuronation/sulfation and/or fucosylation. GlcNAc, Fuc and Man can be linked to the protein at serine indirectly, via a phosphate diester bridge. Glycosaminoglycan (GAG) (synonym: mucopolysaccharide) chains are linked to proteoglycans via a β-linked xylose to serine or threonine. The major challenge for the structural analysis of GAG samples is the determination of chain size, which is difficult due to the high number of carboxylate and sulfate functional groups. Such samples usually contain diverse cations in tight ion pairs, as illustrated for MALDI analysis with basic amines of known structure. In intact proteoglycans different types of GAG chains of length dispersity could be present, as well as other types of glycosylation. In the collagen-like modules the hydroxylated lysine can be glycosylated by a β-linked galactose via an O-glycosidic linkage.

6.3.1
Release of O-Glycans from the Parent Protein

Mass spectrometry can be applied to the analysis of purified components or complex mixtures of O-glycans released from the parent protein. In this case, the analysis can be performed in the same way as for N-glycans. Traditionally, O-glycans are detached from the parent protein by β-elimination using strong bases, such as sodium hydroxide or hydrazine. Depending on the reaction conditions used, both N- and O-glycans can be released during hydrazinolysis [135], where selective isolation of O-glycans from proteins – which carry both N- and O-glycosylation – is not always achieved. Classical β-elimination by sodium hydroxide under reductive conditions [136,137] offers additional options for chemical derivatizations by tagging the reduced end. The non-reductive release of O-linked oligosaccharides from mucin glycoproteins by β-elimination is also possible, and is frequently carried out as a first step in the synthesis of neoglycolipids, substrates carrying the defined glycan structure, covalently linked to a defined lipid portion, as used in the case of E-selectin ligand identification [138]. The β-elimination of O-linked glycans takes place during the permethylation reaction of intact glycoproteins at high pH values. This procedure is of practical value for determining the overall O-glycosylation status of both glycoproteins and glycopeptides, as applied to human glycopeptides from urine to confirm the type of glycan present [139]. In order to obtain short oligosaccharide blocks from the starting glycoproteins with long oligosaccharide chains, other type of hydrolases, endoglycosidases, can be used; this approach is used for the eliminative cleavage of GAGs by chondroitinase ABC from proteoglycans.

6.3.2
MALDI-TOF MS of Released O-Glycans

With regard to the choice of MALDI matrix for native and derivatized samples, the same preferences can be applied as for N-glycan analysis (see Section 6.2). For compositional analysis, the ratio of Hex and HexNAc moieties can indicate the type of the core. In standard preparations of O-glycan mixtures obtained by β-elimination from glycoproteins, the m/z values in the MALDI-TOF molecular ion maps are used for this compositional analysis calculation. This general concept was applied to the mucin-type O-glycan mixture from the frog *Rana temporaria* oviduct glycoprotein, in which molecular species were detected in preparations obtained by alkaline borohydride release, ion-exchange chromatography, HPLC separation, and permethylation [140].

An instrumental advantage in carrying out high-sensitivity analysis at accurate mass determination was demonstrated on a FTICR analyzer using positive ion MALDI desorption of the highly heterogeneous neutral O-glycan mixture from the frog *Xenopus laevis*, where 12 previously unknown molecular species were detected in the mixture [141]. Negative ion mode MALDI-TOF is the method of choice for mapping sialylated O-glycan mixtures, which could be derivatized at the reduced end. A more complex set-up was designed to detect single components in the complex mixture of heparin-like GAGs by adding defined basic peptides to the GAG mixtures for the formation of non-covalent complexes prior to the MALDI-TOF analysis [142,143].

6.3.3
Determination of O-Glycosylation Sites by PSD-MALDI-TOF MS

PSD-MALDI-TOF fragment ion analysis under standard conditions in α-cyano-4-hydroxycinnamic acid as matrix renders rather complex pictures, providing enough ions for the assignment of O-glycosylation sites, if prolines as preferable cleavage sites are present in the chain. However, the amount of deglycosylated fragment ions, neutral losses and internal peptide chain cleavages is significant. The formation of b and y peptide fragment ions in TAP25 MUC1 mono-, di- and triglycosylated species was dominant, although a significant degree of neutral losses (CO, H_2O, NH_3) was also observed [144]. Fragmentation analysis by PSD-MALDI-TOF was studied in the O-glycosylation of MUC4 peptides with blocked N- and C-termini. Fragmentation patterns of the glycosylated MUC4 peptides were compared to those of the non-glycosylated peptides to study the influence of sugar constituents on the fragmentation behavior and their relative ion abundance. The fragmentation patterns obtained were similar to those obtained with low-energy CID in the ion trap analyzer [145].

6.3.4
Determination of O-Glycosylation Status by MALDI-TOF MS

In many cases it is necessary to determine the GalNAc O-glycosylation status in proteins or peptides. In order to study the epigenetic regulatory mechanisms ruling the initiation of O-glycosylation *in vitro,* the incorporation of GalNAc moieties into peptides containing up to 25 amino acids, prone to folding, and five to six potential glycosylation sites by the recombinant GalNAc-transferase, was monitored by MALDI-TOF. Here, the sequence of utilization of single glycosylation sites was determined by Edman sequencing [146]. Of practical importance is the MALDI-TOF O-glycosylation status analysis, which is carried out in parallel in the linear and reflectron mode, and is where the difference in mass in these two experiments can indicate the extent of deglycosylation; for example, $\Delta m = 203$ for HexNAc, 162 for Hex, and 146 for dHex.

6.3.5
Identification of O-Glycosylation in Proteins by In-Gel Alkylaminylation

The first step in the analysis of proteomes is usually two-dimensional gel electrophoresis, followed by in-gel proteolysis of the excised protein bands and mass spectrometric peptide mapping. N- or/and O-glycosylated proteins – in particular those with a high carbohydrate content – are difficult to detect in gels, due either to low levels of staining or to the diffuse appearance of spots in the gel. The other aspect is the steric hindrance at proteolysis of glycoproteins, notably within domains with clustered O-carbohydrate chains, where the number of mismatched proteolytic cleavage sites can be rather high, as in mucins. Modified β-elimination under mild conditions has been tested by different research groups, using ammonia to introduce the amino group to the glycosylation site [147], or ethylamine for improved yields of the detached sugar portion [15]. Alkylamines (RNH_2) as reagents are superior to the use of ammonia due to the labeling of the former glycosylation site: $\Delta m = RNH - 1$ is the mass increment for the glycosylation marker pro occupied glycosylation site, instead of $\Delta m = -1$ mass increment for sub-

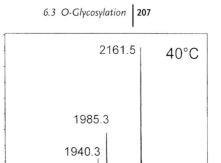

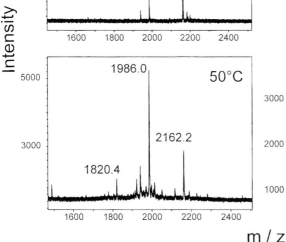

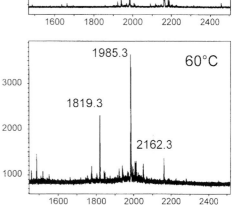

m / z

Fig. 6.14 Temperature dependency of β-elimination/ethylaminylation. The glycopeptide at $m/z = 2162.2$ (average mass) has been submitted to the reaction with 70% aq. ethylamine, rendering the ethylaminated product at $m/z = 1986.0$ (average mass). Aliquots were concentrated and acidified with 0.1% aq. TFA prior to MALDI-TOF analysis in the positive ion mode. (Reprinted with permission from Ref. [148].)

stitution of the hydroxy group on serine or threonine by an amino group (Fig. 6.14) [148]. The aminoalkyl site-specific marker does not interfere with the enzymatic proteolysis for protein identification, does not compromise the ionization of modified peptides in complex proteolytic mixtures during the MALDI desorption, and the modified peptide is identified by the mass difference induced by the reaction.

6.3.6
O-Glycoforms in Urine: Identification of O-Glycosylated Amino Acids in Mixtures by MALDI-MS and Computer Assignment

Urine is a rich source of glycans, frequently containing O-glycosylated amino acids. The content of these compounds is particularly high, and also highly complex, in the case of patients suffering from hereditary diseases, such as congenital disorder of glycosylation (CDG). In order to develop standard protocols for the glycoscreening of complex mixtures containing unknown N- and O-glycans and their metabolites of high dynamic range, MALDI MS experiments were carried

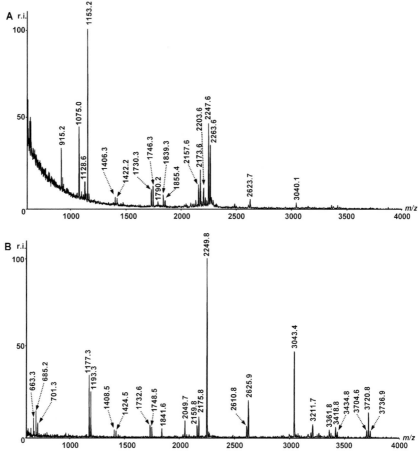

Fig. 6.15 (a) Negativea and (b) positive ion MALDI-TOF
analysis of a fraction from the urine of a patient diagnosed with
CDG. (Reprinted with permission from Ref. [149].)

out in both positive- and negative-ion modes using the reflectron mode for higher resolution and less overlapping of molecular ions. To obtain a good identification coverage of molecular species, DHB was chosen as a most effective matrix for neutral carbohydrates in the positive ion mode, and ATT for acidic glycans in the negative ion mode. For automatic assignment of m/z values to a particular carbohydrate composition, a computer algorithm has been developed. To model the composition of the respective glycoconjugate ions, different combinations of glycan building blocks (dHex, Hex, HexNAc and NeuAc), amino acids (Ser, Thr) and possible sugar modifications such as phosphates (P) and sulfates (S) at various types of cationization – that is, protonation as well as attachment of sodium and potassium – were considered.

Figure 6.15 illustrates negative- (a) and positive- (b) ion MALDI-TOF MS of a single fraction from the urine of a CDG patient. A computer algorithm calculation indicated that the majority of ions were to be assigned to oligosaccharides, some of them being modified by sulfate or phosphate, but none of them to a Ser- or Thr-linked O-glycan structure. In the positive ion mode, however, three distinct ions were found to fit the mass of glycans attached to Ser, Thr and Ser-Thr, at m/z 1067.6, 1119.5, and 1322.7; these corresponded to NeuAc$_2$HexHexNAc-Thr, Hex$_5$HexNAc-Ser and NeuAc$_2$dHexHexHexNAc-SerThr, respectively [149]. This approach of combined MALDI mapping and computerized compositional assignment is suitable for high-throughput glycoscreening.

6.3.7
Identification of the Non-Covalent O-Glycoprotein Dimer by Blot IR-MALDI-MS

Glycophorin A (GpA), an integral transmembrane protein from the human erythrocyte membrane, contains within its N-terminal 72 amino acids stretch 15 O- and one N-glycosylation sites, expressing glycans with different degree of sialylation. In the case of maximal glycosylation, the molecular mass of the monomeric glycoprotein is calculated to be 30 kDa. An experiment to analyze a dimeric protein as a non-covalent complex under MALDI conditions was designed using IR-MALDI-MS as a "softer" desorption technique than the UV-MALDI, and succinic acid as an IR-MALDI matrix [150]. The dimeric GpA, which is very stable under the conditions of the SDS-PAGE separation, electroblotting and desorption/ionization, was identified directly from the blot in parallel by MS, and in Western blot using a monoclonal antiGpA antibody. Molecular ions with one, two, three, and four positive charges were detected with declining abundance. For MALDI-MS, a laboratory-built TOF instrument at the University of Münster using 12-keV ion energy and a single-stage reflection followed by a secondary-electron multiplier (EMI 9643; Electron Tube Ltd., Ruislip, UK) equipped with a conversion dynode using 15 keV postacceleration was used, equipped with an Er-YAG laser (SEO 123; Schwartz Electrooptics, Orlando, FL, USA; 2.94 μm, 150-ns pulse duration). The molecular mass of the stained GpA band was approximated by comparison to commercial standards set at 66 kDa. The membrane was incubated in the solution of matrix, and the GpA band cut from the blot and glued to the target. In the IR-MALDI-MS the molecular $[M + H]^+$ ion was found at $m/z = 59\,860 \pm 50$ (Fig. 6.16) [150].

6.3.8
Determination of Overall O-Glycosylation of a Glycoprotein by Permethylation

By making use of the alkaline lability of the O-glycosidic linkage to a protein backbone, a rapid protocol has been developed for detaching all O-glycans from the protein or peptide by β-elimination and permethylation as a one-pot reaction [138]. The reaction is performed in a slurry of powdered NaOH in dimethyl sulfoxide, with methyl iodide as a methylating agent [40]. The products are obtained by chloroform extraction, and salts are removed by washing with water. The original

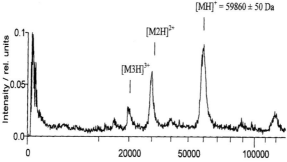

Fig. 6.16 IR-MALDI-MS of the glycophorin A desorbed directly from the Western blot. A band at 66 kDa has been stained by the monoclonal antibody directed against the N-terminus of the glycoprotein in the Western blot analysis, indicating the identity of glycophorin A. (Reprinted with permission from Ref. [150].)

reaction can be scaled down to the 1–5 μg starting material, if a carrier saccharide (mostly glucose) is added. The mixture of permethylated *O*-glycans can be submitted to MALDI-MS for rapid mapping, and is claimed by some authors to provide the possibility for rough quantification of species in the mixture.

References

1 J.B. Lowe, J.D. Marth, *Annu. Rev. Biochem.* **2003**, *72*, 643–691.

2 A. Varki, *Glycobiology* **1993**, *3*, 97–130.

3 E. Ioffe P. Stanley, *Proc. Natl. Acad. Sci. USA.* **1998**, *91*, 728–732.

4 D.H. Dube, C.R. Bertozzi, *Nat. Rev. Drug Discov.* **2005**, *4*, 477–488.

5 T. Marquardt, J. Denecke, *Eur. J. Pediatr.* **2003**, *162*, 359–379.

6 J.C. Paulson, *Trends Biochem. Sci.* **1989**, *14*, 272–276.

7 R. Kornfeld, S. Kornfeld, *Annu. Rev. Biochem.* **1985**, *54*, 631–664.

8 P.M. Rudd R.A. Dwek, *Crit. Rev. Biochem. Mol. Biol.* **1997**, *32*, 1–100.

9 S. Takasaki, T. Mizuochi, A. Kobata, *Methods Enzymol.* **1982**, *83*, 263–268.

10 T. Patel, J. Bruce, A. Merry, C. Bigge, M. Wormald, A. Jaques, R. Parekh, *Biochemistry* **1999**, *32*, 679–693.

11 T. Mizuochi, *Methods Mol. Biol.* **1993**, *14*, 55–68.

12 R.N. Iyer, D.M. Carlson, *Arch. Biochem. Biophys.* **1971**, *142*, 101–105.

13 M. Oheda, E. Tominaga, Y. Nabuchi, T. Matsuura, N. Ochi, M. Tamura, S. Hase, *Anal. Biochem.* **1996**, *236*, 369–371.

14 R. Geyer, H. Geyer, *Methods Mol. Biol.* **1993**, *14*, 131–142.

15 W. Chai, T. Feizi, C.T. Yuen, A.M. Lawson, *Glycobiology* **1997**, *7*, 861–872.

16 Y. Huang, T. Konse, Y. Mechref, M.V. Novotny, *Rapid Commun. Mass Spectrom.* **2002**, *16*, 1199–1204.

17 A.L. Tarentino, R.B. Trimble, T.H. Plummer, Jr., *Methods Cell Biol.* **1989**, *32*, 111–139.

18 F. Maley, R.B. Trimble, A.L. Tarentino, T.H. Plummer, Jr., *Anal. Biochem.* **1989**, *180*, 195–204.

19 O. Belgacem, A. Buchacher, K. Pock, D. Josic, C. Sutton, A. Rizzi, G. Allmaier, *J. Mass Spectrom.* **2002**, *37*, 1118–1130.

20 B. Kuster, M. Mann, *Anal. Chem.* **1999**, *71*, 1431–1440.

21 J. Gonzalez, T. Takao, H. Hori, V. Besada, R. Rodriguez, G. Padron, Y. Shimonishi, *Anal. Biochem.* **1995**, *205*, 151–158.

22 V. Tretter, F. Altmann, L. Marz, *Eur. J. Biochem.* **1991**, *199*, 647–652.

23 I.B. Wilson, R. Zeleny, D. Kolarich, E. Staudacher, C.J. Stroop, J.P. Kamerling, F. Altmann, *Glycobiology* **2001**, *11*, 261–274.

24 S.M. Haslam, G.C. Coles, E.A. Munn, T.S. Smith, H.F. Smith, H.R. Morris, A. Dell, *J. Biol. Chem.* **1999**, *271*, 30561–30570.

25 J. Charlwood, J.M. Skehel, P. Camilleri, *Anal. Biochem.* **2000**, *284*, 49–59.

26 B. Kuster, S.F. Wheeler, A.P. Hunter, R.A. Dwek, D.J. Harvey, *Anal. Biochem.* **1997**, *250*, 82–101.

27 D. Sagi, P. Kienz, J. Denecke, T. Marquardt, J. Peter-Katalinic, *Proteomics* **2005**, *5*, 2689–2701.

28 A.L. Tarentino, F. Maley, *Biochem. Biophys. Res. Commun.* **1975**, *67*, 455–462.

29 T. Tai, K. Yamashita, S. Ito, A. Kobata, *J. Biol. Chem.* **1987**, *252*, 6687–6694.

30 A.L. Tarentino, T.H. Plummer, Jr., *Methods Enzymol.* **1994**, *230*, 44–57.

31 R.B. Trimble, A.L. Tarentino, *J. Biol. Chem.* **1991**, *266*, 1646–1651.

32 M. Nimtz, E. Grabenhorst, H.S. Conradt, L. Sanz, J.J. Calvete, *Eur. J. Biochem.* **1999**, *265*, 703–718.

33 N.H. Packer, M.A. Lawson, D.R. Jardine, J.W. Redmond, *Glycoconj. J.* **1998**, *15*, 737–747.

34 R. Marusyk, A. Sergeant, *Anal. Biochem.* **1980**, *105*, 403–404.

35 H. Gorisch, *Anal. Biochem.* **1988**, *173*, 393–398.

36 G.S. Kansas, O. Spertini, L.M. Stoolman, T.F. Tedder, *J. Cell Biol.* **1991**, *114*, 351–358.

37 J.C. Rouse, J.E. Vath, *Anal. Biochem.* **1996**, *238*, 82–92.

38 N.G. Karlsson, N.L. Wilson, H.J. Wirth, P. Dawes, H. Joshi, N.H. Packer, *Rapid Commun. Mass Spectrom.* **2004**, *18*, 2282–2292.

39 S. Hakomori, *J. Biochem. (Tokyo)* **1964**, *55*, 205–208.

40 I. Ciucanu, F. Kerek, *Carbohydr. Res.* **1984**, *131*, 209–217.

41 H. Perreault, C.E. Costello, *J. Mass Spectrom.* **1999**, *34*, 184–197.

42 S. Handa, K. Nakamura, *J. Biochem. (Tokyo)* **1984**, *95*, 1323–1329.

43 A.K. Powell, D.J. Harvey, *Rapid Commun. Mass Spectrom.* **1996**, *10*, 1027–1032.

44 S. Hase, *J. Chromatogr. A* **1996**, *720*, 173–182 (1996).

45 F.N. Lamari, R. Kuhn, N.K. Karamanos, *J. Chromatogr. B. Analyt. Technol. Biomed. Life Sci.* **2003**, *793*, 15–36.

46 D.J. Harvey, *J. Am. Soc. Mass Spectrom.* **2000**, *11*, 900–915.

47 T. Takao, Y. Tambara, A. Nakamura, K.-I. Yoshino, H. Fukuda, M. Fukuda, Y. Shimonishi, *Rapid Commun. Mass Spectrom.* **1996**, *10*, 637–640.

48 G. Okafo, L. Burrow, S.A. Carr, G.D. Roberts, W. Johnson, P. Camilleri, *Anal. Chem.* **1996**, *68*, 4424–4430.

49 P. Camilleri, D. Tolson, H. Birrell, *Rapid Commun. Mass Spectrom.* **1998**, *12*, 144–148.

50 M. Okamoto, K. Takahashi, T. Doi, Y. Takimoto, *Anal. Chem.* **1997**, *69*, 2919–2926.

51 S. Hase, T. Ibuki, T. Ikenaka, *J. Biochem. (Tokyo)* **1984**, *95*, 197–203.

52 J.C. Bigge, T.P. Patel, J.A. Bruce, P.N. Goulding, S.M. Charles, R.B. Parekh, *Anal. Biochem.* **1995**, *230*, 229–238.

53 K.R. Anumula, S.T. Dhume, *Glycobiology* **1998**, *8*, 685–694.

54 J. Lemoine, M. Cabanes-Macheteau, M. Bardor, J.C. Michalski, L. Faye, P. Lerouge, *Rapid Commun. Mass Spectrom.* **2000**, *14*, 100–104.

55 H. Suzuki, O. Müller, A. Guttman, B.L. Karger, *Anal. Chem.* **1997**, *69*, 4554–4559.

56 J.F. Cipollo, A.M. Awad, C.E. Costello, C.B. Hirschberg, *J. Biol. Chem.* **2005**, *280*, 26063–26072.

57 K.O. Börnsen, M.D. Mohr, H.M. Widmer, *Rapid Commun. Mass Spectrom.* **1995**, *9*, 1031–1034.

58 B. Stahl, M. Steup, M. Karas, F. Hillenkamp, *Anal. Chem.* **1991**, *63*, 1463–1466.

59 D.J. Harvey, *Rapid Commun. Mass Spectrom.* **1993**, *7*, 614–619.

60 M. Karas, H. Ehring, E. Nordhoff, B. Stahl, K. Strupat, F. Hillenkamp, M. Grehl, B. Krebs, *Mass Spectrom.* **1993**, *28*, 1476–1481.

61 M.D. Mohr, K.O. Bornsen, H.M. Widmer, *Rapid Commun. Mass Spectrom.* **1995**, *9*, 809–814.

62 P. Chen, A.G. Baker, M.V. Novotny, *Anal. Biochem.* **1997**, *244*, 144–151.

63 S.F. Wheeler, D.J. Harvey, *Anal. Chem.* **2000**, *72*, 5027–5039.

64 S.F. Wheeler, D.J. Harvey, *Anal. Biochem.* **2001**, *296*, 92–100.

65 J.O. Metzger, R. Woisch, W. Tuszynski, R. Angermann, *Fresenius J. Anal. Chem.* **1994**, *349*, 473–474.

66 B. Stahl, S. Thurl, J. Zeng, M. Karas, F. Hillenkamp, M. Steup, G. Sawatzki, *Anal. Biochem.* **1994**, *223*, 218–226.

67 D.I. Papac, A. Wong, A.J. Jones, *Anal. Chem.* **1996**, *68*, 3215–3223.

68 B. Stahl, A. Linos, M. Karas, F. Hillenkamp, M. Steup, *Anal. Biochem.* **1915**, *246*, 195–204.

69 N. Xu, Z.-H. Huang, J.T. Watson, D.A. Gage, *J. Am. Soc. Mass Spectrom.* **1997**, *8*, 116–124.

70 A. Pfenninger, M. Karas, B. Finke, B. Stahl, G. Sawatzki, *J. Mass Spectrom.* **1999**, *34*, 98–104.

71 J. Zaia, *Mass Spectrom. Rev.* **2004**, *23*, 161–227.

72 R. Orlando, C.A. Bush, C. Fenselau, *Biomed. Environ. Mass Spectrom.* **1990**, *19*, 747–754.

73 L.C. Ngoka, J.F. Gal, C.B. Lebrilla, *Anal. Chem.* **1994**, *66*, 692–698.

74 M.T. Cancilla, S.G. Penn, J.A. Carroll, C.B. Lebrilla, *J. Am. Chem. Soc.* **1996**, *118*, 6736–6745.

75 J. Lemoine, B. Fournet, D. Despeyroux, K.R. Jennings, R. Rosenberg, E. Dehoffmann, *J. Am. Soc. Mass Spectrom.* **1993**, *4*, 197–203.

76 M.T. Cancilla, A.W. Wong, L.R. Voss, C.B. Lebrilla, *Anal. Chem.* **1999**, *71*, 3206–3218.

77 D.J. Harvey, *Mass Spectrom. Rev.* **1999**, *18*, 349–450.

78 H. Egge, J. Peter-Katalinić, *Mass Spectrom. Rev.* **1987**, *6*, 331–393.

79 A. Dell, J. Thomas-Oates, *Analysis of Carbohydrates by GLC MS.* CRC Press.

80 D. Garozzo, G. Impallomeni, G. Montaudo, E. Spina, *Rapid Commun. Mass Spectrom.* **1992**, *6*, 550–552.

81 B. Domon, D.R. Müller, W.J. Richter, *Biomed. Environ. Mass Spectrom.* **1990**, *19*, 390–392.

82 R.A. Laine, E. Yoon, T.J. Mahier, S. Abbas, B. de Lappe, R. Jain, K. Matta, *Biol. Mass Spectrom.* **1991**, *20*, 505–514.

83 B. Domon, C.E. Costello, *Glycoconj. J.* **1988**, *5*, 397–409.

84 D.J. Harvey, *J. Mass Spectrom.* **2000**, *35*, 1178–1190.

85 G.E. Hofmeister, Z. Zhou, J.A. Leary, *J. Am. Chem. Soc.* **1991**, *113*, 5964–5970.

86 T.J. Naven, D.J. Harvey, J. Brown, G. Critchley, *Rapid Commun. Mass Spectrom.* **1997**, *11*, 1681–1686.

87 D.J. Harvey, T.J. Naven, B. Kuster, R.H. Bateman, M.R. Green, G. Critchley, *Rapid Commun. Mass Spectrom.* **1995**, *9*, 1556–1561.

88 B. Spengler, D. Kirsch, R. Kaufmann, J. Lemoine, *J. Mass Spectrom.* **1995**, *30*, 782–787.

89 N. Viseux, C.E. Costello, B. Domon, *J. Mass Spectrom.* **1999**, *34*, 364–376.

90 F. Zal, B. Kuster, B.N. Green, D.J. Harvey, F.H. Lallier, *Glycobiology* **1998**, *8*, 663–673.

91 T. Yamagaki, H. Nakanishi, *Proteomics* **2001**, *1*, 329–339.

92 J.C. Rouse, A.M. Strang, W. Yu, J.E. Vath, *Anal. Biochem.* **1998**, *256*, 33–46.

93 J. Lemoine, F. Chirat, B. Domon, *J. Mass Spectrom.* **1996**, *31*, 908–912.

94 T. Yamagaki, H. Nakanishi, *Glycoconj. J.* **1999**, *16*, 385–389.

95 D.J. Harvey, R.H. Bateman, R.S. Bordoli, R. Tyldesley, *Rapid Commun. Mass Spectrom.* **2000**, *14*, 2135–2142.

96 S. Hanrahan, J. Charlwood, R. Tyldesley, J. Langridge, R. Bordoli, R. Bateman, P. Camilleri, *Rapid Commun. Mass Spectrom.* **2001**, *15*, 1141–1151.

97 J.W. Gauthier, T.R. Trautman, D.B. Jacobson, *Anal. Chim. Acta* **1991**, *246*, 211–225.

98 D.P. Little, J.P. Speir, M.W. Senko, P.B., O'Connor, F.W. McLafferty, *Anal. Chem.* **1994**, *66*, 2809–2815.

99 R.A. Zubarev, N.L. Kelleher, F.W. McLafferty, *J. Am. Chem. Soc.* **1998**, *120*, 3265–3266.

100 S.G. Penn, M.T. Cancilla, C.B. Lebrilla, *Anal. Chem.* **1996**, *68*, 2331–2339.

101 Y. Xie, C.B. Lebrilla, *Anal. Chem.* **2003**, *75*, 1590–1598.

102 C.S. Creaser, J.C. Reynolds, D.J. Harvey, *Rapid Commun. Mass Spectrom.* **2002**, *16*, 176–184.

103 C.E. Von Seggern, P.E. Zarek, R.J. Cotter, *Anal. Chem.* **2003**, *75*, 6523–6530.

104 C.E. Von Seggern, S.C. Moyer, R.J. Cotter, *Anal. Chem.* **2003**, *75*, 3212–3218.

105 Y. Mechref, M.V. Novotny, C. Krishnan, *Anal. Chem.* **2003**, *75*, 4895–4903.

106 Y. Mechref, P. Kang, M.V. Novotny, *Rapid Commun. Mass Spectrom.* **2006**, *20*, 1381–1389.

107 A. Kobata, *Anal. Biochem.* **1979**, *100*, 1–14.

108 C.J. Edge, T.W. Rademacher, M.R. Wormald, R.B. Parekh, T.D. Butters, D.R. Wing, R.A. Dwek, *Proc. Natl. Acad. Sci. USA* **1992**, *89*, 6338–6342.

109 C.W. Sutton, J.A. O'Neill, J.S. Cottrell, *Anal. Biochem.* **1994**, *218*, 34–46.

110 B. Stahl, T. Klabunde, H. Witzel, B. Krebs, M. Steup, M. Karas, F. Hillenkamp, *Eur. J. Biochem.* **1994**, *220*, 321–330.

111 S. Liedtke, R. Geyer, H. Geyer, *Glycoconj. J.* **1997**, *14*, 785–793.

112 B. Kuster, T.J. Naven, D.J. Harvey, *J. Mass Spectrom.* **1996**, *31*, 1131–1140.

113 H. Geyer, S. Schmitt, M. Wuhrer, R. Geyer, *Anal. Chem.* **1999**, *71*, 476–482.

114 Y. Mechref, M.V. Novotny, *Anal. Chem.* **1998**, *70*, 455–463.

115 K.K. Mock, M. Davy, J.S. Cottrell, *Biochem. Biophys. Res. Commun.* **1991**, *177*, 644–651.

116 M.L. Vestal, P. Juhasz, S.A. Martin, *Rapid Commun. Mass Spectrom.* **1995**, *9*, 1044–1050.

117 A. Tsarbopoulos, U. Bahr, B.N. Pramanik, M. Karas, *Int. J. Mass Spectrom. Ion Processes* **1997**, *169*, 251–261.

118 Y. Wada, J. Gu, N. Okamoto, K. Inui, *Biol. Mass Spectrom.* **1994**, *23*, 108–109.

119 T. Nakanishi, N. Okamoto, K. Tanaka, A. Shimizu, *Biol. Mass Spectrom.* **1994**, *23*, 230–233.

120 J. Peter, C. Unverzagt, W.-D. Engel, D. Renauer, C. Seidel, W. Hösel, *Biochim. Biophys. Acta* **1998**, *1380*, 93–101.

121 U. Bahr, J. Stahl-Zeng, E. Gleitsmann, M. Karas, *J. Mass Spectrom.* **1997**, *32*, 1111–1116.

122 C. Sottani, M. Fiorentino, C. Minoia, *Rapid Commun. Mass Spectrom.* **1997**, *11*, 907–913.

123 Y.J. Kim, A. Freas, C. Fenselau, *Anal. Chem.* **2001**, *73*, 1544–1548.

124 T. Denzinger, H. Diekmann, K. Bruns, U. Laessing, C.A. Stuermer, M. Przybylski, *J. Mass Spectrom.* **1999**, *34*, 435–446.

125 S. Berkenkamp, C. Menzel, M. Karas, F. Hillenkamp, *Rapid Commun. Mass Spectrom.* **1997**, *1*, 1399–1406.

126 B. Garner, A.H. Merry, L. Royle, D.J. Harvey, P.M. Rudd, J. Thillet, *J. Biol. Chem.* **2001**, *276*, 22200–22208.

127 E. Mortz, T. Sareneva, S. Haebel, I. Julkunen, P. Roepstorff, *Electrophoresis* **1996**, *17*, 925–931.

128 K. Mills, A.W. Johnson, O. Diettrich, P.T. Clayton, B.G. Winchester, *Tetrahedron: Asymmetry* **2000**, *11*, 75–93.

129 N.L. Wilson, B.L. Schulz, N.G. Karlsson, N.H. Packer, *J. Proteome Res.* **2002**, *1*, 521–539.

130 H. Kaji, H. Saito, Y. Yamauchi, T. Shinkawa, M. Taoka, J. Hirabayashi, K. Kasai, N. Takahashi, T. Isobe, *Nat. Biotechnol.* **2003**, *21*, 667–672.

131 P. Hagglund, J. Bunkenborg, F. Elortza, O.N. Jensen, P. Roepstorff, *J. Proteome Res.* **2004**, *3*, 556–566.

132 H. Zhang, X.J. Li, D.B. Martin, R. Aebersold, *Nat. Biotechnol.* **2003**, *21*, 660–666.

133 M.R., Larsen, P. Hojrup, P. Roepstorff, *Mol. Cell. Proteomics* **2005**, *4*, 107–119.

134 H. Egge, J. Peter-Katalinić, *Mass Spectrom. Rev.* **1987**, *6*, 331–391.

135 T. Patel, J. Bruce, A. Merry, C. Bigge, M. Wormald, A. Jaques, R. Parekh, *Biochemistry* **1993**, *32*, 679.

136 D.M. Carlson, *J. Biol. Chem.* **1968**, *243*, 616.

137 F.G. Hanisch, H. Egge, J. Peter-Katalinić, G. Uhlenbruck, C. Dienst, R. Fangmann, *Eur. J. Biochem.* **1985**, *152*, 343–351.

138 W. Chai, T. Feizi, C.T. Yuen, A.M. Lawson, *Glycobiology* **1997**, *7*, 861.

139 H.-U. Linden, R.A. Klein, H. Egge, J. Peter-Katalinić, J. Dabrowski, D. Schindler, *Biol. Chem. Hoppe-Seyler* **1989**, *370*, 661.

140 A. Coppin, E. Maes, W. Morelle, G. Strecker, *Eur. J. Biochem.* **1999**, *266*, 94.

141 K. Tseng, L.L. Lindsay, S. Penn, L.J. Hedrick, B.C. Lebrilla, *Anal. Biochem.* **1997**, *250*, 18.

142 P. Juhasz, K. Biemann, *Proc. Natl. Acad. Sci. USA* **1994**, *91*, 4333.

143 G. Venkataraman, Z. Shiver, R. Raman, R. Sasisekharan, *Science* **1999**, *286* (5439) 537.

144 S. Goletz, B. Thiede, F.-G. Hanisch, M. Schultz, J. Peter-Katalinić, S. Müller, O. Seitz, U. Karsten, *Glycobiology* **1997**, *7*, 881.

145 K. Alving, R. Körner, H. Paulsen, J. Peter-Katalinić, *J. Mass Spectrom.* **1998**, *33*, 1124.

146 F.-G. Hanisch, S. Müller, H. Hassan, H. Clausen, N. Zachara, A.A. Gooley, H. Paulsen, K. Alving, J. Peter-Katalinić, *J. Biol. Chem.* **1999**, *274*, 9946.

147 G.J. Rademaker, S.A. Pergantis, L. Blok-Tip, J.I. Langridge, A. Kleen, J. Thomas-Oates, *Anal. Biochem.* **1998**, *257*, 149.

148 F.-G. Hanisch, M. Jovanović, J. Peter-Katalinić, *Anal. Biochem.* **2001**, *290*, 47.

149 S. Vakhrushev, *PhD Thesis*, University of Muenster, Germany, July **2006**.

150 I. Meisen, *PhD Thesis*, University of Bonn, Germany, **1997**; I. Meisen, K. Strupat, J. Peter-Katalinić, unpublished results.

151 J. Peter-Katalinić, *Methods Enzymol.* **2005**, *405*, 139–171.

7
MALDI-MS of Lipids

Jürgen Schiller

7.1
Introduction

In this chapter, before the applications of MALDI-TOF MS are discussed in greater detail, some comments on the special structural properties of lipids will be provided. Special emphasis will be placed on why interest in lipids in general – and in their analysis in particular – has greatly increased during the past decade.

7.1.1
Why are Lipids of Interest?

Nowadays, there is immense interest in proteins and DNA, as these biomolecules are often considered to be the most important in living organisms. Moreover, it is accepted that gene defects and protein deficiencies or overexpressions are important causes of diseases, and possess considerable diagnostic relevance.

In contrast, interest in lipids has been much lower, mainly because in the past they have been considered primarily to be relevant only as an energy source (fat-rich food), as the storage form of energy (triacylglycerols), and as the "packaging material" of more relevant components [e.g., the phospholipids (PLs) of biomembranes, or lipoproteins] [1]. Due to the simultaneous presence of both polar and apolar moieties, PLs are able to form a characteristic "bilayer" structure [2].

Additionally, lipids such as (poly)-phosphoinositides, phosphatidic acids, lysophospholipids (that lack one fatty acid residue in comparison to the corresponding PL) and diacylglycerols (Fig. 7.1) are important messenger molecules, or serve as their precursors in signal transduction processes [1,3]. Highly unsaturated fatty acids released from lipids, especially arachidonic acid (20:4), are precursors of bioactive molecules such as prostaglandins or leukotrienes [4]. Finally, cholesterol is a precursor not only of glycocorticoids and sexual hormones, but also of vitamin D [1].

In animal cells, the primarily occurring phospholipids include phosphatidylcholine (PC), phosphatidylserine (PS), phosphatidylethanolamine (PE),

MALDI MS. A Practical Guide to Instrumentation, Methods and Applications.
Edited by Franz Hillenkamp and Jasna Peter-Katalinić
Copyright © 2007 Wiley-VCH Verlag GmbH & Co. KGaA, Weinheim
ISBN: 978-3-527-31440-9

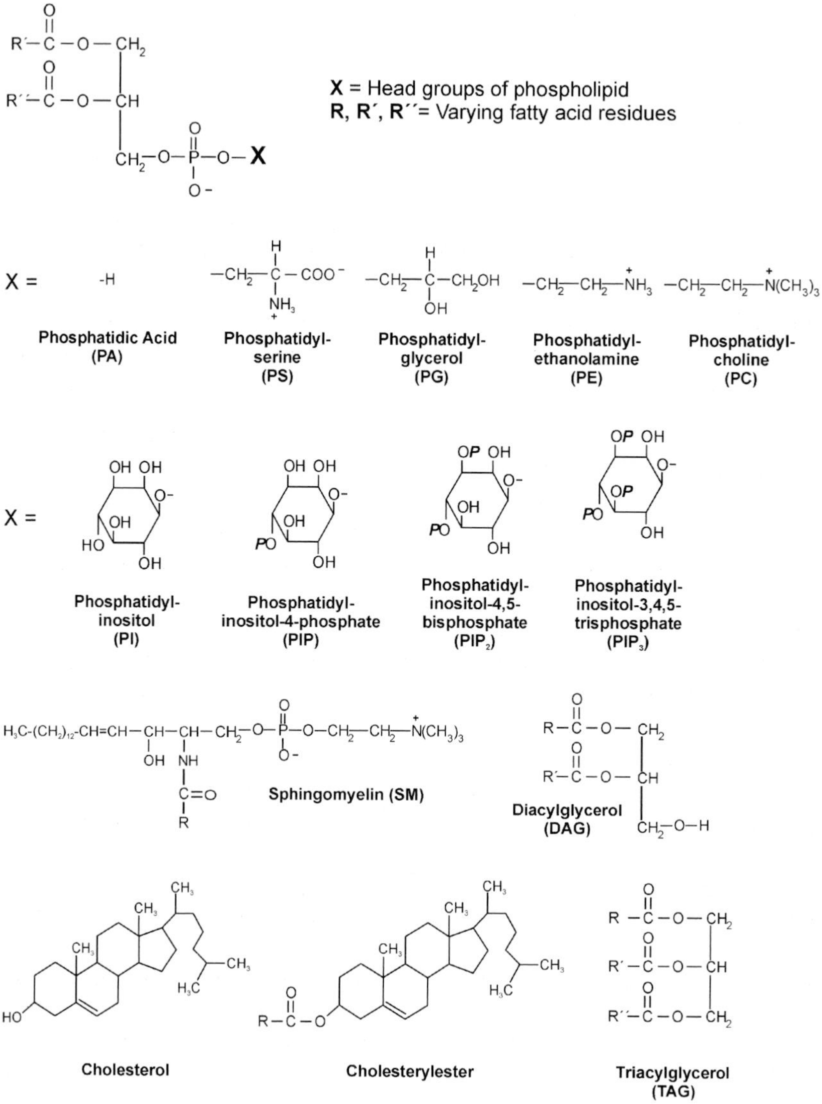

Fig. 7.1 Chemical structures of important naturally occurring lipids and phospholipids. X represents the varying headgroups of glycerophospholipids, whereas R, R' and R" represent varying fatty acid residues. Lipids are sorted into glycerophospholipids (SM is an exception) and further lipids. P indicates the presence of an organic phosphate group.

sphingomyelin (SM), and cholesterol (Fig. 7.1). Acidic PLs such as phosphatidic acid (PA), phosphatidylglycerol (PG), phosphatidylinositol (PI) or its phosphorylated derivatives, are present only in much smaller amounts, but are of paramount interest [1].

In the majority of animal tissues and cells, fatty acids are linked to the glycerol backbone via acyl linkages – that is, by ester bonds. However, it should be noted that lipids may also be generated from fatty aldehydes [5], under which conditions there are no acyl linkages, but rather alkyl or alkenyl linkages. These lipids are referred to as "etherlipids" or "plasmalogens", respectively.

7.1.2
Problems in Lipid Analysis: A Brief Comparison of Methods

For the above-mentioned reasons – namely, the differences in fatty acid composition and the linkage type – it can be imagined that lipid analysis might be a challenging task. A comprehensive discussion of the advantages and drawbacks of individual methods is clearly beyond the scope of this chapter, and only a few comments will be provided here and for further information the interested reader should consult a comprehensive textbook [6].

Traditionally, the analysis of lipids is based on chromatographic techniques such as thin-layer chromatography (TLC) [7] or high-performance liquid chromatography (HPLC) [8]. Although well established, these methods often provide only limited resolution [6]. An additional drawback is that the detectability of a lipid depends on its fatty acid composition.

Besides chromatography, a number of different spectroscopic techniques have been used in lipid analysis in the past, and continue to be used today. Among these, nuclear magnetic resonance (NMR) spectroscopy is capable of providing the most detailed information on lipid composition of a mixture. Although the most frequently applied methods, ^{1}H and ^{13}C [9] NMR fail to analyze complex lipid mixtures because of the overlap of individual lipid classes. One especially convenient method in PL research is that of ^{31}P NMR, which permits the detection of PL in the 100-μg range [10]. By using a suitable solvent or detergent system, highly resolved NMR spectra are obtained even from crude mixtures within a few minutes. These spectra allow differentiation to be made between all major PL classes, and even a coarse differentiation of the fatty acid residues [11]. Additionally, positional fatty acid analysis is possible [12], which in turn means that isomers of, for example, lysophosphatidylcholines (1- and 2-LPC) can be differentiated. One disadvantage of the NMR approach is that it requires larger amounts of PL than either chromatography or mass spectrometry (MS) [10]. Another disadvantage is that only PL – and not lipids – are in general detectable by ^{31}P NMR.

7.1.3
Analysis of Lipids by MS

Almost all currently available ionization techniques of MS, including electron impact (EI) [13,14], field desorption (FD) [15], plasma desorption (PD) [16], chemical ionization (CI) [17], and fast atom bombardment (FAB) [18,19], have been successfully applied to lipids. The individual advantages and disadvantages of these approaches are discussed in [20].

Despite the limitations of EI, the coupling between MS and gas chromatography (GC/MS) is still widely used [21,22]. However, the lipid sample must be hydrolyzed prior to analysis by GC/MS, and the resulting free fatty acids converted into the corresponding trimethylsilyl or methyl esters to enhance their volatility [22]. However, as the measured quantity is provided by the released free fatty acids, information on the fatty acid composition of individual lipids is lost.

7.1.4
Capabilities and Limitations of MALDI-TOF MS in the Lipid Analysis

Currently, two important "soft-ionization" techniques are available that induce only negligible fragmentation of the analyte, namely electrospray ionization (ESI) and MALDI (matrix-assisted laser desorption and ionization) MS. Both methods have revolutionized MS, and the inventors of the techniques were each awarded the Nobel prize for Analytical Chemistry in 2002.

Although ESI MS has to date been used primarily in lipid analysis, this chapter will be dedicated exclusively to MALDI-TOF MS. Those readers interested in the ESI MS of lipids should consult the excellent review by Pulfer and Murphy [23].

Although most commonly used in that combination, it should be noted that there is no absolute need to combine "MALDI" as the ion source with "TOF" as the detection system. However, "MALDI-TOF" will be discussed exclusively here as these devices are extremely widespread.

It is surprising how little attention MALDI-TOF MS has yet attracted in lipid analysis. Indeed, among almost 10 000 reports published to date in the context of MALDI, only 2–4% have dealt with the analysis of lipids. A short survey indicating to what extent the individual lipid classes have yet been investigated by MALDI-TOF MS is provided in Figure 7.2. It is clear that PCs were primarily investigated, whereas other lipids, for example diacylglycerols (DAGs), were rather

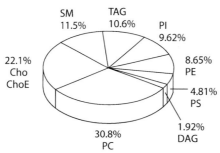

Fig. 7.2 Distribution of the individual lipid classes characterized to date by MALDI-TOF MS. Numbers in the pie correspond to the available references. Data were taken from the "Web of Science" database. Please note that there may be a considerable overlap between the individual topics. For details of abbreviations, see list (p. 239–240).

neglected. The reason for such interest in PCs, besides their vast abundance in biological systems [6], also relates to the excellent commercial availability of PCs with strongly varying fatty acid composition.

The as-yet limited interest in the application of MALDI-TOF MS to lipids is especially surprising when the high reproducibility is considered, this being due to the even co-crystallization of the analyte and matrix (see below) [20]. As both the lipid and the matrix are readily soluble in organic solvents, sample preparations do not require solvent mixtures (e.g., CH_3CN and H_2O). Another important advantage of lipids is the scarce solubility of salts in organic solvents; therefore, for the majority of applications, desalting of the lipid extract prior to MALDI-TOF MS analysis is unnecessary, even if concentrated buffers have been used.

The most serious drawback of MALDI-TOF MS of lipids is that the quantitative estimation of two different PL classes is rather difficult because the detection limits are different for each PL class. This fact will be outlined more comprehensively in the following sections.

7.1.5
Choosing an Appropriate Matrix

The matrix has two important tasks: (i) to absorb the laser energy; and (ii) to separate the analyte molecules from each other in order to avoid cluster formation. Therefore, the matrix and sample are mixed with an approximately 100- to 1000-fold molar excess of the matrix. This excess of matrix also helps to avoid direct laser "hits" on the analyte that would lead to marked fragmentation. Accordingly, the most important requirements of a "good" matrix are its absorption properties at the laser wavelength (very often 337 nm) and good mixing properties with the analyte to provide homogeneous co-crystallization. Although many different compounds have been suggested as MALDI matrices, the analysis of lipids with their rather low molecular mass requires the matrix to possess special properties. First, it is important that the yield of matrix ions generated by laser irradiation is as low as possible; this is the reason why compounds that undergo photoreactions (especially matrices derived from cinnamic acid such as sinapinic acid) should not be used, in order to avoid saturation of the detector.

So far, 2,5-dihydroxybenzoic acid (DHB) seems to be the most appropriate matrix for lipids [24], and only very few attempts have been made to use other matrices [25]. The application of DHB offers the additional advantage that positive- as well as negative-ion mass spectra can be recorded from the same sample [24]. One recent approach has been to use of graphite [26] to minimize the background, while very recently ionic liquid matrices were also proposed as the matrices of choice for lipid analysis because the homogeneity of the lipid-matrix co-crystals is excellent under these conditions [27]. A survey of typical matrices used in lipid analysis by MALDI-TOF MS is provided in Table 7.1.

Especially when DHB is used as matrix, it is advisable to evaporate the solvent rapidly in order to obtain rapid crystallization, leading to smaller and more homogeneous matrix/analyte co-crystals. Fast solvent evaporation can be achieved

Table 7.1 Typical matrix compounds used to date in lipid and phospholipid analysis by MALDI-TOF MS. For details, see text.

Chemical structure	Common name	Remarks
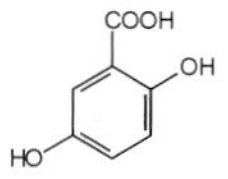	2,5-Dihydroxy-benzoic acid ("Gentisic acid")	To date, the most frequently used matrix for lipids [24,43,44]. Provides positive- and negative-ion spectra. The lithium salt of DHB was postulated to be more useful than the free acid when apolar lipids are of interest [64].
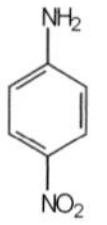	p-Nitroaniline (PNA)	Useful for the detection of PE in mixtures with PC. In the presence of PNA, PE may be detected as negative ion, whereas PC is only detected as positive ion. Enables quantitative analysis of PL [25,72].
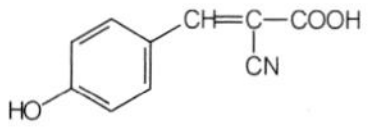	α-Cyano-hydroxy-cinnamic acid	Provides good sensitivity when positive-ions spectra of negatively charged PL are recorded. Gives comparatively strong background [43].
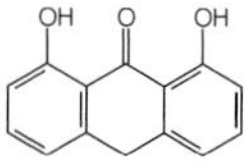	Dithranol	Not very often used. Gives strong matrix background [65].
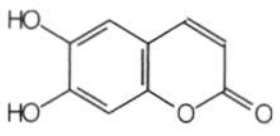	6,7-Dihydroxy-coumarin (Esculetine)	Seldom used. Provides good results with PC. Not very effective for other PL [43].
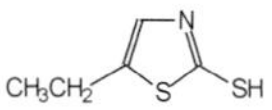	5-Ethyl-2-mercapto-thiazole (EMT)	Works well with apolar compounds, e.g., TAG. Gives only one significant peak in positive- ($m/z = 289$) and negative- ($m/z = 144$) ion spectra. Not yet commercially available [65].
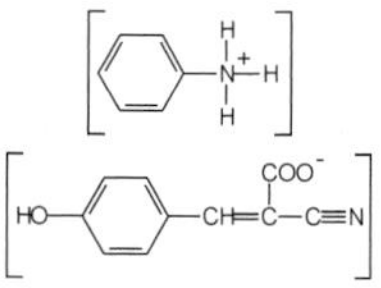	Ionic liquid matrices	Very recently developed (one selected structure is shown to the left). Very homogeneous co-crystals between matrix and analyte. One serious drawback is that these compounds are not yet commercially available [27].
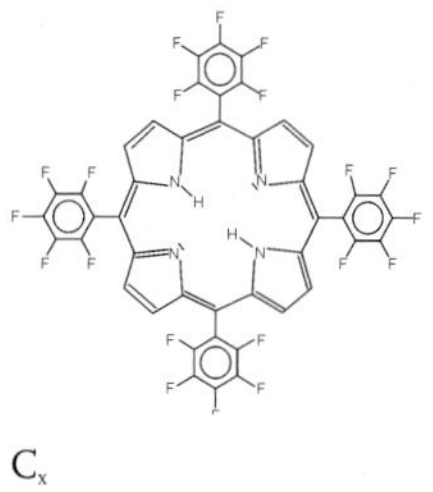	Meso-tetrakis-(pentafluorophenyl)-porphyrin	This matrix compound has a relatively large MW (974.6) and does not give any peaks in the lower mass range. Suggested to be the matrix of choice for small molecules, e.g., free fatty acids. Seems not to work properly with unsaturated lipids (oxidation?) [53,54].
C_x	Graphite	Seems very useful for apolar lipids and analysis of free fatty acids. Does not provide matrix background [26].

in the simplest case by treating the sample with a warm stream of air from a hair-dryer. In our experience, no changes of the lipid sample occur under these conditions, even if highly unsaturated lipids are analyzed [28].

Since the spectra of lipids are recorded primarily in the positive-ion detection mode, cationizing agents are often added in order to increase the ion yield. Although biological samples do always contain alkali metal cations (mainly Na^+ or K^+), even after extraction with organic solvents, trifluoroacetic acid (TFA) is typically used as an additive to the matrix in a concentration of about 0.1 vol%. Therefore, H^+, Na^+ and (with lower intensity) K^+ adducts are simultaneously detected [24]. However, an equilibrium between these adducts exists that can be influenced by the addition of further ions, that are not naturally occurring (e.g., Cs^+). This approach is helpful if there is a need to shift peaks selectively in order to avoid peak-overlap (which results from differences in the fatty acid composition and different adducts, and is more comprehensively discussed below [29]).

It should also be noted that the shape of the laser desorption spectra of the pure matrix depends significantly on the ion content that influences peak intensities [20].

7.1.6
Sample Preparation, Extraction and Purification

The best spectra are obtained when lipids are handled in organic solvents. Accordingly, lipids should be extracted from the biological material of interest, for example, with mixtures of methanol and chloroform [30]. After separation of the organic and aqueous layers, the organic phase can be used without further purification for MALDI-TOF MS, as impurities such as salts and the majority of water-soluble components remain in the aqueous layer. Of course, samples such as blood can also be analyzed without previous extraction, although spectra exhibit lower resolution and require enhanced laser intensities.

Although the extraction of lipids from aqueous samples sounds simple, it is actually a science in its own right, and the user should be aware of the potential loss of lipids during extraction [6]. Generally, higher charged lipids are more difficult to extract than neutral PLs.

Detergents such as Triton are often used instead of organic solvent mixtures to solubilize the lipids from biological tissues [31]. This approach is not advisable when the lipids are subsequently to be analyzed by MS, however, as Triton itself produces very intense signals that make the detection of peaks of interest difficult. The complete removal of Triton is almost impossible, whether by extraction or by using specially developed kits for Triton removal [32]. When a detergent is absolutely required, sodium cholate rather than Triton is to be preferred as it has a defined molecular mass that results in a single peak. It should also be noted that chloroform solutions are best not handled in plastic vessels as the $CHCl_3$ releases plasticizers from the plastic material, and these produce strong peaks in the mass spectra [20].

7.2
Analysis of Individual Lipid Classes and their Characteristics

The MALDI-TOF MS characteristics of some physiologically important lipid classes will be discussed in the next sections. It must be highlighted at this point that spectral quality may vary between individual MALDI-TOF devices. For example, the applied laser intensity has the most significant impact on spectral quality and should be kept as low as possible. A suitable measure of the appropriate laser strength is the isotopic distribution and the contribution of fragment ions as well as dimers.

It should also be noted that, depending upon the instrumental settings, the intensity ratios of compounds in a mixture may be altered [33]. This is especially true when lipid mixtures are being investigated using a reflectron in combination with delayed ion extraction. Thus, all measurements should be performed with highly defined settings.

7.2.1
The Apolar Lipids: Diacylglycerols, Triacylglycerols, Cholesterol and Cholesteryl Esters

Diacylglycerols (DAG) and triacylglycerols (TAG) represent rather simple structured lipids that are, however, of great relevance in signal transduction [3] and nutrition, respectively. Because of the considerable commercial interest, much more attention was paid to TAG than DAG [32], and consequently TAG spectra were among the very first lipid MALDI-TOF mass spectra to be analyzed quantitatively.

TAG yield exclusively the corresponding alkali metal adducts, whereas proton adducts are never detected [34]. Although the ion affinities of individual lipid classes have not yet been investigated in detail, two different opinions exist. The first indicates that all lipids that do not contain nitrogen give exclusively alkali adducts [35]. The second opinion explained the observed differences of adduct formation by the different tendencies of Na^+ and H^+ to bind to the glycerol backbone or the polar headgroup, respectively [32].

Different matrices, for example α-cyanohydroxycinnamic acid, DHB (gentisic acid) or dithranol, were tested in TAG analysis. Among these, DHB gave the best reproducibility and a three orders of magnitude higher sensitivity. Zollner et al. [36] suggested the use of $K_4[Fe(CN)_6]$ in glycerol as the matrix of choice because of the achievable gain in sensitivity. However, when using this matrix other authors were not able to produce any ions [34] and, therefore, the advantages of this system seem equivocal.

Very recently, it was also reported that TAG can be even analyzed in the absence of a UV-absorbing matrix [37] by using standard stainless-steel targets. Although the ionization process is not yet clarified, it was stated that the spectra of TAG are highly reproducible under these conditions and can be accurately quantified as there is no interference with matrix peaks [37].

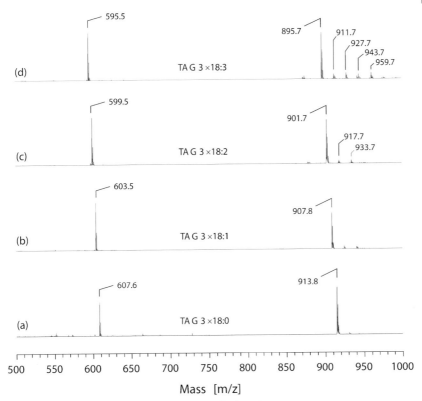

Fig. 7.3 Positive-ion MALDI-TOF mass spectra of selected triacylglycerols. (a) TAG 3 × 18:0; (b) TAG 3 × 18:1; (c) TAG 3 × 18:2 and TAG 3 × 18:3 (d). All spectra were recorded using 0.1 μg TAG and 0.5 mol L⁻¹ DHB in CH₃OH as matrix. All peaks are marked according to their *m/z* ratio. Note the increasing content of oxidation products in the spectra of higher unsaturated TAG. For details, see text.

DAG and TAG yield very stable ions. The most pronounced, laser-intensity-dependent fragmentation of DAG is the loss of sodium hydroxide under the generation of the corresponding carbenium ion [28], whereas the loss of the sodium salt of a fatty acid residue predominates in TAG spectra [34]. Presumably due to peroxidation, different fatty acid residues provide specific fragmentation patterns that are helpful for identification of fatty acid residues [32].

The positive-ion MALDI-TOF mass spectra of four selected TAGs are shown in Figure 7.3. These TAGs were chosen because they differ in their content of double bonds: Each TAG is detectable as the sodium adduct, and the cleavage of one sodiated fatty acid residue is clearly detectable. For instance, the mass difference in trace (a) is 306, which corresponds to the loss of sodium stearate.

It should be noted that the content of oxidation products increases with the number of double bonds. For example, trilinolenoyl glycerol (TAG 3 × 18:3) yields the most intense oxidation products that are easily detectable by the characteristic mass difference of 16 Da corresponding to the addition of oxygen. This oxidation process is not caused by the MALDI ionization process, but occurs during storage of the TAG (lipids become "rancid" when stored over a longer period).

In contrast to DAG, cholesterol is not detectable as the H^+ or Na^+ adduct, but only subsequent to water elimination leading to $m/z = 369.3$ ($M+H^+-H_2O$) [38]. Although only rarely investigated to date, cholesteryl esters produce similar peak patterns as DAG or TAG, and are detected exclusively as Na^+ adducts [39]. The spectra of cholesteryl esters always contain signals of cholesterol ($m/z = 369.3$), with cleavage of the fatty acid residue being the most prominent fragmentation reaction [40]. It must be emphasized here that cholesteryl esters and TAG are detected in biological samples with lower sensitivity in comparison with other PLs. This difference is not derived from the ionization process, but rather is caused by a reduced extraction efficiency of apolar lipids in comparison to PLs [39].

7.2.2
Zwitterionic Phospholipids: Sphingomyelin, Phosphatidylcholine and Phosphatidylethanolamine

These PLs represent the prime constituents of the membranes of animal cells. Accordingly, the majority of MALDI-TOF investigations are targeted at these PL classes, though with special emphasis being placed on phosphatidylcholine (PC) (see Fig. 7.2).

PC, sphingomyelin (SM) and phosphatidylethanolamine (PE) are easily detectable by MALDI-TOF MS, although PE is detected with reduced sensitivity [33]. PC and SM each possess the choline head group – a quaternary ammonium group with a permanent positive charge – that leads to a high yield of positive ions [24].

PC is frequently investigated by MALDI-TOF MS [24,41,42], and was one of the first PLs to be studied using this technique [43,44]. As the pioneer in this field, Harvey [43], in 1995, investigated different PLs and which matrix would provide the best spectral quality. Among others, α-cyano-4-hydroxycinnamic acid, 6,7-dihydroxy-coumarin and DHB produced the best results.

Some basic concepts regarding the shape of lipid spectra are provided in Figure 7.4, which shows the MALDI-TOF mass spectra of PC 16:0/16:0 (a), PE 16:0/16:0 (b), and PG 16:0/16:0 (c,d). It should be noted that PG is a negatively charged PL. Spectra (a–c) represent positive-ion spectra, whereas (d) is a negative-ion spectrum. The PC spectrum (a) (M = 733.6 Da) shows the expected peaks at $m/z = 734.6$ and 756.6 corresponding to the H^+ and Na^+ adducts, respectively. In contrast, the PE spectrum (b) (M = 691.5 Da) is characterized by a more complex peak pattern caused by the ammonium group that contains exchangeable protons. Peaks at $m/z = 692.5$ and 714.5 can be explained analogously to PC. The peak at $m/z = 736.5$, however, is caused by the exchange of an H^+ by Na^+ in the neutral state of the PE ($M-H^++Na^+$). The resulting neutral molecule (M = 713.5 Da) is subsequently cationized by the addition of H^+ or Na^+. A similar mechanism holds for the PG that was used as the sodium salt (M = 744.5): Spectrum (c) can be explained

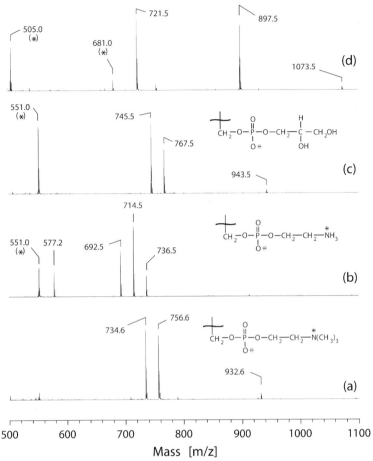

Fig. 7.4 Positive-ion MALDI-TOF mass spectra of selected phospholipids (PL). (a) PC 16:0/16:0; (b) PE 16:0/16:0, and PG 16:0/16:0 (c). In (d) the negative-ion spectrum of PG 16:0/16:0 is also shown. All spectra were recorded using 0.1 μg PL with 0.5 mol L^{-1} DHB in CH$_3$OH as matrix. All peaks are marked according to their m/z ratio, and typical matrix peaks are labeled with asterisks. The characteristic headgroups of the individual PLs are also shown. Note the peaks of cluster ions between DHB and analyte at the right end of the spectra.

by the addition of H$^+$ or Na$^+$ (m/z = 745.5 and 767.5). The negative-ion peak at m/z = 721.5 of PG (d) is caused by the loss of one Na$^+$ from the neutral PG molecule.

It should be noted that, in addition to some DHB peaks (marked with asterisks [33]), there are also peaks at higher m/z values. These are most pronounced in the negative-ion spectrum and correspond to cluster ions of the matrix with the analyte. As one selected example, the peak at m/z = 897.5 may be explained by the addition of the sodium salt of DHB (molecular weight 176) to the negatively charged PG (m/z = 721.5).

To the present author's knowledge, the parameters that determine the yield of such cluster ions have not yet been comprehensively studied. Although all PL were used in the same amounts, the comparison of PL signal intensities with the typical matrix peak at $m/z = 551$ (positive ion) clearly [33] indicates that individual PL are detected with different sensitivities: PC is most sensitively detected (detection limit ca. 0.5 pmol, corresponding to ca. 0.4 ng PC on the target) [43]. This is of paramount interest when PL mixtures are investigated: PLs containing quaternary ammonium groups – such as PC, LPC or SM – may prevent the detection of further species in a lipid mixture [33]. The fragmentation pathways of PC were also studied by the post-source decay (PSD) technique [41], and stability differences of the H+ and the Na+ adducts were monitored. The protonated PC produced only one pronounced fragment corresponding to the loss of the head group ($m/z = 184$). In contrast, the Na+ adduct of PC is less stable, and a loss of the fatty acid residue occurs primarily at the *sn*-1 position. Fragmentations of the head group, namely M+Na+-59, M+Na+-183 and M+Na+-205 corresponding to the loss of trimethylamine, choline phosphate and sodiated choline phosphate, respectively, also occurred [41]. Besides the effect of the applied laser strength, there are also indications of an influence of the sample preparation on the yield of fragmentation products. This especially holds for the water content of the sample [43].

In addition to differences in the PL head group structure, the fatty acid composition also influences detectability. PL with lower molecular masses are more sensitively detected than those with higher molecular masses, whereas the degree of saturation has only a minor influence. A similar behavior was also reported for TAGs [34]. This is a significant difference in comparison to ESI MS, where the degree of saturation strongly influences the ion yield of PLs [45].

Unfortunately, potential applications are impaired by: (i) the overlap between peaks resulting from differences in the fatty acid composition; and (ii) the superposition of H+ and Na+ adducts. One of the first attempts to overcome this problem was to digest the PC with phospholipase C (PLC). This enzyme produces the corresponding DAGs, that are exclusively detected as Na+ adducts and, therefore, peak overlap is reduced [41]. This is an alternative approach to the application of an excess of Cs+ [25,29].

Para-nitroaniline (PNA) was recently suggested as matrix of choice for PL mixture analysis [25], the main advantage here being that PE becomes detectable as the negative ion. As PC is not detectable as a negative ion, problems arising from the suppression of PE by the presence of PC can be overcome [25]. This was also proven for the case of complex lipid mixtures [46].

MALDI-TOF MS has also shown great promise with regard to the determination of enzyme activities. For example, it was shown that the activity of phospholipase A2 (PLA2) can be easily monitored as both, the substrate (PC) and the reaction product (LPC), are simultaneously detectable [47]. Because of its lower molecular mass, the sensitivity of detection is greater for LPC than for PC, and this is a considerable advantage when only low-level enzyme activities are under investigation. Of course, the most important advantage of this assay is that there is no need to label the substrate, and all PC species can be used independently of their fatty acid composition.

Similar studies on sphingolipids are also available and have provided good results for ceramide monohexosamide and especially SM [48].

7.2.3
Acidic Phospholipids: Phosphatidic Acid, Cardiolipin, Phosphatidylglycerol, Phosphatidylserine, Inositol and Higher Phosphoinositides and their Lysolipids

All of these PLs are negatively charged at physiological pH (7.4), and therefore yield intense negative-ion signals. Singly charged, negative ions [24,33] are generated by the abstraction of H^+ or Na^+ from the neutral molecule and, therefore, the m/z values of the negative ions are shifted by 2, 24, or 46 mass units in comparison to the positive ions (cf. traces (c) and (d) in Fig. 7.4).

PS, PI and PG possess similar molecular masses and are characterized by comparable detection limits [43]. In contrast, PLs with a higher negative charge density are more difficult to detect and have been investigated only rarely. For example, to date, the spectra of PA and its reduced detectability in comparison to other PL have been described only once [33]. An interesting report was made, however, on the detection of lysophosphatidic acid (LPA) by MALDI-TOF MS [49]. These authors used the characteristic mass shift resulting from the total charge change (from −2 to +1) due to 1:1 complexation of LPA (2-) with a dinuclear zinc (II) complex that exhibits three positive charges at physiological pH. The resulting complex with LPA could be detected as a single positively charged ion, whereas no other adducts were detectable. The detection limit of LPA using this approach was about 100 fmol. These studies [49] were especially remarkable as the authors did not use "standard" DHB, but rather the 2,3-isomer. This is surprising because the absorbance at 337 nm of this isomer is far lower than that of the 5-isomer.

The required amounts of poly-phosphoinositides (e.g., PIP$_2$ and PIP$_3$) are relatively high because these compounds differ from standard PLs in three points: (i) the charge state; (ii) the polarity; and (iii) the enhanced molecular mass. While PI is characterized by only one negative charge, the charge of PIP is −3, of PIP$_2$ is −5, and of PIP$_3$ is −7, at pH 7.4 [50]. Accordingly, PIP and PIP$_2$ can be detected as positive and negative ions, whereas PIP$_3$ is exclusively detectable as the negative ion [24]. Therefore, the detection thresholds for PPIs are much higher in comparison to those of other PLs. While PC is detectable in amounts of approximately 3 pmol, 700 pmol of PIP$_3$ are required to produce sufficient signal intensities [50]. Some typical data to indicate this fact are shown in Figure 7.5. It should be noted that defining the signal-to-noise (S/N) ratio (= 5) as the detection limit was only an arbitrary decision.

Shadyro et al. [51] used MALDI-TOF MS in combination with TLC to monitor the effects of γ-irradiation on cardiolipin (one of the largest PLs), and were able to monitor the reaction products of phosphatidic acid and diacylphosphatidyl-hydroxyacetone [51].

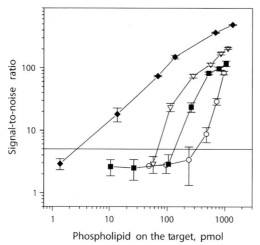

Fig. 7.5 Signal-to-noise (S/N) ratios as a function of the phospholipid amount on the target for MALDI-TOF positive-ion mass spectra of phosphatidylcholine (DPPC, ◆), PI (▽), PIP (■), and PIP$_2$ (○). 2,5-Dihydroxybenzoic acid (0.5 mol L^{-1} in methanol) containing 0.1 vol.% TFA was used as matrix. Data points represent the mean of six measurements (two different preparations with three measurements each), error bars represent the S.D. The line at S/N = 5 represents the threshold defined as the detection limit. (Reprinted with permission from *Chem. Phys. Lipids* **2001**, *110*, 151–164, with permission from Elsevier.)

A variety of glycocardiolipins from bacteria has also been further characterized by MALDI-TOF MS [52] (for further details, see Chapter 6).

7.2.4
Free Fatty Acids

MALDI-TOF MS studies of fatty acids are hampered by their small molecular mass and the resulting overlap with matrix peaks.

One promising approach has, however, been suggested by Ayorinde and coworkers, who used *meso*-tetrakis(pentafluorophenyl)porphyrin [53] as matrix that does not produce any peaks in the mass region of fatty acids. Fatty acids were obtained by alkaline saponification of different TAG mixtures [54]. As an excess of sodium acetate was added, the Na$^+$ adducts of the sodium salts of fatty acids were detected exclusively as positive ions [54]. The results obtained were in good agreement with data obtained using other methods [55].

Unfortunately, this approach is only useful when saturated fatty acids are investigated. In the presence of even small amounts of unsaturated fatty acids, all peaks – even that of saturated fatty acids – are shifted by 14 Da to higher *m/z* values, though the reasons for this surprising behavior are, to date, unknown [54].

Graphite has also been used successfully as a matrix for fatty acid detection [26].

7.3
MALDI-TOF MS of Typical Lipid Mixtures

In this section we will review all currently available data relating to MALDI-TOF MS of lipid mixtures. Although attention will be focused on the analysis of lipid extracts from cell suspensions and body fluids, analyses of plant oils and of the oxidation products of PLs will also be discussed.

7.3.1
Vegetable Oils and Frying Fats

It has been reported by several authors that TAG mixtures from plant oils [34] as well as cod-liver oil [56] can be easily characterized by MALDI-TOF MS, without any time-consuming work-up. The fatty acid composition of a given oil can be calculated directly from the spectrum since TAGs yield exclusively Na$^+$ adducts [57,58]. Of course, it is not possible to determine under these conditions the position of a certain fatty acid residue, but only the overall composition of the TAG mixture. Therefore, it is very common to use the following nomenclature. When a TAG is referred to "TAG 54:3", this indicates that the three fatty acids consist in total of 54 carbon atoms with three double bonds.

Comprehensive MALDI studies of TAGs are available. For example, it was shown that olive, sunflower, safflower, walnut and linseed oils yield characteristic mass spectra, thereby allowing an estimation to be made of the degree of saturation [59]. In this way, a good correlation between the MALDI-TOF MS and GC/MS data [55] could be established.

Jakab et al. [60] used sophisticated mathematical models in order to interpret the MALDI-TOF mass spectra of different plant oils. MALDI-TOF MS data were compared with data by HPLC/atmospheric pressure chemical ionization (APCI) MS, and a good agreement between both techniques was obtained [60]. This indicated clearly that MALDI represented a potential alternative analytical technique.

As vegetable oils are often used to fry food, the changes in TAG upon heating were also investigated [61]. The data obtained indicated that the addition of atmospheric oxygen to unsaturated TAG, as well as the subsequent β-scission under the generation of the corresponding carbonyl compound, could be monitored. Interestingly, thermally induced changes in completely saturated oils (e.g., coconut oil) could also be monitored [62].

7.3.2
Plant and Tissue Extracts

When biological tissues are extracted with organic solvents, the extracts generally represent complex lipid mixtures. Thus, it is normally not possible to provide a complete compositional analysis without any previous separation of the mixture into the individual lipid classes. Unfortunately, direct coupling between

MALDI-TOF MS and chromatography is not possible in this situation because solid samples are being used. Although this is a clear drawback of MALDI-TOF MS, some reports have been made describing the direct use of TLC plates for MALDI-TOF MS analysis [63].

To date, very few reports have been made dealing with plant lipids. For example, the fatty acid composition of PC from potatoes was quantitatively determined by digesting the PC with phospholipase C (PLC) in order to convert the PCs into the corresponding DAGs. These are more simple to analyze because they produce (exclusively) the Na$^+$ adduct, but not a mixture of H$^+$ and Na$^+$ adducts [41].

It was also reported recently that lipids from leaves as well as high-molecular mass hydrocarbons of plants can be analyzed using lithium 2,5-dihydroxybenzoate as matrix [64]. A similar investigation was undertaken with coconut oil and 5-ethyl-2-mercaptothiazole as a matrix that shows a very intense absorbance at the laser wavelength, and produces only small peaks of its own [65]. Finally, lipid second messengers such as LPC and DAG could be quantified in growing plant cells [66] when MALDI-TOF MS was used subsequent to TLC separation. Among the reports made to date, by far the largest majority have related to animal tissues rather than to plant tissues; consequently, investigations into the PLs of the brain and eye lens tissue are discussed as representative examples in the following sections.

7.3.2.1 Brain Phospholipids

Brain tissue is highly complex, and contains almost all known PL classes, to a point where lipids constitute approximately 12% of the brain (as wet weight). It is an additional challenge that, besides diacyl-lipids, there are also alkyl-acyl and alkenyl-acyl (plasmalogens) lipids present in brain. Notably, brain tissue is characterized by its content of highly unsaturated fatty acids, such as docosahexaenoic acid (22:6).

Because of the enhanced detection of molecules with quaternary ammonium groups, the positive ion MALDI-TOF MS spectra of brain are dominated by PC, LPC and SM, whereas PI is the most abundant negative-ion PL. In order to be able to detect further lipid species, a previous separation of the total brain extract into the individual lipid classes is necessary [33].

Because of the rapid performance and the simple equipment required, TLC can be used advantageously only when small amounts of lipids are available [67]. The most critical step is staining of the TLC plate. Staining procedures which are accompanied by changes in the lipids' molecular masses must not be used when further MS analysis is planned. In the present author's opinion, primuline (Direct Yellow 59) is the dye of choice for lipids, as it binds non-covalently to the apolar fatty acid residues and individual lipids are visualized as violet spots upon irradiation by UV light (254 nm). Spots of interest can be scratched from the TLC plate, and the lipids eluted and evaluated by MS. Primuline is simply released from the lipid by the addition of solvent [68].

One interesting feature of brain is its considerable content of alkyl- and alkenyl-ether-linked PLs. These are detectable with a sensitivity comparable to that for diacyl PL, but this leads to characteristic mass differences. The extreme sensitivity of plasmalogens to even traces of acids is the reason why lysolipids are always detectable when TLC separation of brain lipids was performed. The acidic groups on the TLC plate suffice to induce hydrolysis of plasmalogens [67].

The brain PL composition was also determined by ion-mobility MS [69], whereupon specific interactions between drugs and PLs could be monitored. One recent study [70] suggested direct investigation of the TLC sample plate using FT-MALDI-TOF MS. Consequently, under optimized conditions there was only a small reduction in resolution compared to that seen for lipids in organic solvents [71].

7.3.2.2 Eye Lens Phospholipids

Interesting studies of the PL composition of eye lenses have been conducted by the group of Yappert [72], with the PLs of eye lenses being characterized by a high content of highly unsaturated fatty acids. MALDI-TOF MS could be used to characterize the cholesterol-enriched fractions [73], to estimate the glycerophospholipids content compared to sphingolipid content [74], and to monitor radical-induced lipid peroxidation [75].

7.3.2.3 Other tissues

The PL composition of thin tissue slices of dystrophic mouse leg muscle was also investigated recently [76].

7.3.3
Important Body Fluids and Cells of the Human Body

Although, in the majority of PL analyses, organic extracts of body fluids and cells are utilized, some procedures will require the use of aqueous extraction. The most commonly used methods are those of Folch [30] and Bligh and Dyer [77], both of which employ a solvent system comprising chloroform/methanol/water, albeit in varying volume ratios. Although the Bligh and Dyer method has been suggested as more suitable for body fluids, and the Folch method for tissues with a lower water content, both procedures can be used principally in all cases. Further details about lipid extraction are also available on the Internet [78].

One further extraction method employs an isopropanol/hexane mixture [79]. This mixture is useful because it not only provides an almost complete extraction of apolar lipids but also uses less-aggressive solvents. There is also the considerable advantage that plastic materials (Eppendorff tubes, pipette tips, etc.) can be used, in contrast to methods which utilize $CHCl_3$ [20].

Recently, a number of reported methods have attempted to avoid extraction. Although reports on "whole-cell" lipid analysis have been published [80–82], only one report has been made on "whole-animal" analysis. *Daphnia galeata*, a type of water flea, with a body diameter of a few millimeters, were analyzed by pulling the insects to pieces and placing them directly onto the target [83]. Although the reproducibility of this technique is questionable, the approach has the advantage that it cannot be influenced by extraction efficiency. On the other hand, however, the high salt content of biological samples may lead to rather poor S/N ratios.

7.3.3.1 Blood

Blood or its cell-free analogue, blood plasma, belong to the most important body fluids of diagnostic interest. Surprisingly, few MALDI studies have been conducted on the lipid composition of blood, and in the majority of cases only one specific metabolite was investigated [84]. An example is the identification in blood of sphingosine-1-phosphate and sphingophosphocholine.

In a recent study, Fuchs et al. [85] used the PC/LPC ratio as a measure of the inflammatory state in plasma sampled from patients suffering with rheumatoid arthritis. Here, the PC/LPC ratio was shown to increase when the patient was treated with anti-rheumatic drugs; consequently, MALDI may play a role in assessing the success of drug therapies.

In the context of arteriosclerosis, the lipids of the individual lipoprotein fractions of human blood plasma were more comprehensively investigated [39,40,86], and differences in the lipid composition of LDL and HDL could be monitored. In the presence of an excess of Cs^+ ions, problems with overlap between different adducts on the one hand and differences in the fatty acid composition on the other hand, could be minimized [29] and unambiguous peak assignment was possible.

The extracts of high-density lipoprotein (HDL) and low-density lipoprotein (LDL) (Fig. 7.6) differ considerably as LDL shows a significantly higher SM content ($m/z = 703.6$ and 725.6), whereas the content of LPC is higher in HDL (data not shown). This is in good agreement with known compositional differences [87]. Additionally, monitoring the influence of reactive oxygen species (ROS) is also possible (see below) [86], and the effects of enzymatic treatments can also be assessed [47]. For example, information relating to the activity of the enzyme cholesteryl esterase and its selective inhibition could be obtained by means of MALDI-TOF MS [40].

7.3.3.2 Spermatozoa and Neutrophils

These cells were chosen as representative examples because their analysis is aggravated by the following properties. Spermatozoa contain significant amounts of highly unsaturated fatty acid residues, especially docosahexaenoic acid (22:6). This is required to provide the spermatozoa cell with high membrane flexibility that facilitates sperm-oocyte fusion. In contrast, neutrophils represent an important component of the immune system and undergo very rapid changes of their PL composition upon stimulation [24].

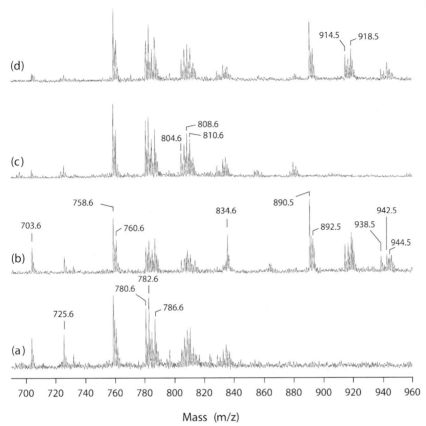

Fig. 7.6 Positive-ion MALDI-TOF mass spectra of chloroform/methanol extracts of LDL (a and b) and HDL (c and d). Spectra (a) and (c) were recorded in the absence of CsCl, whereas for (b) and (d) the chloroform layers were washed with aqueous 1 M CsCl. (Reprinted with permission from *Chem. Phys. Lipids* **2001**, *113*, 123–131, with permission from Elsevier.)

As shown in Figure 7.7, the spectra of lipid extracts of human spermatozoa are dominated by PC 16:0/22:6 ($m/z = 806.6$ and 828.6), SM 16:0 ($m/z = 703.6$ and 725.6) and cholesterol ($m/z = 369.3$) [38]. In contrast to spermatozoa, the seminal plasma lacks the PC peaks.

It is also evident from Figure 7.7 that further abundant PLs (e.g., PE and PS) are not detectable [38]. This clearly indicates that spectra of complex lipid mixtures must be carefully interpreted in order to avoid data misinterpretation.

Although these results can hardly be called a detailed lipid analysis, the sensitive detection of PC and SM, as well as the corresponding lysophospholipids, may provide important information. For example, subsequent to a freezing–thawing cycle, intense peaks of lysophospholipids – namely LPC and ceramide – appeared in the lipid extracts of spermatozoa. These changes could be explained by

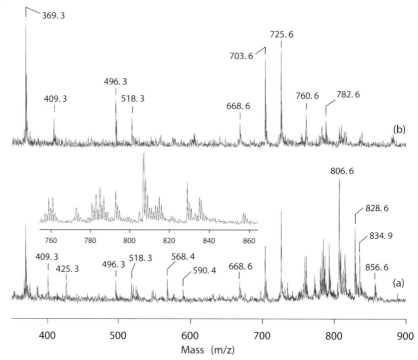

Fig. 7.7 Positive-ion matrix-assisted laser desorption and ionization time-of-flight (MALDI-TOF) mass spectra of a typical sample of human spermatozoa (a) and the corresponding seminal plasma (b). Peaks attributable to fixed products are marked with their mass. The insert shows the expansion of the mass region between 755 and 865 mass units of the spermatozoa extract. (Reprinted with permission from *Chem. Phys. Lipids* **2000**, *106*, 145–156, with permission from Elsevier.)

activities of the enzymes PLA$_2$ and sphingomyelinase, both of which are released when the spermatozoan membrane loses its integrity by the freeze–thaw process [38].

Another important application is to evaluate sperm quality. One common clinical approach is the "annexin-V-cell-viability assay", which is based on the high affinity of the protein annexin V for PS. PS is normally located exclusively in the inner leaflet of the membrane, and hence no binding is observed when the spermatozoan membrane is intact. A correlation between the annexin-V-cell-viability assay and the intensities of the LPC peaks could be established, with annexin-V-positive spermatozoa in all cases being characterized by enhanced LPC moieties [88]. This indicated that the LPC concentration (and, accordingly, PLA$_2$ activity) is a useful measure of sperm cell viability.

One characteristic property of some animal (e.g., bull or boar) spermatozoa is their significant plasmalogen content. These PLs are extremely sensitive towards acids [89], and therefore the routine use of trifluoroacetic acid (TFA) as a cationizing reagent [24] must be regarded with great caution [90,91] as small amounts of TFA can cause plasmalogen destruction. When a sample contains unexpectedly high concentrations of LPC, the measurement should be repeated in the absence of TFA in order to check whether the LPC is from the sample or is simply an artifact of the plasmalogen hydrolysis. Such hydrolysis may even occur under conditions used for TLC, due to the acidic groups on the silica surface [89].

Neutrophilic granulocytes (neutrophils) are very interesting cells which are assumed to contribute to the development of arteriosclerosis or rheumatoid arthritis [92]. Neutrophils must be activated in order to leave the bloodstream, the activation process being accompanied by changes in the neutrophils' lipid composition. Thus, via lipid analysis, MALDI-TOF MS offers the possibility of determining the extent of stimulation, for example with chemotactic tripeptides [24,93]. Accordingly, the lysophospholipid concentration within the cells is an important measure of stimulation [94,95]. Additionally, the dominant role of PA in stimulation was recently confirmed by MALDI-TOF MS [96], while similar investigations were also carried out with physiologically active DAG [97]. Finally, a recent investigation was conducted into the extraction efficiency of cells by different solvents; however, this was performed not with neutrophils (which cannot be grown in culture) but rather with HL-60 cells that can be differentiated into neutrophils [98].

7.3.4
Characterization of Typical Oxidation Products of Lipids

Many studies have aimed to investigate the modification of PLs by selected ROS. The focus was on hypochlorous acid (HOCl) that is generated under *in-vivo* conditions from H_2O_2 and Cl^- in the presence of the enzyme myeloperoxidase (MPO).

The effects of HOCl on unsaturated PLs were first described many years ago; HOCl is known to be added to the olefinic residues of the PL with the generation of corresponding chlorohydrins and (in lower yields) the glycols and dichlorides as reaction products.

By using MALDI-TOF MS, it could be shown that PC with highly unsaturated fatty acid residues (20:4 or 22:6) yield primarily the corresponding LPC. Under the influence of HOCl, the unsaturated fatty acid residue is cleaved completely from the PL backbone [99,100]. Based on the results of these experiments, it became clear that LPC may not be generated exclusively by the enzyme PLA₂ but also by HOCl. Surprisingly, only a single report was made which described the generation of LPC by the treatment of lipoproteins with physiologically relevant concentrations of HOCl [86].

An interesting study on the egg tempera paints which were commonly used several hundred years ago, was performed by van den Brink and coworkers [101]. These authors used high-resolution MALDI FT MS to determine changes in the products of PC and TAG. The long-term exposure of egg tempera paints to light in the presence of atmospheric oxygen yields oxygenated PC and TAG, which can easily be differentiated by MS. As the yield of oxidation products is heavily dependent on light exposure as well as the NO_x and SO_2 concentrations, these authors suggested that such an MS approach might be used as a "paint dosimeter" for the environmental monitoring of museum display conditions [101].

7.4
Quantitative Aspects of MALDI-TOF MS of Lipids

Although our knowledge of the ion-generation process in MALDI is steadily increasing [102], there remains an ongoing debate as to whether MALDI-TOF mass spectra might be analyzed quantitatively, and what level of accuracy might be achievable.

Reports available to date have indicated that the quantitative analysis of smaller molecules (and, accordingly, lipids) is possible [103]. In the case of lipid mixtures, it is difficult to derive the absolute amounts of any individual lipid, as even if correction factors were used to adjust for the different detectabilities of the individual PL classes, the determination of PL present in very small amounts would not be accurate.

In contrast, determination of the fatty acid distribution of a certain PL class, or the PL/LPL ratio, could be carried out in a more rapid and convenient manner than by using other MS methods.

7.4.1
Approaches Used To Date

The most important prerequisite for the quantitative evaluation of MALDI-TOF mass spectra is a highly controlled sample preparation, together with homogeneous crystallization of the matrix and analyte. Ideally, this "solid solution" between matrix and analyte should yield exactly the same amount of ions by each laser shot. It is a great advantage of lipids that both, matrix and analyte, are readily soluble in the same organic solvent, as this leads to very homogeneous matrix/sample mixtures [24]. This is especially true when the matrix and analyte are "premixed", because under these conditions the ratio of both is exactly defined. In contrast, the so-called "add-on" sample preparation technique [104], where first the matrix and afterwards the analyte solution is deposited onto the MALDI target, should not be used because of the reduced homogeneity.

To date, the following approaches have been used to quantify MALDI-TOF mass spectra: (i) the peak intensity of the compound of interest was related to the intensity of a known internal standard [28]; (ii) the S/N ratio was determined [50,105]; or (iii) the intensity of the peak of the lipid of interest was compared with a known matrix peak [33].

So far, method (i) has been the most frequently used and has provided convincing results for DAG [28] and TAG [34] analysis. A linear correlation between peak intensity on the one hand and lipid concentration on the other hand over about one order of magnitude can be obtained on a regular basis [28]. The added standard should be of the same lipid class as the lipid of interest, and also have a comparable molecular mass. Lipids containing fatty acids with odd carbon numbers that are commercially available may be advantageously used because these compounds do not occur under natural conditions.

In method (ii), the peak intensities of the standard and analyte can be compared directly subsequent to subtraction of the noise level. This method provides the most accurate results when the intensities of the peaks of the standard and the lipid are comparable. Therefore, some prior knowledge of the sample composition is required in order to avoid adding an excess or an insufficient amount of the standard. To date, method (ii) has been used for the analysis of LPC [105], the determination of PLA_2 activities [47], and for the quantitative analysis of acidic phospholipids (e.g., phosphoinositides) [50].

Method (iii) was successfully applied to the analysis of TAG mixtures, and shown to be superior when compared to the other methods [34]. A special advantage of this approach is that the sample is not altered at all. Unfortunately, this method does not work well for polar lipids, one potential reason being the ion content of the applied solutions. The intensity of the matrix peaks depends on the ion content, especially the Na^+ concentration [33]. The solutions of negatively charged PL normally contain higher amounts of ions than neutral lipids, and therefore the presence of acidic lipids may alter the intensity of matrix peaks and make the quantitative analysis ambiguous.

7.5
Problems, and their Potential Solution

7.5.1
"Small" Molecules and Fragmentation Products

Clearly, MALDI-TOF MS was originally developed for the analysis of compounds with high molecular masses. Such compounds do not interfere with matrix peaks and, therefore, a clear distinction can be made: the lower the molecular mass of the compound of interest, the more marked is the influence of matrix peaks and fragmentation products of the analyte.

The following suggestions might serve as useful ways around this problem when compounds with molecular masses less than about 500 Da are be analyzed:

- Omission of the matrix: This is only possible when the analyte itself gives a strong absorption at the laser wavelength. This does normally not hold for lipids. Very recently it was shown that, by using stainless-steel targets, reasonable spectra could be recorded in the absence of a matrix, though only at very high laser intensities [37]. Surprisingly, no pronounced fragmentation of the analyte occurred and the results were comparable to spectra recorded in the presence of a matrix. Unfortunately, this method has so far only been used in TAG analysis, and it is still to be tested to determine it is generally applicable.
- Choosing a matrix with a higher molecular mass than the analyte of interest. Here, a "porphyrin-like matrix" – *meso*-tetrakis(pentafluorophenyl)-porphyrin – has been used. This matrix [54] has a molecular mass of $974.6 \, g \, mol^{-1}$ and does not produce peaks at smaller m/z values. However, whether it can be generally applied is not yet known.

- Choosing a matrix that does not give any signals in the mass range of interest. Here, inorganic matrices (e.g., graphite or nanotubes) were suggested and some application examples provided. A similar technique termed "Dios-(Desorption ionization on porous silicon)-MS" was also recently suggested [106]. Attempts have also been made to couple AP-MALDI with ion trap detection devices [107] in order to increase the accuracy of quantitative MALDI data. A more comprehensive discussion of small molecule analysis is provided in Chapter 4.

7.5.2
Auxiliary Reagents

In the present author's opinion there are only two relevant useful auxiliary chemicals for MALDI-TOF MS analysis of lipids. The first of these is CsCl, and the second is the enzyme PLA$_2$. Both are very useful and allow the identification of very complex mixtures, even on very simple devices.

For example, a peak at m/z = 918.5 corresponds to the Cs$^+$ adduct of PC 36:2, but if *only* this information is available it is simply a matter of speculation as to whether it is PC 18:1/18:1 or PC 18:0/18:2. It is also not clear which fatty acid is located in which position. Digestion of the PC sample with PLA$_2$ would rapidly provide this information, since in the first case LPC 18:1 (m/z = 522.3) would be the only product, whereas in the second case LPC 18:0 (m/z = 524.3) is the only product when the saturated fatty acid is located in *sn*-1 position, while LPC 18:2 (m/z = 520.3) is generated when the unsaturated fatty acid is located in *sn*-2 position. Performing peak assignment in such a manner is often more straightforward than using more sophisticated MS methods that are not available on all devices.

7.6
Summary and Outlook

There is growing evidence that MALDI-TOF MS is a very useful technique in lipid research. Although in the past the relevance of lipids has been neglected in favor of proteins and molecular biology, the interest in lipids and their analysis is steadily increasing. Today, in addition to terms such as "proteomics" or "metabonomics", lipidomics has entered the scientific vocabulary and is now in common use [108].

Separation into individual fractions appears to be a prerequisite for the unambiguous mass assignment of complex mixtures, and as ESI may be more easily coupled with chromatography, ESI MS has become the preferred approach.

When MALDI is used without previous separation, the spectra are dominated by PLs with quaternary ammonium groups (i.e., PC, LPC and SM), whereas the other PLs are suppressed. This is especially problematic for PE as this PL does not provide significant amounts of negative ions, whereas PS – for example – can easily be detected as a negative ion without interference from PC or SM. It should also be noted, however, that positive- and negative-ion detection modes are not independent – that is, large amounts of PC also reduce the detection limits of the negative-ion mode. Nevertheless, for some medical applications it may be sufficient to have information available on a special PL class (e.g., PC) as this may allow conclusions to be drawn about metabolic processes and the activities of related enzymes. Finally, the routine coupling of MALDI-TOF MS with TLC provides a number of advantages. Typically, separation by TLC is very fast, with rapid identification of lipids by spraying with a suitable dye. Less-resolved samples might require re-application to the TLC separation. Moreover, as TLC plates are discarded after use, then memory effects are invalid. Further advantages include the possibility of analyzing several samples simultaneously, the option of re-analyzing the analyte spots, and last – but not least – the minimal instrumentation required. Thus, it is the present author's opinion that coupling between TLC and MALDI will in time become a highly favorable method and help to overcome the current problems related to lipid analysis by MALDI-TOF MS.

Acknowledgments

The author thanks all colleagues and friends who helped in the writing of this chapter. Special thanks go to R. Süß for performing the majority of MALDI-TOF MS experiments and TLC separations. Many fruitful discussions with Dr. Beate Fuchs were very helpful, and these are gratefully acknowledged. Finally, the authors thanks Prof. Dr. Klaus Arnold for the continuous support of these investigations.

These studies were supported by the Deutsche Forschungsgemeinschaft (Schi 476/5-1) and the Federal Ministry of Education and Research (Grant "MS CartPro", 982000-041).

Abbreviations

APCI	atmospheric pressure chemical ionization
DAG	diacylglycerols
DE	delayed extraction
DHB	2,5-dihydroxybenzoic acid
DIOS	desorption ionization on porous silicon
EI	electron impact
ESI	electrospray ionization
FAB	fast atom bombardment
FD	field desorption
FI	field ionization
FT	Fourier transform
GC/MS	gas chromatography/mass spectrometry
HDL	high-density lipoprotein

HPLC	high-performance liquid chromatography
LD	laser desorption
LDL	low-density lipoprotein
LPA	lyso-phosphatidic acid
LPC	lyso-phosphatidylcholine
LPL	lyso-phospholipid
M	molecular mass
MALDI	matrix-assisted laser desorption and ionization
MS	mass spectrometry
NMR	nuclear magnetic resonance
PA	phosphatidic acid
PC	phosphatidylcholine
PD	plasma desorption
PE	phosphatidylethanolamine
PI	phosphatidylinositol
PL	phospholipid
PPI	(poly)-phosphoinositides
PS	phosphatidylserine
PG	phosphatidylglycerol
PLA$_2$	phospholipase A$_2$
PLC	phospholipase C
ROS	reactive oxygen species
SM	sphingomyelin
sn	stereospecific numbering
TAG	triacylglycerol
TFA	trifluoroacetic acid
TLC	thin-layer chromatography
TOF	time-of-flight

References

1 A. Nicolaou, G. Kokotos. *Bioactive Lipids*, Oily Press, UK, **2004**.

2 G.I. Siegel. *Basic Neurochemistry: Molecular, Cellular and Medical Aspects*, Lippincott-Raven Publishers, USA, **1999**.

3 M.N. Hodgkin, T.R. Pettitt, A. Martin, R.H. Mitchell, A.J. Pemberton, M.J. Wakelam. *Trends Biochem. Sci.* **1998**, *23*, 200–204.

4 B. Samuelsson, M. Goldyne, E. Granstrom, M. Hamberg, S. Hammarstrom, C. Malmsten. *Annu. Rev. Biochem.* **1978**, *47*, 997–1029.

5 D. Riendeau, E. Meighen. *Experientia* **1985**, *41*, 707–713.

6 W.W. Christie. *Lipid Analysis*, Oily Press, UK, **2003**.

7 J.C. Touchstone. *J. Chromatogr. B* **1995**, *671*, 169–195.

8 R.H. McCluer, M.D. Ullman, F.B. Jungalwala. *Adv. Chromatogr.* **1986**, *25*, 309–353.

9 P. Pollesello, O. Eriksson, K. Höckerstedt. *Anal. Biochem.* **1996**, *236*, 41–48.

10 J. Schiller, K. Arnold. *Med. Sci. Monitor* **2002**, *8*, MT205–MT222.

11 J.M. Pearce, R.A. Komoroski. *Magn. Reson. Med.* **1993**, *29*, 724–731.

12 J.M. Pearce, R.A. Komoroski. *Magn. Reson. Med.* **2000**, *44*, 215–223.

13 R.A. Klein. *J. Lipid Res.* **1971**, *12*, 628–634.

14 R.A. Klein. *J. Lipid Res.* **1971**, *12*, 123–131.

15 G.W. Wood, P.Y. Lau. *Biomed. Mass Spectrom.* **1974**, *1*, 154–155.

16 P.A. Demirev. *Biomed. Environ. Mass Spectrom.* **1987**, *14*, 241–246.

17 C.G. Crawford, R.D. Plattner. *J. Lipid Res.* **1983**, *24*, 456–460.

18 N.J. Jensen, K.B. Tomer, M.L. Gross. *Lipids* **1986**, *21*, 58–588.

19 K.A. Kayganich-Harrison, R.C. Murphy. *Anal. Biochem.* **1994**, *221*, 16–24.

20 J. Schiller, K. Arnold. in: *Encyclopedia of Analytical Chemistry*, Wiley, USA, **2000**, pp. 559–585.

21 W.W. Christie. *Lipids* **1998**, *33*, 343–353.

22 H.Y. Kim, N. Salem, Jr. *Prog. Lipid Res.* **1993**, *32*, 221–245.

23 M. Pulfer, R.C. Murphy. *Mass Spectrom. Rev.* **2003**, *22*, 332–364.

24 J. Schiller, J. Arnhold, S. Benard, M. Müller, S. Reichl, K. Arnold. *Anal. Biochem.* **1999**, *267*, 46–56.

25 R. Estrada, M.C. Yappert. *J. Mass Spectrom.* **2004**, *39*, 412–422.

26 K.H. Park, H.J. Kim. *Rapid Commun. Mass Spectrom.* **2001**, *15*, 1494–1499.

27 Y.L. Li, M.L. Gross, F.F. Hsu. *J. Am. Soc. Mass Spectrom.* **2005**, *16*, 679–682.

28 S. Benard, J. Arnhold, M. Lehnert, J. Schiller, K. Arnold. *Chem. Phys. Lipids* **1999**, *100*, 115–125.

29 J. Schiller, R. Süß, M. Petković, N. Hilbert, M. Müller, O. Zschörnig, J. Arnhold, K. Arnold. *Chem. Phys. Lipids* **2001**, *113*, 123–131.

30 J. Folch, M. Lees, G.H.S. Stanley. *J. Biol. Chem.* **1957**, *226*, 497–509.

31 S. Schuck, M. Honsho, K. Ekroos, A. Shevchenko, K. Simons. *Proc. Natl. Acad. Sci. USA* **2003**, *100*, 5795–5800.

32 J. Schiller, R. Süß, J. Arnhold, B. Fuchs, J. Leßig, M. Müller, M. Petković, H. Spalteholz, O. Zschörnig, K. Arnold. *Prog. Lipid Res.* **2004**, *43*, 449–488.

33 M. Petković, J. Schiller, M. Müller, S. Benard, S. Reichl, K. Arnold, J. Arnhold. *Anal. Biochem.* **2001**, *289*, 202–216.

34 G.R. Asbury, K. Al-Saad, W.F. Siems, R.M. Hannan, H.H. Hill. *J. Am. Soc. Mass Spectrom.* **1999**, *10*, 983–991.

35 K.A. Al-Saad, V. Zabrouskov, W.F. Siems, N.R. Knowles, R.M. Hannan, H.H. Hill, Jr. *Rapid Commun. Mass Spectrom.* **2003**, *17*, 87–96.

36 P. Zollner, E.R. Schmid, G. Allmaier. *Rapid Commun. Mass Spectrom.* **1996**, *10*, 1278–1282.

37 C.D. Cosima, F. Palmisano, C.G. Zambonin. *Rapid Commun. Mass Spectrom.* **2005**, *19*, 1315–1320.

38 J. Schiller, J. Arnhold, H.-J. Glander, K. Arnold. *Chem. Phys. Lipids* **2000**, *106*, 145–156.

39 J. Schiller, O. Zschörnig, M. Petković, M. Müller, J. Arnhold, K. Arnold. *J. Lipid Res.* **2001**, *42*, 1501–1508.

40 O. Zschörnig, M. Pietsch, R. Süß, J. Schiller, M. Gütschow. *J. Lipid Res.* **2005**, *46*, 803–811.

41 K.A. Al-Saad, W.F. Siems, H.H. Hill, V. Zabrouskov, N.R. Knowles. *J. Am. Soc. Mass Spectrom.* **2003**, *14*, 373–482.

42 V. Zabrouskov, K.A. Al-Saad, W.F. Siems, H.H. Hill, N.R. Knowles. *Rapid Commun. Mass Spectrom.* **2001**, *15*, 935–940.

43 D.J. Harvey. *J. Mass Spectrom.* **1995**, *30*, 1333–1346.

44 J.A. Marto, F.M. White, S. Seldomridge, A.G. Marshall. *Anal. Chem.* **1995**, *67*, 3979–3984.

45 M. Koivusalo, P. Haimi, L. Heikinheimo, R. Kostiainen, P. Somerharju. *J. Lipid Res.* **2001**, *42*, 663–672.

46 R. Estrada, M.C. Yappert. *J. Mass Spectrom.* **2004**, *39*, 1531–1540.

47 M. Petković, J. Müller, M. Müller, J. Schiller, K. Arnold, J. Arnhold. *Anal. Biochem.* **2002**, *308*, 61–70.

48 T. Fujiwaki, S. Yamaguchi, M. Tasaka, M. Takayanagi, M. Isobe, T. Taketomi. *J. Chromatogr. B* **2004**, *806*, 47–51.

49 T. Tanaka, H. Tsutsui, K. Hirano, T. Koike, A. Tokumura, K. Satouchi. *J. Lipid Res.* **2004**, *45*, 2145–2150.

50 M. Müller, J. Schiller, M. Petković, W. Oehrl, R. Heinze, R. Wetzker, K. Arnold, J. Arnhold. *Chem. Phys. Lipids* **2001**, *110*, 151–164.

51 O.I. Shadyro, I.L. Yurkova, M.A. Kisel, O. Brede, J. Arnhold. *Int. J. Radiat. Biol.* **2004**, *80*, 239–245.

52 C. Schaffer, A.I. Beckedorf, A. Scheberl, S. Zayni, J. Peter-Katalinić, P. Messner. *J. Bacteriol.* **2002**, *184*, 6709–6713.

53 F.O. Ayorinde, P. Hambright, T.N. Porter, Q.L. Keith. *Rapid Commun. Mass Spectrom.* **1999**, *13*, 2474–2479.

54 F.O. Ayorinde, K. Garvin, K. Saeed. *Rapid Commun. Mass Spectrom.* **2000**, *14*, 608–615.

55 F. Guyon, C. Absalon, A. Eloy, M.H. Salagoity, M. Esclapez, B. Medina. *Rapid Commun. Mass Spectrom.* **2003**, *17*, 2317–2322.

56 F.O. Ayorinde, Q.L. Keith, L.W. Wan. *Rapid Commun. Mass Spectrom.* **1999**, *13*, 1762–1769.

57 G. Stubinger, E. Pittenauer, G. Allmaier. *Phytochem. Anal.* **2003**, *14*, 337–346.

58 C. Hlongwane, I.G. Delves, L.W. Wan, F.O. Ayorinde. *Rapid Commun. Mass Spectrom.* **2001**, *15*, 2027–2034.

59 J. Schiller, R. Süß, M. Petković, K. Arnold. *J. Food Lipids* **2002**, *9*, 185–200.

60 A. Jakab, K. Nagy, K. Heberger, K. Vekey, E. Forgacs. *Rapid Commun. Mass Spectrom.* **2002**, *16*, 2291–2297.

61 J. Schiller, R. Süß, M. Petković, K. Arnold. *Eur. Food Res. Technol.* **2002**, *215*, 282–286.

62 J. Schiller, R. Süß, M. Petković, G. Hanke, A. Vogel, K. Arnold. *Eur. J. Lipid Sci. Technol.* **2002**, *104*, 496–505.

63 H. Hayen, D.A. Volmer. *Rapid Commun. Mass Spectrom.* **2005**, *19*, 711–720.

64 J. Cvacka, A. Svatos. *Rapid Commun. Mass Spectrom.* **2003**, *17*, 2203–2207.

65 N.P. Raju, S.P. Mirza, M. Vairamani, A.R. Ramulu, M. Pardhasaradhi. *Rapid Commun. Mass Spectrom.* **2001**, *15*, 1879–1884.

66 K. Viehweger, B. Dordschbal, W. Roos. *Plant Cell* **2002**, *14*, 1509–1525.

67 J. Schiller, R. Süß, B. Fuchs, M. Müller, O. Zschörnig, K. Arnold. *Chromatographia* **2003**, *57*, S297–S302.

68 T. White, S. Bursten, D. Federighi, R.A. Lewis, E. Nudelman. *Anal. Biochem.* **1998**, *258*, 109–117.

69 S.N. Jackson, H.Y.J. Wang, A.S. Woods. *J. Am. Soc. Mass Spectrom.* **2005**, *16*, 133–138.

70 V.B. Ivleva, Y.N. Elkin, B.A. Budnik, S.C. Moyer, P.B. O'Connor, C.E. Costello. *Anal. Chem.* **2004**, *76*, 6484–6491.

71 S.N. Jackson, H.-Y.J. Wang, A.S. Woods. *Anal. Chem.* **2005**, *77*, 4523–4527.

72 M. Rujoi, R. Estrada, M.C. Yappert. *Anal. Chem.* **2004**, *76*, 1657–1663.

73 M. Rujoi, J. Jin, D. Borchman, D. Tang, M.C. Yappert. *Invest. Ophthalmol. Vis. Sci.* **2003**, *44*, 1634–1642.

74 M.C. Yappert, M. Rujoi, D. Borchman, I. Vorobyov, R. Estrada. *Exp. Eye Res.* **2003**, *76*, 725–734.

75 E.M. Oborina, M.C. Yappert. *Chem. Phys. Lipids* **2003**, *123*, 223–232.

76 D. Touboul, H. Piednoel, V. Voisin, S. De La Porte, A. Brunelle, F. Halgand, O. Laprevote. *Eur. J. Mass Spectrom.* **2004**, *10*, 657–664.

77 E.G. Bligh, W.J. Dyer. *Can. J. Biochem. Physiol.* **1959**, *37*, 911–917.

78 http://www.cyberlipid.org/extract/extr0002.htm#top

79 A. Hara, N.S. Radin. *Anal. Biochem.* **1978**, *90*, 420–426.

80 J.J. Jones, M.J. Stump, R.C. Fleming, J.O. Lay, C.L. Wilkins. *Anal. Chem.* **2003**, *75*, 1340–1347.

81 J.J. Jones, M.J. Stump, R.C. Fleming, J.O. Lay, C.L. Wilkins. *J. Am. Soc. Mass Spectrom.* **2004**, *15*, 1665–1674.

82 Y. Ishida, A.J. Madonna, J.C. Rees, M.A. Meetani, K.J. Voorhees. *Rapid Commun. Mass Spectrom.* **2002**, *16*, 1877–1882.

83 Y. Ishida, O. Nakanishi, S. Hirao, S. Tsuge, J. Urabe, T. Sekino, M. Nakanishi, T. Kimoto, H. Ohtani. *Anal. Chem.* **2003**, *75*, 4514–4518.

84 K. Liliom, G.P. Sun, M. Bunemann, T. Virag, N. Nusser, D.L. Baker, D.A. Wang, M.J. Fabian, B. Brandts, K. Bender, A. Eickel, K.U. Malik, D.D. Miller, D.M. Desiderio, G. Tigyi, L. Pott. *Biochem. J.* **2001**, *355*, 189–197.

85 B. Fuchs, J. Schiller, U. Wagner, H. Häntzschel, K. Arnold. *Clin. Biochem.* **2005**, *38*, 925–933.

86 O. Zschörnig, C. Bergmeier, R. Süß, K. Arnold, J. Schiller. *Lett. Org. Chem.* **2004**, *1*, 381–390.

87 B.W. Shen, A.M. Scanu, F.J. Kezdy. *Proc. Natl. Acad. Sci. USA* **1977**, *74*, 837–841.

88 H.J. Glander, J. Schiller, R. Süß, U. Paasch, S. Grunewald, J. Arnhold. *Andrologia* **2002**, *34*, 360–366.

89 R.C. Murphy. *Mass spectrometry of phospholipids: tables of molecular and product ions.* Illuminati Press, USA, **2002**.

90 J. Schiller, K. Müller, R. Süß, J. Arnhold, C. Gey, A. Herrmann, J. Leßig, K. Arnold, P. Müller. *Chem. Phys. Lipids* **2003**, *126*, 85–94.

91 J. Leßig, C. Gey, R. Süß, J. Schiller, H.-J. Glander, J. Arnhold. *Comp. Biochem. Physiol. B* **2004**, *137*, 265–2677.

92 J. Schiller, B. Fuchs, J. Arnhold, K. Arnold. *Curr. Med. Chem.* **2003**, *10*, 2123–2145.

93 B.H. Kim, Y.S. Chang, B.D. Lee, S.H. Ryu, D.H. Shin. *Microchem. J.* **1999**, *63*, 3–8.

94 J. Müller, M. Petković, J. Schiller, K. Arnold, S. Reichl, J. Arnhold. *Luminescence* **2002**, *17*, 141–149.

95 J. Müller, M. Petković, J. Schiller, J. Arnhold. *Z. Naturforsch.* **2001**, *56*, 1150–1156.

96 M. Petković, A. Vocks, J. Schiller, J. Arnhold. *Physiol. Res.* **2005**, *54*, 105–113.

97 J. Arnhold, S. Benard, U. Kilian, S. Reichl, J. Schiller, K. Arnold. *Luminescence* **1999**, *14*, 129–137.

98 M. Petković, A. Vocks, M. Müller, J. Schiller, J. Arnhold. *Z. Naturforsch.* **2005**, *60c*, 143–151.

99 J. Arnhold, A.N. Osipov, H. Spalteholz, O.M. Panasenko, J. Schiller. *Free Radic. Biol. Med.* **2001**, *31*, 1111–1119.

100 J. Arnhold, A.N. Osipov, H. Spalteholz, O.M. Panasenko, J. Schiller. *Biochim. Biophys. Acta* **2002**, *1572*, 91–100.

101 O.F. van den Brink, J.J. Boon, P.B. O'Connor, M.C. Duursma, R.M. Heeren. *J. Mass Spectrom.* **2001**, *36*, 479–492.

102 M. Karas, R. Kruger. *Chem. Rev.* **2003**, *103*, 427–440.

103 L.H. Cohen, A.I. Gusev. *Anal. Bioanal. Chem.* **2002**, *373*, 571–586.

104 T.R. Klein, D. Kirsch, R. Kaufmann, D. Riesner. *Biol. Chem.* **1998**, *379*, 655–666.

105 M. Petković, J. Schiller, J. Müller, M. Müller, K. Arnold, J. Arnhold. *Analyst* **2001**, *126*, 1042–1050.

106 Z. Shen, J.J. Thomas, C. Averbuj, K.M. Broo, M. Engelhard, J.E. Crowell, M.G. Finn, G. Siuzdak. *Anal. Chem.* **2001**, *73*, 612–619.

107 V.V. Laiko, S.C. Moyer, R.J. Cotter. *Anal. Chem.* **2000**, *72*, 5239–5243.

108 M.R. Wenk. *Nat. Rev. Drug Discov.* **2005**, *4*, 594–610.

8
MALDI-MS for Polymer Characterization

Liang Li

8.1
Introduction

With the development of new polymerization chemistry, catalysis and formulation processes, a great number of polymeric materials with diverse properties can be produced. A detailed characterization of these materials is important to relate their chemical structure and composition to their functions. For example, modification of the end groups of a polymer can significantly alter its characteristics, such as chemical reactivity, solubility, and miscibility with other chemicals. Polymer characterization is not a simple task and often involves the use of multiple analytical techniques, with each generating a piece of useful information that is necessary to provide a comprehensive interrogation of the polymer. A number of analytical techniques, including chromatographic methods, spectroscopy, and mass spectrometry (MS), have been developed and applied to study areas such as polymer structure, polymer composition, molecular mass and molecular mass distribution, bulk and surface properties and impurity content [1–4].

During recent years, MS has become an indispensable tool for polymer characterization [5,6], and today can produce rich chemical information that is highly specific for polymer structural analysis. MS is also very sensitive, allowing the detection and identification of minor polymer components or impurities in a composed polymeric material, and any byproducts of polymerization reactions of a desired polymer. In addition, MS can potentially provide quantitative information required to determine the average molecular mass and molecular mass distribution of a polymer, or to characterize the relative amounts of the different components of a polymer mixture. Some forms of MS, such as secondary ion mass spectrometry (SIMS), can also be used to characterize polymer surfaces [7].

Many different types of MS technique have been used for polymer characterization [5]. Before the introduction of matrix-assisted laser desorption ionization (MALDI) and electrospray ionization (ESI) during the 1980s, MS was limited to the analysis of relatively low mass polymers of less than 3000 Da. Desorption techniques such as SIMS, fast atom bombardment (FAB), and laser desorption/ionization (LDI) could ionize a polymer with mass of up to 10000 Da.

MALDI MS. A Practical Guide to Instrumentation, Methods and Applications.
Edited by Franz Hillenkamp and Jasna Peter-Katalinić
Copyright © 2007 Wiley-VCH Verlag GmbH & Co. KGaA, Weinheim
ISBN: 978-3-527-31440-9

However, this type of analysis was limited to some favorable polymeric systems, such as poly(ethylene glycol)s. Moreover, even for these polymers the analysis was not routine and usually conducted by an experienced researcher.

MALDI-MS transforms the practice of polymer characterization [8–12], and today has become a widely used technique for the analysis of a huge variety of polymers. There are several unique attributes of MALDI-MS which, together, make it a powerful technique for polymer characterization. In this chapter, these attributes will be discussed, along with many technical issues related to the use of MALDI-MS for polymer characterization. A few selected applications of MALDI-MS and MS/MS will also be outlined in order to illustrate the power of the technique in solving practical problems. This chapter does not aim to survey all published studies in the area of MALDI polymer characterization; rather, it attempts to provide an overview on the technique, attributes, and current limitations of MALDI-MS for polymer analysis.

8.2
Technical Aspects of MALDI-MS

MALDI can be used to generate ions from many different types of polymers. In general, a polymer sample is mixed with a suitable matrix and a cationization reagent, such as AgNO$_3$. The mixture is deposited onto a MALDI target which is then inserted into a mass spectrometer for MALDI. For many polymers, the individual oligomer is ionized during the MALDI process by the attachment of a cation (cationization) which, in most cases, takes place in the gas phase [13–16].

Unlike biopolymers, where the sample amount is often limited, the sample size of an industrial polymer is generally not a problem for MS analysis. It is, in fact, a relatively easy task to generate a mass spectrum from a polymer sample. However, it is certainly less easy to produce the correct mass spectrum that reflects the true oligomer distribution and composition of a polymer sample. Thus, an understanding of the technical issues related to the entire process of MALDI mass spectrum acquisition and data processing becomes very important. It can assist in optimizing the experimental procedure for producing a correct spectrum, as well as evaluating the results obtained under any given condition. There are many issues governing the success of polymer analysis by MALDI-MS, and the most important of these are grouped into three categories, namely sample preparation, instrumentation, and data processing (Fig. 8.1). These issues are discussed in detail in the following sections.

8.2.1
Sample Preparation Issues

Sample preparation plays the key role in producing a reliable MALDI mass spectrum of a polymer [17–24]. In sample preparation, depending on the nature of a polymer sample, the type of matrix, cationization reagent, and solvent or solvent

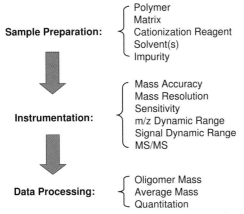

Sample Preparation:
{
Polymer
Matrix
Cationization Reagent
Solvent(s)
Impurity
}

Instrumentation:
{
Mass Accuracy
Mass Resolution
Sensitivity
m/z Dynamic Range
Signal Dynamic Range
MS/MS
}

Data Processing:
{
Oligomer Mass
Average Mass
Quantitation
}

Fig. 8.1 Major issues related to MALDI analysis of polymers.

mixture needs to be properly selected. (The solvent-free sample preparation method is discussed in Section 8.2.1.4.) A number of matrices, including 4-hydroxy-α-cyanocinnamic acid (HCCA) [8–12], 2,5-dihydroxybenzoic acid (DHB) [8–12], 2-(4-hydroxyphenylazo)benzoic acid (HABA) [25], trans-3-indoleacrylic acid (IAA) [26], 9-nitroanthracene [27], 1,8,9-trihydroxyanthracene (dithranol) [27,28], all-trans retinoic acid (RA) [29], pentafluorobenzoic acid [30,31], pentafluorocinnamic acid [30], 7,7,8,8-tetracyanoquinodimethane [32], and others [27,33–37], have been found to be useful for polymer MALDI-MS.

A typical experiment of analyzing a polymer sample starts with screening and selecting a suitable matrix from a list of known matrices that have been shown to be effective for the same or similar type of polymer. This usually involves four steps. The first step is to identify a solvent system that will dissolve both the polymer and the matrix; the second step is to find a suitable cationization reagent dissolved in a solvent that is compatible with the polymer/matrix solvent system. Next, about $1\,\mu L$ of the mixture is taken and deposited onto a MALDI target and, after drying, the crystal morphology is carefully examined. Finally, the target is inserted into a mass spectrometer to acquire MALDI mass spectra using optimal instrumental conditions.

8.2.1.1 Matrix

Because there is great diversity among polymers in terms of their chemical structure and composition, there is no single MALDI matrix that can be universally applied for mass analysis. For some polymers, several different matrices can be used to generate the MALDI mass spectra from the same polymer sample. However, not all matrices will produce the same results. In some cases, the effect of matrix type on the quality of MALDI spectra can be very dramatic. Figure 8.2 shows an example of analyzing poly(ethylene glycol) 15000 by MALDI time-of-flight (TOF) MS using different matrices [38]. This figure is separated into two mass ranges: the top of the figure shows the region between m/z 0 to 10000, and the

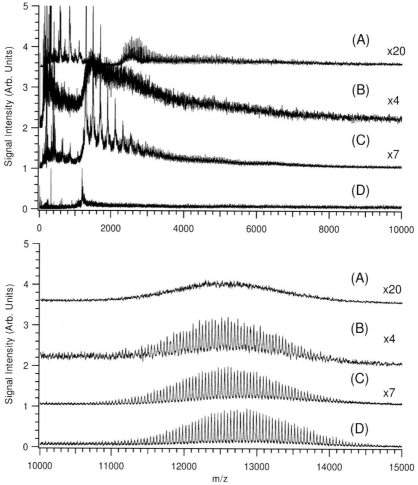

Fig. 8.2 MALDI mass spectra of poly(ethylene glycol) 15 000 obtained by operating the instrument in (A) continuous extraction mode with HABA as the matrix and sodium cationization, and (B–D) time-lag focusing extraction mode with (B) DHB, (C) IAA, and (D) HABA as the matrices and sodium cationization. Reprinted from Ref. [38] by permission of American Chemical Society; © American Chemical Society, 1997.

bottom between m/z 10 000 and 15 000. Figure 8.2(A) was obtained using HABA as matrix and continuous extraction; individual oligomers are not resolved above 9000 Da using the 1-meter linear TOF instrument. In contrast, the individual oligomers are well resolved in Figure 8.2(B–D), where they were collected using time-lag focusing (TLF) or delayed extraction (DE) with DHB, IAA, and HABA as matrices, respectively. Figure 8.2 shows that the type of matrix used can significantly affect the apparent mass resolution in TLF MALDI. At 12 700 Da, the resolution is 700, 1060, and 1400 fwhm for Figure 8.2(B–D), respectively. Lower mass oligomers are not resolved with DHB under the DE conditions used, whereas oligomers are resolved across the whole spectrum using IAA and HABA as matrices. In addition, the overall detection sensitivity, judged from the signal to baseline-noise (S/N) ratios, is higher with HABA or IAA than with DHB. HABA provides a flat baseline, extending to the low-mass region, while the baselines elevate in the low-mass region for the spectra obtained with DHB and IAA. Intense matrix clusters are observed with IAA extending to m/z 3000, obscuring polymer peaks in this mass range. The extent of clustering observed with IAA could be reduced with a reduction in laser power, but this would result in a complete loss of the polymer signal.

The reason for the difference in resolution for polymer spectra obtained with different matrices is not obvious. All of the matrix preparations gave densely packed microcrystals on the probe surface. Well-packed microcrystals give improved mass accuracy, resolution, and shot-to-shot reproducibility over preparations that produce large crystals [39,40]. DHB required twice the laser power used for HABA and IAA, and the shot-to-shot reproducibility was significantly poorer; lower sensitivity and resolution can be expected. However, the sensitivity and shot-to-shot reproducibility of HABA and IAA are similar. It can be speculated that the high abundance of matrix clusters observed with IAA indicates a higher charge density in the desorption plume, leading to increased Coulombic repulsion and an unanticipated spatial distribution that cannot be compensated with TLF – that is, there is no longer an exact relationship between an ion's position and its velocity at the time the extraction pulse is applied and slight peak broadening results.

The example shown in Figure 8.2 illustrates that the selection of a proper matrix is important to generate a MALDI spectrum reflective of the polymer sample composition. In this case, if only DHB were used, the MALDI spectrum produced would not reveal the low mass oligomers (m/z <10 000) actually present in this sample. As only a handful of matrices are found to be practically useful for polymer analysis, it is often worthy spending the time to test these matrices on a given sample to identify the best matrix that provides good sensitivity, mass resolution, and reproducibility over a broad mass range.

8.2.1.2 Cationization Reagent

Unlike protein analysis by MALDI, in which analyte ionization is generally achieved by protonation, the ionization of industrial or synthetic polymers is achieved usually through other monovalent cations such as Na^+, K^+, and Ag^+. Selection of the correct cationization reagent is crucial to allow for a large mass range of analysis and uniformity of detection from oligomer to oligomer [15,16,18,41–55]. An example of the effect of cationization reagent on MALDI analysis is shown by the spectra for polyisoprene 3280, obtained by using $AgNO_3$ and $Cu(NO_3)_2$ as cationization reagents and all-trans-retinoic acid as matrix [42] (Fig. 8.3). Here, the S/N ratio is significantly greater in the analysis with copper ions, allowing for a more precise determination of oligomer mass and average molecular mass. With $Cu(NO_3)_2$ as the cationization reagent, the determination of

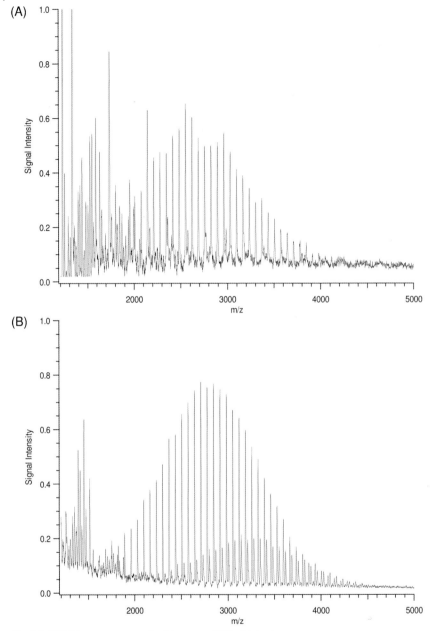

Fig. 8.3 MALDI mass spectra of polyisoprene 3280 obtained using *all-trans*-retinoic acid with (A) AgNO$_3$ and (B) Cu(NO$_3$)$_2$ as the cationization reagents. Reprinted by permission of Elsevier from Ref. [42]; © American Society for Mass Spectrometry, 1997.

individual oligomer masses indicates that the principal distribution is generated from Cu$^+$ ion attachment. The generation of [polyisoprene+Cu]$^+$ species is most likely induced upon desorption into the gas phase. Laser irradiation of the solid sample would generate a gas plume containing free electrons, and radical forms of the matrix capable of reducing Cu^{2+} to Cu$^+$. The Cu$^+$ ion forms a complex with the double bond(s) in the polyisoprene chain. Interestingly, AgNO$_3$ is a much better cationization reagent than Cu(NO$_3$)$_2$ for polystyrene analysis by MALDI. The aromatic rings in the polystyrene chain provide a favorable attachment site for the silver ion during the desorption/ionization process.

The need for a cationization reagent in MALDI analysis of polymers can also create some complications in mass spectral interpretation [42,56]. For example, the spectrum in Figure 8.3(B) shows a secondary distribution of lower intensity in addition to the principal distribution. This secondary distribution could be due to cation adduction with different ionic species and/or the presence of other polymeric species with different end-group structures. In this case, the secondary distribution has oligomer mass shifts of +22.4 Da from the nearest oligomer of lower mass in the principal distribution. This is consistent with the generation of salt cluster complexes, similar to what has been observed in the ESI of polystyrene [57]. For polyisoprene, it takes the form of [polyisoprene+Cu(copper retinoate)]$^+$. This amounts to an actual mass shift of +363.0 Da with respect to the principal distribution of [polyisoprene+Cu]$^+$. This is consistent with the observed mass shift of 22.4 Da plus six repeat units of 64.2 Da.

It is also possible for a polymer to form adducts involving the counterion of the added inorganic salt. Figure 8.4 illustrates the comparative adduction behavior of polyisoprene 3280 resulting from different cationization reagents. In Figure 8.4(A) there are four peaks in the secondary distribution arising from cationization with Cu(NO$_3$)$_2$, with an apparent mass shift of 22.4 Da from the principal distribution. Doping molar equivalents of KCl and Cu(NO$_3$)$_2$ into the polymer/matrix solution gives rise to a spectrum showing two secondary distributions (Fig. 8.4(B)). One distribution is isobaric with the secondary distribution of Figure 8.4(A). The other, with an apparent mass shift of 30.9 Da, is isobaric with the secondary distribution of Figure 8.4(C). This latter spectrum was obtained by using CuCl$_2$ as the cationization reagent. With an apparent mass shift of 30.9 Da, this is consistent with a structural assignment of [polyisoprene+Cu(CuCl)]$^+$, in that the actual mass shift generated from such adduction behavior is 99.0 Da. Figure 8.4(B) strongly suggests that both [polyisoprene+Cu(copper retinoate)]$^+$ and [polyisoprene+Cu(CuCl)]$^+$ are present. A further implication of this series of spectra is that chloride ion competes effectively with the retinoate anion in the adduction process.

It is clear that selecting the correct cationization reagent and understanding the adduction behavior of the reagent used are important in order to generate high-quality MALDI mass spectra that can be correctly interpreted.

8.2.1.3 Solvent

The type and quality of solvents can influence the MALDI analysis of polymer samples. For example, the dryness and purity of tetrahydrofuran (THF) used to

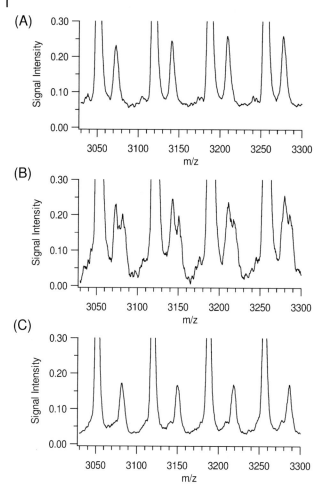

Fig. 8.4 The effect of cationization reagent on the appearance of the secondary distribution in polyisoprene 3280, using *all-trans*-retinoic acid with (A) Cu(NO$_3$)$_2$, (B) Cu(NO$_3$)$_2$ with KCl and (C) CuCl$_2$. Reprinted by permission of Elsevier from Ref. [42]; © American Society for Mass Spectrometry, 1997.

prepare polymer samples play a central role in the success of detecting high-molecular mass polymers [29]. The solvent system used can affect analyte incorporation and distribution in matrix crystals. As has been shown in MALDI biopolymer analysis, analyte distribution in matrix crystals can significantly affect the signal reproducibility, detection sensitivity, and relative intensities of individual components in a mixture [40]. However, unlike biopolymer analysis – where a common solvent can be often found to dissolve both the analyte and matrix – the choice of a solvent system for polymer analysis by MALDI is much more critical [17,20,31,58–65]. In particular, solvents used to dissolve polymers may not be compatible with the matrix or cationization reagent. The current (and still recommended) practice is that, whenever possible, a single solvent system should

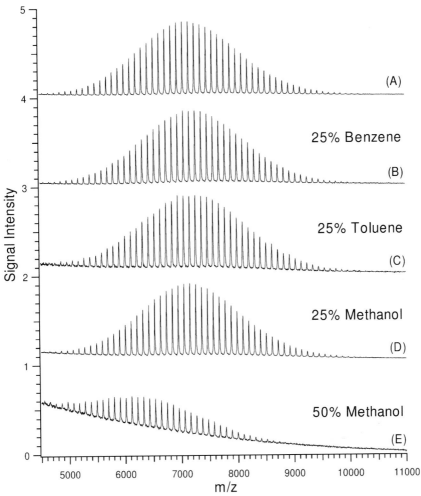

Fig. 8.5 MALDI mass spectra of polystyrene 7000 obtained by using different solvent systems for sample preparation. (A) 99.5% THF:0.5% ethanol; (B) 25% benzene:74.5% THF:0.5% ethanol; (C) 25% toluene:74.5% THF:0.5% ethanol; (D) 25% methanol:74.5% THF:0.5% ethanol; and (E) 50% methanol: 49.5% THF:0.5% ethanol. *All-trans*-retinoic acid was used as the matrix and silver nitrate as the cationization reagent. Reprinted by permission of Elsevier from Ref. [60]; © American Society for Mass Spectrometry, 1998.

be sought to prepare the polymer/matrix sample. However, for a number of applications, the use of a solvent mixture cannot be avoided. In this case, the choice of solvents becomes an important issue in the development of a useful sample/ matrix preparation protocol.

The effect of solvent or solvent mixture on MALDI analysis of polymers appears to follow a systematic pattern [60]. A solvent mixture containing a polymer non-solvent gives rise to poor reproducibility and erroneous average molecular mass results. For example, Figure 8.5 shows five MALDI mass spectra of polystyrene

(PS) 7000 obtained by using different solvents for sample preparation [60]. The solvent system used for producing the spectrum shown in Figure 8.5(A) consisted of 0.5% ethanol and 99.5% THF. For the spectra shown in Figure 8.5(B–D), all experimental conditions were the same as those used for Figure 8.5(A) except that the solvent systems consisted of 0.5% ethanol, 74.5% THF, and 25% benzene, or toluene, or methanol. The average masses (M_n and M_w) and polydispersity (PD) values obtained for these samples by MALDI are listed in Table 8.1. The relative differences in average masses obtained from different solvent systems are also shown. The average molecular mass values obtained from the gel permeation chromatography (GPC) for this polymer are $M_n = 6770$, $M_w = 6962$, and PD = 1.03. It should be noted that, whilst the MALDI results are close to the GPC data in this case, any direct comparison of values generated from these two different techniques should be made with care, even for narrow-polydispersity polymers [66–69]. There are a number of examples showing significant differences between the GPC and MALDI data [70–74], and this is particularly true in cases where polymer standards do not exist for calibrating GPC. For example, GPC data of a novel polymer with different structure or composition from the polymer standards are not reliable due to the retention difference between the analyte and the standards. However, in MALDI, no polymer standards are required to calibrate the instrument, thus potentially offering accurate mass measurement of a polymer.

The data in Table 8.1 indicate that using a solvent mixture containing 25% benzene, toluene or methanol does not affect the measurement of M_n and M_w of PS 7000. The differences are well within the statistical errors at the 99% confidence limit. However, a different spectrum was obtained in the case of using 50% methanol as the third solvent (see Fig. 8.5(E)). This figure shows a severe mass discrimination at the high mass region of the polymer distribution. As the data of Table 8.1 illustrate, the M_n and M_w values are reduced by 9.6% and 9.7%, respectively, from those obtained by using the solvent system containing 0.5% ethanol and 99.5% THF. The precisions for M_n and M_w measurements (as indicated by the standard deviations in Table 8.1) are also reduced. When 50% benzene or toluene is used as the third solvent, similar spectra to those shown in Figure 8.5(B,C) were obtained, and the average mass results were not affected (see Table 8.1). The effect of water addition on mass spectral patterns of PS 7000 was even more pronounced. Typically, when a small amount of water (5%) was used in sample preparation, the MALDI mass spectra could not be readily reproduced, resulting in poor precision (see Table 8.1). In addition, large differences in M_n from those obtained by other solvents were observed.

The results shown in Figure 8.5 and Table 8.1 can be explained by considering the solubility property of the solvent used in the sample preparation. An ideal polymer non-solvent is characterized by its inability to dissolve any amount of polymer at any temperature under atmospheric pressure. For example, water or methanol is a non-solvent for PSs, and it can be readily observed that the addition of an excess amount of water or methanol to the THF sample solution of PS can cause turbidity. One possible misunderstanding of the solvent effect is that, when a solvent mixture is used – particularly with the use of a small portion of non-

Table 8.1 MALDI results for the analysis of polystyrene 7000 using different solvent systems for sample preparation.[a,b,c]

Solvent system[d]	M_n	M_w	PD	Difference [%][e]
THF:ethanol (99.5:0.5)	7057 ± 22 (7029–7085)	7174 ± 19 (7157–7203)	1.017	—
Methanol:THF:ethanol (25:74.5:0.5)	7085 ± 21 (7054–7107)	7206 ± 23 (7175–7223)	1.017	0.4
Toluene:HF:ethanol (25:74.5:0.5)	7073 ± 29 (7035–7107)	7196 ± 19 (7178–7225)	1.017	0.2
Benzene:THF:ethanol (25:74.5:0.5)	7063 ± 24 (7036–7098)	7196 ± 16 (7162–7206)	1.019	0.1
Methanol:THF:ethanol (50:49.5:0.5)	6377 ± 140 (6196–6512)	6476 ± 145 (6287–6624)	1.016	−9.5
Toluene:THF:ethanol (50:49.5:0.5)	7056 ± 23 (7017–7075)	7179 ± 27 (7138–7199)	1.017	0.0
Benzene:THF:ethanol (50:49.5:0.5)	7057 ± 28 (7029–7097)	7174 ± 33 (7142–7224)	1.017	0.0
Water:THF:ethanol (5:94.5:0.5)	6103 ± 1464 (3375–7340)	6455 ± 1072 (4334–7573)	1.058	−13.5

[a] See Section 8.2.3 for definitions of tabulated quantities.
[b] GPC data provided by the supplier: M_n = 6770; M_w = 6962; PD = 1.03.
[c] M_n, M_w, and standard deviations calculated from five trials except for the solvent system containing water, where results were from nine trials. The ranges for M_n and M_w are shown in parentheses.
[d] Solvent systems expressed as vol.%.
[e] % Difference in M_n, compared to that obtained by using THF:ethanol (99.5:0.5) as solvent.

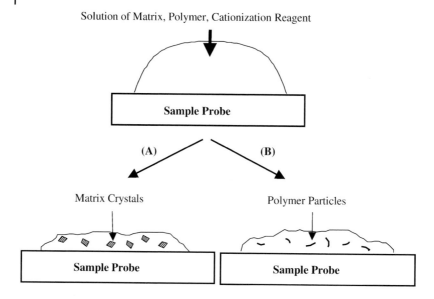

(A). Solvent composition in the drying solution favors matrix crystallization;
(B). Solvent composition in the drying solution favors polymer precipitation.

Fig. 8.6 Effect of solvent composition on MALDI sample preparation. (A) Polymer incorporation into matrix crystals; (B) polymer precipitation.

solvent – the initial sample solution may be transparent, which may lead one to believe that the solvent system used is adequate. The overall solvent effect can be explained by considering Figure 8.6, where the competing processes of matrix crystallization and polymer precipitation are illustrated.

As Figure 8.6 shows, in MALDI sample preparation, solvent evaporation takes place after the sample solution is deposited onto the probe. If a solvent mixture is used for preparing the initial sample solution, the solvent composition is expected to change during the solvent evaporation process because of differences in their volatility. This can lead to a change in the polymer's solubility. If a solvent mixture consists of a solvent and a non-solvent, and the non-solvent is less volatile, its content in the final sample solution on the probe prior to the formation of matrix crystals would be expected to be much higher than in the initial solution. In this case (Fig. 8.6(B)) there is a good possibility that the polymer might precipitate before matrix crystal formation [60]. Because the precipitation of polymer is often a function of molecular mass, mass discrimination can occur during the sample preparation stage. Specifically, any polymer ions detected in MALDI are from the oligomers incorporated into the matrix crystals. The relative contents of the oligomers in the polymer distribution may be altered because of the mass-dependent precipitation. In Figure 8.6, the term "polymer precipitation" refers to precipitation of the polymer before the formation of matrix crystals, whereas the term "polymer incorporation" refers to the event that involves the co-crystallization of polymer, matrix, and cationization reagent (where applicable). It is clear that, in MALDI, it is desirable that polymer incorporation takes place prior to polymer precipitation (Fig. 8.6(A)).

It can be concluded that the selection of a correct solvent system for polymer/matrix sample preparation must be carefully considered. Any solvent conditions that favor polymer precipitation during the sample preparation stage – including the drying of the sample solution on the MALDI probe – will result in possible errors in average molecular mass measurement. It is important to recognize that, when a clear, dilute stock solution is made with a particular solvent system, it does not guarantee that the polymer is still well dissolved at the onset of matrix crystal formation. The use of polymer non-solvent in a solvent system must be carried out with great care.

8.2.1.4 Solvent-Free Sample Preparation

Conventional MALDI-MS analysis of polymers requires the use of a suitable solvent for preparing the matrix/analyte sample. One interesting technique that has been shown to be useful for analyzing some polymeric systems is that of "solvent-free" sample preparation. In this method, a polymer sample is directly mixed with the solid matrix, without using an extraneous solvent. There are several ways of mixing matrix and polymer. For example, Skelton et al. used mechanical mixing of the analyte and matrix by mortar and pestle and applied the pressed pellet to a MALDI target for analysis [75]. Low molecular-mass polyamides (<5000 Da) which are difficult to solubilize in conventional solvents were ionized using this method. Later, Meier and colleagues used a vapor deposition method to deposit matrix onto a poly(ethylene glycol) 6000 sample without using a solvent, and generated a good-quality MALDI spectrum [76–79]. Przybilla et al. applied a mechanical mixing of sample and matrix with a ball mill and then transferred the dry powder onto the MALDI target by using a spatula [80]. MALDI spectra were also obtained for insoluble large polycyclic aromatic hydrocarbons. Several other methods including vortexing [81], mini-ball mill [82] and on-target grinding [81] have been developed for solvent-free sample preparation. This solvent-free method was found to be useful for generating MALDI spectra for a variety of polymers [32,82–86]. An example is shown in Figure 8.7 for poly(dimethylsiloxane) [82], whereby the MALDI spectrum was obtained by using a sample prepared simply by mixing the oily polymer with a premixed DHB:silver powder (2.5:1 mass ratio) in a ratio of 1:15 via mortar and pestle treatment. In contrast, conventional solvent-based MALDI-MS was unable to generate a mass spectrum of the polymer in the mass range shown in Figure 8.7.

Whilst the solvent-free sample preparation is useful for generating MALDI spectra for some polymers, the mechanism underlying the success of the method seems to contradict the conventional wisdom that, in MALDI, analyte molecules should be well incorporated into matrix crystals. It has been shown recently,

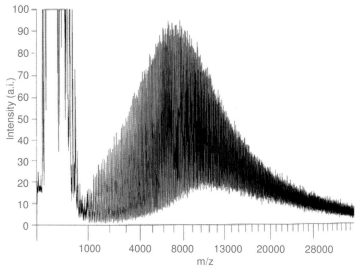

Fig. 8.7 Solvent-free MALDI mass spectrum of a
poly(dimethylsiloxane) sample. Reprinted by permission of
Elsevier from Ref. [82]; © American Society for Mass
Spectrometry, 2006.

however, that in favorable cases such as large accessible surface areas – as would
be generated in samples consisting of layers of many small matrix crystals – the
adsorption of analyte at the matrix crystal surface suffices for the generation of
MALDI spectra [87]. However, at present this method appears to be limited to the
analysis of relatively low mass polymers (i.e., <30 000 Da).

8.2.2
Instrumental and Measurement Issues

8.2.2.1 Mass Resolution and Accuracy

At present, MALDI polymer characterization is commonly carried out by using
TOF MS. TOF mass spectrometers equipped with reflectron and TLF or delayed
extraction provide good mass-resolving power and mass measurement accuracy
(for instrumental details, see Chapter 2). Typically, a low-mass oligomer
(<5000 Da) can be resolved isotopically with a mass measurement accuracy of better than 100 ppm using external mass calibration. Better accuracy (i.e., <30 ppm)
can be achieved by using internal mass calibration with a suitable calibrant added
to the polymer sample. As the polymer mass increases, the isotope peaks of an
oligomer become difficult to resolve by TOF. However, if oligomer resolution can
be achieved, it can be very useful for structural and compositional analysis of a
polymeric material [88,89].

Figure 8.8(A–C) shows mass spectra of narrow polydispersity PS 18 700, 32 660,
and 45 000 obtained with TLF MALDI-TOF [38]. The pulse voltage was set to

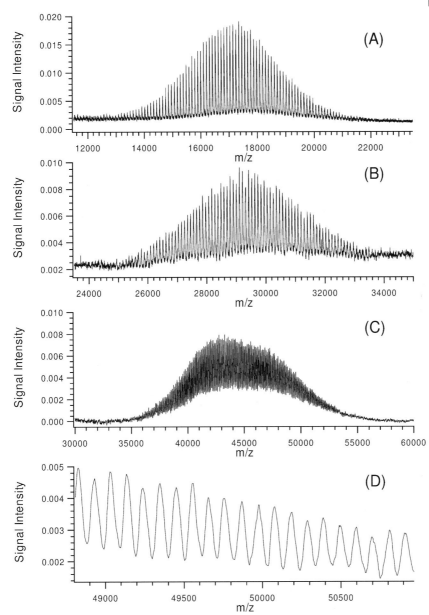

Fig. 8.8 MALDI mass spectra of: (A) poly(styrene) 18 700;
(B) poly(styrene) 32 660; (C) poly(styrene) 45 000; and (D) the
expanded spectrum of poly(styrene) 45 000 using retinoic acid
as the matrix with silver cationization. Reprinted by permission
of American Chemical Society from Ref. [38]; © American
Chemical Society, 1997.

optimize the resolution at the center of the distribution of each compound. In all cases, oligomer ions were well resolved, with the mass resolution for PS 18700 being between 725 and 905 fwhm, for PS 32660 between 870 and 1065 fwhm, and for PS 45000 between 550 and 670 fwhm. Figure 8.8(D) is an expansion of Figure 8.8(C) at the high-mass end of the distribution, and illustrates that oligomer peaks, when spaced by a mass of 104 Da, are well resolved at masses beyond 50000 Da. It should be noted that in analyzing these relatively high-mass polymers the use of reflectron does not further improve the mass resolution obtained. Due to losses of high-mass ions in a reflectron TOF – most likely due to metastable fragmentation and/or reduced efficiency of ion focusing – this type of sample is better handled by using a TLF MALDI-TOF instrument operated in a linear mode.

As illustrated in Figure 8.8, the apparent mass resolution gradually decreases as the ion mass increases. The actual value of the upper mass limit where oligomers are resolved is dependent on several factors, including the number of components present in a polymer sample, the mass(es) of the repeat units(s), and the matrix/sample preparation method. For the latter, whenever possible, the formation of two or more cation adducts should be avoided. The addition of a preferred cation suppresses other cationization and can improve the quality of the spectrum by reducing peak overlap.

The advantage of high-resolution and high-accuracy analysis of polymers is illustrated in the MALDI analysis of poly(ethylene glycol) 20000 (Fig. 8.9(A)) [38]. Despite the high mass of the polymer and low mass of the repeat unit (44 Da), oligomer peaks are still well resolved across the entire polymer distribution (Fig. 8.9(B)). The data in Figure 8.9(A) show that this sample has a bimodal mass distribution, with one distribution centered at m/z ~23000 and the other centered at m/z ~18500. Inspection of the low-mass region of Figure 8.9(A) indicates that there is a second type of polymer present in the sample. Figure 8.9(C) is an expansion of a spectrum of poly(ethylene glycol) 20000, obtained by adjusting the extraction pulse potential to achieve optimal resolution at m/z 12000. The peaks at m/z 11142 (calculated 11142) and 11935 (calculated 11935) are from sodium-cationized $HO-(CH_2CH_2O)_n-H$, where n = 252 and 270, respectively. The peaks at m/z 11126 and 11919 likely correspond to $H-(CH_2CH_2O)_n-H$, with calculated masses of 11126 and 11919 Da, respectively. The sodiated form of this distribution was verified, as the addition of potassium chloride also generated oligomer mass increases of 16 Da. This second polymer distribution is observed in the low-mass region of the spectrum. Extensive studies of poly(ethylene glycol) fragmentation patterns suggest that it is not possible to produce predominant fragment ions that arise from the poly(ethylene glycol) parent ion with a loss of 16 or 16+44n Da (where n is an integer number of oligomers) [90]. The MALDI results shown in Figure 8.8 demonstrate that this sample contains two major poly(ethylene glycol) components with different end groups. Since limited information on the synthesis of this sample is provided by the supplier, the exact nature of the end-group for the second distribution cannot be confirmed. Nevertheless, this example illustrates that some structural and compositional information can be obtained for polymers if sufficient mass resolution and mass accuracy can be achieved.

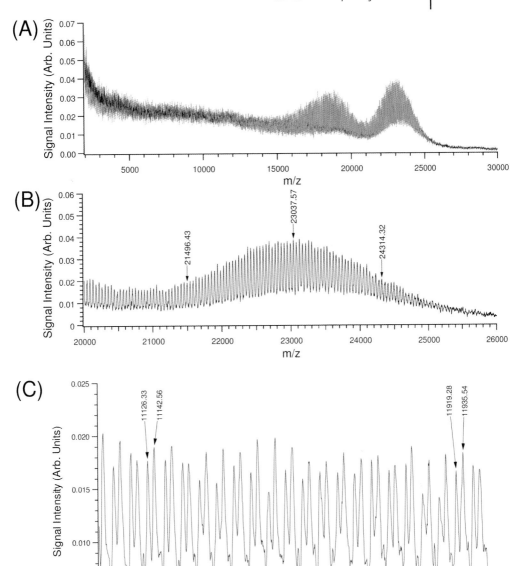

Fig. 8.9 (A) MALDI mass spectrum of poly(ethylene glycol) 20 000 using HABA as matrix with sodium cationization and time-lag focusing set to optimize the resolution of ions at *m/z* 23 000. (B) Expansion of (A) in the high mass region of the spectrum. (C) Expanded MALDI mass spectrum of poly(ethylene glycol) 20 000 using HABA as matrix with sodium cationization and time-lag focusing extraction set to optimize the resolution of ions at *m/z* 12 000. Reprinted by permission of American Chemical Society from Ref. [38]; © American Chemical Society, 1997.

Very high mass resolution and mass measurement accuracy can be obtained in a Fourier-transform (FT) ion-cyclotron resonance (FT-ICR) mass spectrometer [91]. In recognizing the tremendous power of FT-ICR MS for polymer characterization, several research groups have developed this technique in combination with MAL-DI [10,23,92–110]. Earlier studies using MALDI FT-ICR MS for polymer characterization relied on the use of home-built MALDI sources. However, more recently, commercial FT-ICR instruments have been equipped with robust and easily operated MALDI sources. Typically, isotope resolution can be achieved with oligomers of up to 10000 Da and a mass measurement accuracy of several ppm can be expected [91]. With this high performance, the structural and compositional analysis of polymers can be significantly enabled, compared to the use of MALDI-TOF MS.

For example, Jaber and Wilkins very recently reported a study of silver-cationized hydrocarbons with a mass range of up to 12 kDa in a comparative study using an external MALDI 9.4 T FTMS and a reflectron TOF MS [107]. Figure 8.10 shows a MALDI FT-ICR MS spectrum of PS with an average mass of about 10000 Da. Excellent mass measurement accuracy and mass resolution were obtained. In

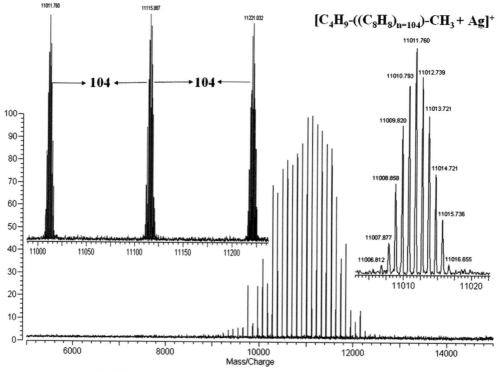

Fig. 8.10 MALDI mass spectrum of poly(styrene) 10000 obtained by using a 9.4 tesla MALDI-FTICR mass spectrometer. Reprinted by permission of Elsevier from Ref. [107]; © American Society for Mass Spectrometry, 2005.

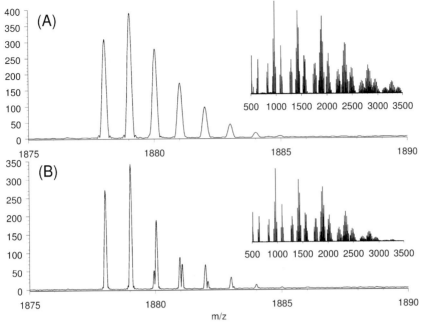

Fig. 8.11 MALDI-FTICR mass spectra of poly[dipropoxylated
bisphenol A-alt-(adipic acid-co-isophtalic acid)] obtained using
(A) high and (B) ultrahigh resolution. Reprinted with
permission from Ref. [110]; © IM Publications, West Sussex,
UK, 2003.

addition, Jaber and Wilkins also addressed the issue of fragmentation as observed
by Chen et al. [111] and Yalcin et al. [112] for polyethylene analysis. The FT-ICR
MS spectra showed no low-mass fragment ions as observed in TOF, this difference being attributed to the different time frames of the measurements [107].

Detailed structural and compositional analysis afforded by FT-ICR MS can be
illustrated in a recent investigation by Mize et al., where homo- and co-polyesters
were characterized [110]. In this case, errors in assigning repeat units from iso-
barically resolved FT-ICR MS spectra were significantly reduced compared to those
from isotopically resolved spectra. For example, Figure 8.11 displays MALDI spectra of poly[dipropoxylated bisphenol A-alt-(adipic acid-co-isophtalic acid)] (U415)
obtained under high and ultra-high mass-resolving conditions. The insets show
the entire mass range for each mass spectrum. Figure 8.11(B) shows the resolution of isobaric isotope peaks from components that differ in molecular mass by
2 Da. Assignments of the cyclic and linear oligomers can be readily carried out at
the higher level of resolution [110].

8.2.2.2 Sensitivity and Dynamic Range

Because the amount of polymer samples available is usually not limited, it is possible to underestimate the sensitivity issue in MALDI polymer characterization.

In reality, the use of a MS instrument that provides high sensitivity and a wide dynamic range of ion detection is pivotal to the success of polymer analysis. This is true not only for the measurement of polymer average mass, but also for the determination of polymer composition [108,110,113–121]. With limited detection sensitivity, some oligomers within a polymer distribution or minor components in a polymer mixture may not be adequately detected, and this may result in measurement errors or misrepresentation of the polymer sample.

Unlike biopolymers such as proteins and peptides, industrial polymers do not have a single exact molecular mass. Rather, they display a distribution of molecular masses, with the molecular mass distribution and average mass of a particular polymeric system depending on the polymerization kinetics and mechanism. In order to measure the average molecular mass accurately, the MALDI method must be able to generate a mass spectrum that reflects the actual oligomer distribution, as well as the relative amounts of all oligomers within the distribution. Sensitivity limitations, background interference, and/or mass discrimination can cause a change in the polymer distribution function, broadening or narrowing of the overall distribution, and/or truncation of detected oligomer peaks within a distribution (i.e., missing low- or high-mass tails). Any one of these variations can result in errors in average mass measurement.

Most commercial MALDI-TOF instruments provide good sensitivity for polymer analysis, although if an instrument is not fine-tuned to its optimal performance, erratic results may be generated. In addition, a less-optimized sample preparation method can lead to low detection sensitivity. The effect of low detection sensitivity on polymer molecular mass measurement is illustrated in Figure 8.12 [120]. This figure shows the mass spectra of PS 7000 obtained under two experimental conditions. In Figure 8.12(A) the spectrum was generated under the optimal condition for PSs, using the retinoic acid/silver nitrate formulation. In Figure 8.12(B), the mass spectrum was obtained using a MALDI preparation where lithium hydroxide was purposely added to the retinoic acid/silver nitrate formulation. It can be seen clearly that the S/N ratio of this spectrum is relatively poor, compared to that shown in Figure 8.12(A). The addition of LiOH reduces the detection sensitivity and increases the background level in the low-mass region.

The spectra shown in Figure 8.12(A,B) were re-plotted as peak area of individual oligomers versus mass-to-charge ratio, and are shown in Figure 8.12(C,D), respectively. The plot of Figure 8.12(C) is quite symmetric and can be well fitted to the Gaussian distribution (note the solid line; the goodness of fit is indicated by $\chi^2 = 0.0201$ with $\chi^2 = 0$ for a perfect fit with unity mass applied to all original data). In contrast, the plot shown in Figure 8.12(D) does not fit well to a Gaussian distribution, particularly for the low-abundance oligomers (the overall fit has $\chi^2 = 0.0522$). It appears to be skewed to the high-mass distribution, most likely due to the uneven elevation of the baseline caused by background noises. For visual comparison, the curve fit of PS 7000 as shown in Figure 8.12(C) (solid line) is included in Figure 8.12(D). Because a Gaussian distribution is expected for this standard polymer sample, the departure from Gaussian distribution suggests that

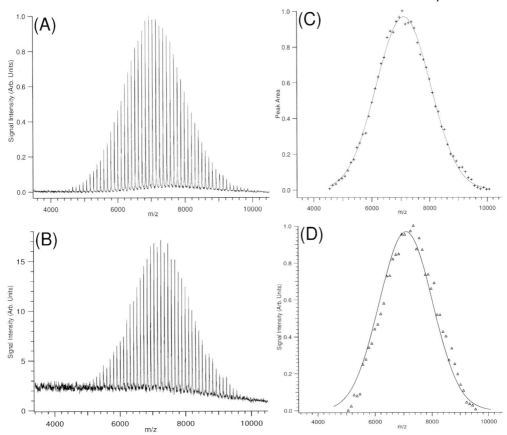

Fig. 8.12 MALDI mass spectra of poly(styrene) 7000 obtained (A) under the optimal condition and (B) under the same conditions except that LiOH was added to the sample preparation. (C) Individual peak areas of oligomers from the MALDI spectrum shown in (A) are plotted as the function of m/z. The result of Gaussian curve fitting is shown as a solid line. (D) Plot of individual peak areas of oligomers as a function of m/z from the spectrum shown in (B) and the Gaussian fit from (C). Reprinted by permission of Elsevier from Ref. [120]; © American Society for Mass Spectrometry, 1998.

there might be an error associated with the MALDI measurement. Under optimal conditions, the average molecular mass values were found to be M_n 6998 (0.4% RSD) and M_w 7132 (0.4% RSD) from five trials. In comparison, using the less-sensitive method, the values were found to be M_n 7166 (0.8% RSD) and M_w 7292 (0.7% RSD). These values were significantly different from those obtained by the more sensitive method (99% confidence limit from t-test). In addition, the precision of the method using the less-sensitive method was also downgraded.

Aside from the uneven baseline elevation, another factor causing error in average molecular mass measurement is related to the truncation of some oligomer signals within the polymer distribution in the mass spectrum obtained by a MALDI method with a limited sensitivity or dynamic range. The relative fraction of the oligomers in a polymer sample can span a wide range. In this context, oligomers with low abundance (e.g., those at the tails of the distribution) may not be detected in a MALDI analysis. The dynamic range of detection can be inferred from examining the polymer spectrum directly. In the case of PS 7000, the most intense peak (100% relative intensity) is the one at m/z 7055 (Fig. 8.12(C)), while the peak at m/z 4555 is the least intense one detectable at the low-mass tail of the distribution, with the S/N ratio 3.9. It has a relative intensity of 1.8%. The least intense peaks (S/N ratio 3.1) from the high-mass tail of the distribution is at m/z 10 075 with a relative intensity of 1.4%. All data were the average values from five trials. In the mass spectrum of PS 7000 obtained by the less-sensitive method (see Figure 8.12(D)), the m/z 7368 peak was the most intense. The least intense peak in the low-mass region of the distribution was at m/z 5076, with a relative intensity of 5.2% and a S/N ratio of 4.0. The peak at m/z 9555 has a relative intensity of 4.7% and a S/N ratio of 4.4. It is clear that the dynamic range of detection is significantly reduced in the less-sensitive method.

In the absence of any information from a more sensitive detection method, Zhu et al. proposed the use of polymer blends to evaluate the dynamic range of detection in MALDI-MS, thus gauging the accuracy of molecular mass measurement [120]. For example, in the case of PS 7000, two bicomponent blends containing PS and a second polymer component of either PS 5050 or PS 11 600 can be prepared. PS 5050 contains oligomers with the same masses as those of oligomers at the low-mass tail of the PS 7000 distribution, whereas PS 11 600 has peak overlap in the high-mass tail of PS 7000. A sensitive method should be able to detect the mass spectral change after a small amount of a second component is added. With the addition of 5% PS 5050 to PS 7000, the optimized method can readily detect the additional oligomer peaks at the low-mass tail of the PS 7000 distribution. The non-optimal method was insufficiently sensitive to extend the detection of oligomers to lower masses beyond those from PS 7000. Similarly, very little signal from PS 11 600 is detected in the blend containing PS 7000 and PS 11 600 in a molar ratio of 100 to 2. In contrast, the sensitive method can detect minor components in the mixture of PS 7000 and PS 11 600, even at a molar ratio of 100 to 1, which suggests that any oligomers with relative contents greater than 1% of the most abundant oligomer in the high-mass tail of the PS 7000 distribution should be detectable in MALDI. For PS 7000, the addition of any peaks with intensities less than 1% of the most abundant peak at both tails of the distribution does not affect the calculated M_n, M_w, and PD values, within the experimental precision. Thus, molecular mass measurement derived via the optimal method should be accurate.

In summary, the dynamic range of detection in MALDI analysis is related to the detection sensitivity and background noise level. Both a decrease in detection sensitivity for the oligomers and an increase in background noise level can result in a reduction of dynamic range. As a consequence, asymmetric distortion of the polymer distribution due to uneven baseline elevations and truncation of the oligomer distribution can be expected. It is worth noting that analyzing a polymer standard such as PS 7000 by MALDI is an excellent way to gauge the sensitivity and detection dynamic range of an instrument. For a given sample preparation protocol, a poorly designed mass spectrometer – or a mass spectrometer that is not optimized for its sensitivity – will generate a low-quality spectrum that is characterized by poor baseline level, asymmetric oligomer distribution and truncation of oligomer peaks at the tails of the distribution. When analyzing polymer mixtures that contain both high- and low-abundance polymers, an instrument with poor sensitivity will result in a failure to detect the low-abundance polymers.

8.2.2.3 Mass Range

The upper mass range of a polymer that can be analyzed by MALDI-MS is dependent on the polymer type. For example, Danis et al. reported the detection of water-soluble poly(styrenesulfonic acid) with a molecular mass just below 400 000 [122], and the detection of a poly(methyl methacrylate) sample with a molecular mass of about 256 000 [17]. Multiply charged ions from a starburst polyamidoamine dendrimer with a molecular mass as high as 1.2 million has been reported by Savickas [123]. Yalcin et al. showed that polybutadienes of narrow polydispersity with masses up to 300 000 Da, and polyisoprenes of narrow polydispersity with masses up to 150 000 Da, can also be analyzed [42].

Schriemer and Li currently hold the record of detecting the highest masses of polymers by MALDI-MS [29]. These authors have shown that accurate molecular mass determinations of up to 1 million can be achieved from the singly charged polymeric species (see Figure 8.13). For a PS with a molecular mass of approximately 1.5 million, signals corresponding to the multiply charged ions of the principal distribution are observed (Fig. 8.13(D)).

The upper mass limit is set by the need to use higher matrix-to-polymer ratios with increasing polymer molecular mass, to a point where the instrument can no longer detect the small quantity of polymer present in the matrix host [29]. Thus, a combination of a highly sensitive MALDI instrument and a highly sensitive sample/matrix preparation method is required to detect high-mass polymers. From an instrumental viewpoint, TOF mass analyzers have no theoretical mass limit and are commonly used for high-mass detection, whereas FT-ICR instruments are applicable only to relatively low-mass polymers (i.e., <20 000 Da) due to the difficulty of trapping the singly charged high-mass ions generated by MALDI in the ICR cell. However, microchannel plate (MCP) detectors are used mostly for detecting the ions in a TOF mass analyzer, which exhibit a strongly decreased detection sensitivity in the high-mass range. This is best illustrated in Figure 8.13(A–C), which shows the mass spectra of three PS samples with nominal molecular masses of 330 000, 600 000 and 900 000 Da.

At the high-mass region shown in Figure 8.13, the adjacent oligomer peaks with a mass difference of 104 Da are unresolved, and consequently the entire oligomer distribution appears as a broad peak. The salient features of these spectra are the appearance of the singly charged molecular ion peaks, as well as peaks from multiply charged ions. For Figure 8.13(B), the low-intensity peak at m/z ~400 000 is from the triply charged dimer of PS 600 000. Likewise, the small peak at m/z

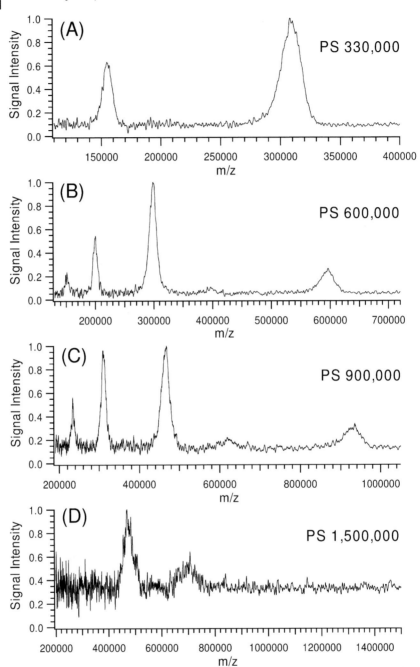

Fig. 8.13 Mass spectra of three poly(styrene) (PS) samples
with nominal molecular masses of (A) 330 000, (B) 600 000,
(C) 900 000, and (D) 1.5 million. Reprinted by permission of
American Chemical Society from Ref. [29]; © American
Chemical Society, 1996.

~620 000 in Figure 8.13(C) is from the triply charged dimer of PS 900 000 (the mass of the dimer should be ~1.8 million Da). As the mass increases, it becomes clear that the multiply charged distributions begin to dominate the spectra. The relative decrease in the intensity of the principal distribution is probably mostly due to a steady fall in the detector efficiency of the MCP, though a shift towards higher charge states with increasing mass has also been observed in MALDI [13]. Future investigations to develop new detectors suitable for high-mass ion detection, such as those based on the detection of charge, the impact energy or sputtered secondary ions, rather than secondary electrons (as discussed in Chapter 2) should enable the analysis of high-mass polymers by MALDI-TOF MS.

Careful sample preparation is also crucial in analyzing high-mass polymers. When preparing polymer samples for MALDI, a general rule of thumb is to use a matrix:analyte molar ratio ranging from 500:1 to 10 000:1. In the MALDI analysis of high-mass polymers, this does not hold, however. When the experimenter experiences poor detection sensitivity, it is intuitively sensible that an increase in the amount of analyte would increase the signal strength, yet just the opposite is the case when analyzing high-mass polymers [29]. For example, the analysis of PS 900 000 was achieved from 10 fmol of total polymer loaded onto the probe tip, with a matrix:analyte ratio of ~8×10^6:1. With the high-mass polystyrenes, the use of molar matrix:analyte ratios ranging from 500:1 to 10 000:1 yielded no ion signals at all. It has been observed that with increasing molecular mass of the analyte, progressively lower molar amounts must be added to the probe tip for a given amount of matrix in order to achieve the highest sensitivity [29]. The exact reason underlying this observation is unknown, but it seems likely that for high-mass analytes the volume ratio of analyte to matrix is the important parameter, rather than the molar ratio. In this way a suitable "solid solution" is obtained, which thereby precludes polymer entanglement and/or the formation of regions of microcrystallinity.

Because it is necessary to reduce the molar amount of polymer loading as the molecular mass increases, this would suggest a practical limit to the mass range for polymer analysis by MALDI. As molecular mass increases, the sensitivity of the instrument is challenged on two fronts: (i) decreased sensitivity due to a loss in detector efficiency; and (ii) decreased sensitivity from the requirement of lower (molar) sample loading. In the analysis of PS, this limit appears to occur at ~1.5 million Da. In order to obtain the MALDI mass spectrum of PS with a nominal mass of 1.5 million Da [29], 5 fmol of total polymer was loaded onto the probe. Even a slight increase in the amount loaded (to 15 fmol, for an identical amount of matrix) resulted in a total, reproducible signal suppression, while loadings of less than 5 fmol proved to be beyond the sensitivity of the instrument. This example illustrates that the correct selection of a matrix:analyte ratio becomes increasingly critical as the molecular mass of the polymer increases. The window of matrix/analyte leading to an observable polymer signal is extremely small as the upper mass limit is approached. In the case of polyisoprene, with a nominal mass of 137 000 Da, this window corresponds to 50 to 500 fmol loaded (M/A =

$1.6 \times 10^6:1$ to $1.6 \times 10^5:1$, with an optimum at 120 fmol loaded ($M/A = 6.7 \times 10^5:1$). An attempted analysis of polyisoprene 144,800 proved unsuccessful. For polybutadiene 315,200, this window corresponds to 15–200 fmol ($M/A = 5.3 \times 10^6:1$ to $4.0 \times 10^5:1$) with an optimum at 70 fmol loaded ($M/A = 1.1 \times 10^6:1$). An attempt to analyze polybutadiene 400 000 failed to generate any signals.

The above discussion indicates that the upper mass limit achieved in MALDI analysis of polymers is determined by the requirement of increasing the M/A ratio as the polymer molecular mass increases. A MALDI instrument providing high detection sensitivity in conjunction with the use of a sensitive sample preparation method can be used to detect very high mass polymers. The studies of high-mass PS analysis by MALDI-TOF MS can be used as a benchmark for future instrumental and method developments. The current upper mass limit of a TOF instrument appears to be about ~1.5 million Da for polymers, and to push beyond this limit would require the development of more sensitive mass spectrometers, preferably with a sensitive detection system tailored for high-mass ions [124]. However, if a sample preparation technique were to be developed that was more sensitive than the method currently used for polyisoprene and polybutadiene, then the upper mass limit might be extended to ~1.5 million Da for these two polymers. Another critical parameter is the ion yield of the analyte species; here, the choice of the best matrix and/or cationization reagent could play a decisive role.

8.2.2.4 MS/MS Capability

Tandem mass spectrometry (MS/MS) is important for structural and compositional analysis of polymers [90,125–129]. With a reflectron TOF instrument, product ions of a mass-selected oligomer can be generated and analyzed for some polymeric systems via a post-source decay (PSD) process (see Chapter 2 for a discussion of this procedure) [126,128,130–139]. For biopolymers, product ions are generally produced from a protonated molecule, but for industrial polymers – as discussed earlier – a cationization reagent is commonly used to produce the molecular ion. The type of cationization reagent used can have a profound effect on the degree of the molecular ion fragmentation and, to some extent, on the type of product ions being formed [90,137,140]. Thus, it is worth the effort and time to try several different reagents or metal cations for a given polymer system. Some metal ions to be tested may not yield the highest MALDI signals, but if the purpose is to generate product ions for structural analysis then these metal ions should still be considered.

Figure 8.14 shows the PSD spectra of several $C_{36}H_{74}$–metal-ion complexes [140]. This non-polar molecule can be considered as a constituent of low-mass polyethylenes [112,141]. Silver ion attachment to this molecule produces the highest signals, but only silver ions are detected in the PSD spectrum of this adduct ion and no fragment ions from the backbone of the long-chain molecule are observed. The use of other transition metal ions can produce reasonably good molecular adduct ions with this molecule. More importantly, fragment ions can be observed, as shown in Figure 8.14. The fragment ion signals provide the information on the repeat units of this molecule. As Figure 8.14 shows, the extent

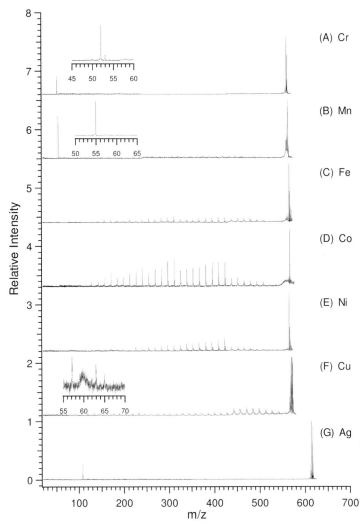

Fig. 8.14 Post-source decay mass spectra of metal ion $C_{36}H_{74}$ adducts. Reprinted by permission of Elsevier from Ref. [140]; © American Society for Mass Spectrometry, 2001.

of fragmentation and the signal strengths of the fragment ions are dependent on the type of metal ions used.

Whilst conventional reflectron TOF instruments can be used to generate PSD spectra, they are not widely used for polymer structural analysis, for reasons such as the difficulty of selecting a narrow mass range of a precursor ion for fragmentation, lack of control on the internal energy of the precursor ions, and relatively poor quality of product ion spectra for unambiguous structural assignments. However, during the past few years major advances in MALDI tandem mass spectrometric instrumentation have taken place. MALDI-MS/MS can now be carried out in several commercially available mass spectrometric platforms, including TOF-TOF, Qq-TOF, ion trap (IT), IT-TOF, and FT-ICR MS (see Chapter 2 for details of these instruments). The availability of these commercial instruments is expected to make MALDI-MS/MS a truly powerful tool for polymer structural characterization. High-quality product ion spectra from the collision-induced dissociation (CID) of a precursor ion can be generated. The ion trap and FT-ICR instruments have the capability of MS^n for detailed structural analysis of a polymer. In addition, other dissociation techniques such as infrared multiphoton dissociation, electron capture dissociation (ECD), and electron transfer dissociation (ETD) are available in some platforms (i.e., FT-ICR). However, for ECD and ETD, one needs doubly or more highly charged precursor ions, which are rarely available in MALDI spectra. An example of high-quality CID spectra of polymers is shown in Figure 8.15, where a Qq-TOF tandem mass spectrometer was used to generate the product ion spectrum of an oligomer ion from a poly(ethylene glycol) sample. Two different metal ions were used as the cationization reagents. No fragment ions were generated from the sodium polymer adduct ions. However, as Figure 8.15(B) shows, the use of lithium ions produces a high-quality product ion spectrum, due to a stronger interaction of lithium ions with the polymer chain, compared to sodium ions. The fragment ions can be assigned to the polymer structures (i.e., $CH_3O-(CH_2CH_2O)_n-H$).

It should be noted that, for most MALDI analyses, the sample plate is inserted into a vacuum or low-pressure region of a mass spectrometer. Yet recent development in atmospheric pressure (AP) MALDI [142,143], where laser desorption/ionization takes place from a plate placed near the entrance of a mass spectrometer, opens the possibility of using existing ESI MS/MS instruments for MALDI-MS/MS. Figure 8.16(A) shows an AP MALDI mass spectrum of Surfynol 465, a telomeric ethoxylated surfactant containing a backbone of 2,4,7,9-tetramethyl-5-decyne-4,7-diol [144]. Figure 8.16(B) shows the AP MALDI ion trap MS/MS spectrum of the m/z 909 ion. Two series of product ions are observed, and their structures can be assigned to the fragment ions of the oligomer ion, which allows for the confirmation of the polymer structure. Hanton et al. further showed that MS^3 product ion spectra could be obtained for this polymer and were valuable in distinguishing between the product ions of the sodiated oligomers and those from the isobaric polymer-matrix adduct ions [144]. The MS^3 experiments helped these authors to resolve unassigned peaks from their previous MALDI PSD experiments of the same sample.

While MALDI-MS/MS is useful for the structural analysis of polymers, it is currently limited to low-mass polymers (i.e., <3000 Da). Higher-mass ions are difficult to dissociate by using CID, but in order to study larger polymers, then technical advances to provide the efficient dissociation of higher-mass ions are required. One possible approach to handling larger polymers might be to degrade the polymer chains into small fragments (either chemically or physically [145–147]) that can then be subjected to MALDI-MS/MS. Although the identification of a degradation process that will not adversely affect the polymer structure may not be trivial, the concept is clearly worthy of future exploration.

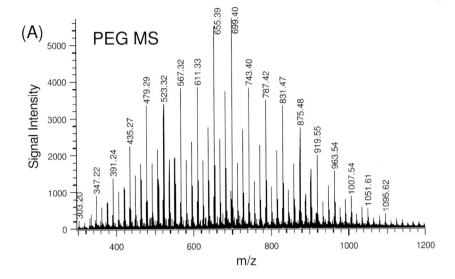

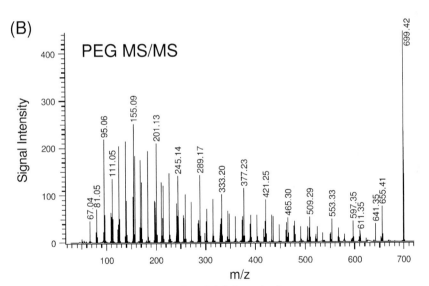

Fig. 8.15 (A) MALDI mass spectrum of a poly(ethylene glycol) (PEG) sample using lithium cationization obtained by quadrupole time-of-flight (QqTOF) MS. (B) MALDI QqTOF product ion spectrum of an oligomer ion at m/z 699.

8.2.3
Data Processing Issues

The determination of polymer molecular mass and distribution by MALDI-MS requires not only accurate mass measurement but also quantitative measurement

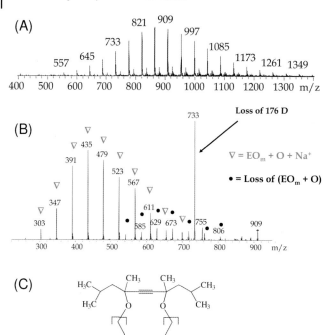

Fig. 8.16 (A) AP MALDI mass spectrum of Surfynol 465 using sodium cationization. (B) AP MALDI ion trap MS/MS spectrum of the *m/z* 909 ion. (C) The chemical structure of the S4XX surfactants. Reprinted by permission of Elsevier from Ref. [144]; © American Society for Mass Spectrometry, 2006.

of ions. The equations used for molecular mass determination and polydispersity (PD) calculation are:

$$M_n = \Sigma(N_iM_i)/\Sigma N_i$$

$$M_w = \Sigma(N_iM_i^2)/\Sigma(N_iM_i)$$

$$PD = M_w/M_n$$

where N_i and M_i represent signal intensity and mass at point *i*. In the case of a polymer spectrum containing well-resolved oligomer peaks, N_i and M_i represent signal intensity in peak area and mass for the oligomer containing *i* monomers, respectively. Numerical integration of the polymer signal was performed in the mass domain. Using a TOF mass spectrometer, correction of the peak intensities is required when converting from a time domain spectrum to a mass spectrum [29,69,148,149]; this is because, in TOF, mass is proportional to t^2. Therefore, calibrating the collected MALDI data and displaying it according to a mass scale can distort the molecular mass distribution function. The necessary correction factor can be expressed as follows:

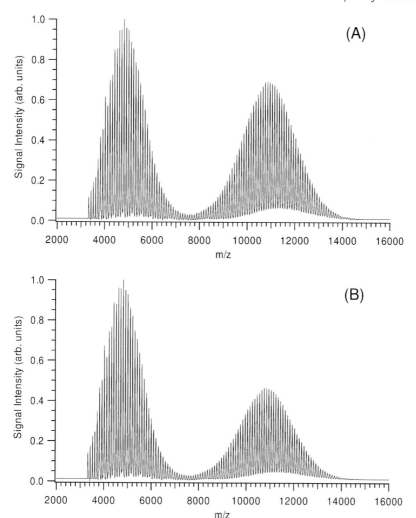

Fig. 8.17 MALDI mass spectra of a two-component blend consisting of polystyrene 5050 and 11 600 (100 pmol each), (A) with no intensity correction applied and (B) with the intensity correction applied. Reprinted by permission of American Chemical Society from Ref. [148]; © American Chemical Society, 1997.

$$q(m) \propto D(t)/(dm/dt)$$

where $D(t)$ is the MALDI detector response as a function of time, t, and $q(m)$ is the corrected number molecular mass distribution as a function of mass, m. The dm/dt term is simply the derivative of the calibration equation. Correction of the data in this fashion allows one to determine the average molecular mass values directly from the mass domain, and also preserves the molecular mass distribution function as determined by MALDI. The effects of such a correction are shown in Figure 8.17 [148]. The two spectra in Figure 8.17 represent the molecular mass distribution of a two-component blend consisting of PS 5050 and 11 600 (100 pmol each). Figure 8.17(A) represents the spectrum without the above correction, while Figure 8.17(B) represents the same spectrum with the correction applied. The difference between the two is readily noticeable, even over this relatively small mass range.

In calculating the average molecular mass, the contribution of the cation attached to the polymer should be corrected. In addition, baseline correction to a MALDI spectrum, if required, should be made with caution, as it can affect the polymer distribution [150].

8.3
Attributes and Limitations of MALDI-MS

MALDI-MS is a powerful tool for polymer characterization. Compared with analytical techniques currently used for polymer analysis, it provides several unique features. In MALDI-MS, molecular mass and molecular mass distribution information can be obtained for polymers of narrow polydispersity with both precision and speed. The accuracy, though difficult to determine due to the lack of well-characterized standards, also appears to be good [151]. The MALDI analysis of polymers does not require the use of polymer standards for mass calibration. Furthermore, this technique uses a minimum amount of solvents and other consumables, which translates into low operational costs. MALDI-MS can also provide structural information, if the instrumental resolution is sufficient to resolve oligomers. In this case, monomer and end-group masses can be deduced from the accurate measurement of the mass of individual oligomers. This is particularly true when a FT-ICR MS is used for polymer analysis. With the use of MALDI-MS/MS, structural characterization can be facilitated. Finally, impurities, byproducts, and subtle changes in polymer distributions can often be detected even for relatively complex polymeric systems such as copolymers.

A number of publications have demonstrated the applications of MALDI-MS for polymer characterization. Of particular interest, Hanton provided a list of publications summarized according to the type of polymers analyzed by MALDI-MS [11]. In addition, an updated web-based resource for polymer/matrix preparation protocols is available from the Polymers Division of the National Institute of Standards and Technology (NIST) homepage (http://polymers.msel.nist.gov/maldirecipes).

In many laboratories, MALDI-MS has become a routine tool for polymer characterization. This is evident from an increasing number of publications in polymer literature (i.e., *Macromolecules*) which indicate the use of MALDI-MS as a tool for characterizing newly grafted or synthesized polymers. In an industry dealing with polymeric materials, MALDI-MS is often combined with other analytical techniques to provide detailed analyses of a polymeric system. In some cases, MALDI-MS is the only technique that can provide the information required to solve a practical problem. One example is in the area of product failure analysis involving four copolymer samples [152]. These ethylene oxide and propylene oxide (EO/PO) copolymer samples provided by the suppliers to our industrial collaborator, Nalco Chemical Co. (Naperville, Illinois), show similar bulk property, but one of them (sample C) shows different surfactant activity, water dispersibility, and stability. The analysis results obtained from traditional techniques, including IR, NMR, GPC, and inductively coupled plasma (ICP) spectrometry, did not reveal much difference among these four samples. These four samples were sent to the author's laboratory for MALDI-MS analysis.

Figure 8.18 shows the MALDI-TOF MS spectra of the samples. There are four main distributions in each spectrum, representing different PO and EO compositions. All peaks are from PO/EO copolymers initiated with the $CH_2=CH-CH_2O-$ group and are Na^+ adduct ions. Figure 8.19 shows the expanded plot of the mass spectral region from m/z 1660 to 1980 for all four samples. The

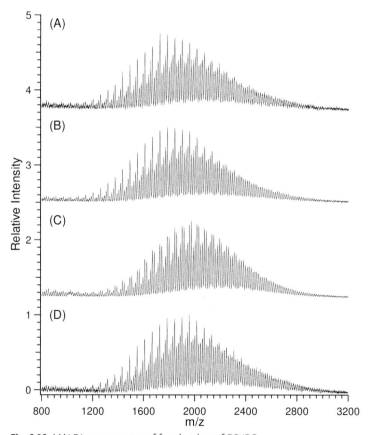

Fig. 8.18 MALDI mass spectra of four batches of EO/PO polymers. Reprinted by permission of Elsevier from Ref. [152]; © American Society for Mass Spectrometry, 2001.

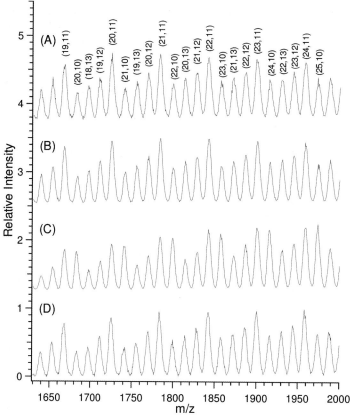

Fig. 8.19 Expanded MALDI mass spectra from Figure 8.18.
Reprinted by permission of Elsevier from Ref. [152];
© American Society for Mass Spectrometry, 2001.

number of EO units in the four distributions in these four samples is determined
by the mass analysis of oligomer peaks to be ranging from 10 to 13, with PO
repeating units ranging from 19 to 25 (see *n, m* values corresponding to each peak
shown in Figure 8.19). The relative amounts of the four adjacent distributions
(between every 58 mass units) are the same for samples A, B, and D, but clearly
different from sample C. For example, the intensity of the mass peak labeled as
(20 PO, 10 EO) in sample C is about twice that of the other three samples.
MALDI-TOF MS results reveal that the composition of sample C is slightly different from the others, suggesting that key differences in copolymer production,
perhaps, are due to process variations that result in a higher 10 EO content in sample C. It transpires that sample C was the one showing abnormal properties. It is
clear that the small compositional variation could not be detected by traditional
analytical techniques within the experimental errors, but causes significant dif-

ferences in the polymer's water dispersibility and activity. By knowing the possible cause revealed by the MALDI-MS data, our collaborator was able to track down the processing problem associated with this particular batch of sample, and an improved quality control mechanism was implemented. It should be noted that, although MALDI-TOF is not quantitative in compositional analysis (see below), this example clearly demonstrates that it is a very useful technique to detect subtle variations among similar samples.

While MALDI-MS is widely used for polymer characterization, it does have certain limitations. The first limitation is that not all narrow-polydispersity polymers can be ionized by MALDI. For example, polyethylene, perfluoropolymers, and polycationic polymers are difficult to analyze by MALDI-MS. Direct laser desorption ionization with the assistance of metal powder can be used to ionize low-molecular mass polyethylene (<5000 Da) [112], but high-mass polyethylene is currently not amenable to MALDI analysis.

The second limitation is that direct analysis of a broad-polydispersity polymer (PD >1.2) is difficult by MALDI-MS since, in many cases, no polymeric signals will be generated. In some cases, only low-mass polymeric ions are detected, but their distribution does not reflect the true polymer distribution [18,113,114,148,153–161]. Mass discrimination in the analysis of polydisperse polymers by MALDI-TOF MS has been investigated by using multicomponent blends to mimic a polydisperse polymer, and also by considering issues related to the sample preparation, desorption/ionization, and TOF instrumentation [18,148]. There are several possible causes for the failure to generate a MALDI spectrum reflective of the true polymer distribution. One is related to the preferential ionization of low-mass polymers. In analyzing a narrow-polydispersity polymer, the mass range of the polymer distribution is small so that the preferential ionization is not noticeable. However, for a wide-polydispersity polymer, the high-mass oligomers can be suppressed. The second reason is that multimer formation favors the high-mass oligomers, which reduces the high-mass ion signals in the principal distribution. The third limitation is that the high-mass ions more readily form multiple charged ions, compared to the low-mass ions. This reduces the high-mass ion signals in the polymer distribution and, moreover, the peaks from the multiple charged ions overlap with those from the low-mass ions in the principal distribution, and this results in an increase in the signals of the low-mass region of the principal distribution. Finally, low-mass ions are detected in TOF with higher sensitivity than high-mass ions. Ion focusing, detector response, and detector saturation bias to the detection of low-mass ions. The combination of these factors, as well as other possibly unknown factors, contributes to the failure of analyzing polymers with wide polydispersity by direct MALDI-MS [18,148].

The analysis of a broad-polydispersity polymer (PD >1.2) can be carried out by combining GPC or size-exclusion chromatography (SEC) with MALDI-MS [162–173]. In this approach, a wide-polydispersity polymer is first separated by GPC and fractions at a defined time interval are collected. The time interval is properly chosen so that the individual fraction would contain only a narrow-polydispersity (PD <1.2) polymer, which can then be analyzed by MALDI-MS for accurate molecular mass determination. The molecular mass information generated from the MALDI analysis of all individual fractions can be used to convert the time domain in the GPC chromatogram into a mass domain. The polymer distribution can be determined from this chromatogram.

This GPC-MALDI-MS technique has several advantages over the traditional method of using polymer standards to calibrate GPC. The MALDI approach does not rely on the availability of polymer standards. This is very important for characterizing newly developed polymeric systems where no standards of accurate masses are available. MALDI-MS should be more accurate in determining the polymer molecular mass to calibrate the GPC chromatogram, compared to the use of polymer standards, unless the polymer standards, if available, are mass-analyzed by MALDI-MS beforehand. In GPC-MALDI-MS, polymer mixtures can be analyzed, because the mass spectral data provide another dimension of differentiating individual polymers based on their mass characteristics.

An example of using GPC-MALDI-MS for the analysis of a functionalized polybutadiene 5000 is illustrated in Figures 8.20–8.22 [174]. Figure 8.20 shows gel-permeation chromatograms of a polybutadiene 5000 before (A) and after (B) epoxidation of its double bonds. The retention time of the functionalized polymer

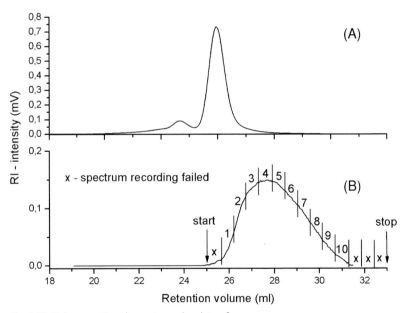

Fig. 8.20 Gel-permeation chromatography plots of polybutadiene 5000 (A) before and (B) after epoxidation using *m*-chloroperbenzoic acid. Fraction marks are indicated in (B). Reprinted by permission of Taylor & Francis Group from Ref. [174]; © Taylor & Francis Group, Abingdon, UK, 2005.

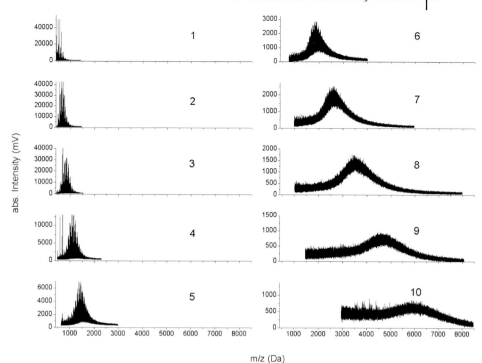

Fig. 8.21 Series of continuously recorded MALDI mass spectra after GPC-MALDI coupling; numbers in the mass spectra are identical with fraction number shown in Figure 8.20. Reprinted by permission of Taylor & Francis Group from Ref. [174]; © Taylor & Francis Group, Abingdon, UK, 2005.

increases, which would indicate a decrease in the molecular mass if the same mass calibration is applied to both chromatograms. However, in reality, the polarity of the polymer after epoxidation increases drastically, resulting in an increase in interactions between the stationary phase and polymer molecules. In this case, mass calibration using polybutadiene standards cannot be applied to the functionalized polymer. GPC fractionation was carried out, and the fractions collected are indicated in Figure 8.20(B). The corresponding mass spectra of these fractions are shown in Figure 8.21. The peak molecular mass information was obtained and used to calibrate the chromatogram shown in Figure 8.20(B); the result is shown in Figure 8.22. Compared to the GPC result obtained by using standard polymers for calibration, the MALDI result is different and should be much more accurate.

Another potential limitation of MALDI-MS for polymer characterization is related to the quantitative analysis of polymer mixtures and copolymers. For a single-component, narrow-polydispersity polymer sample, MALDI-MS can be used to provide accurate relative abundance information on oligomers bearing the same backbone and end-group structures. In this case, each individual oligomer

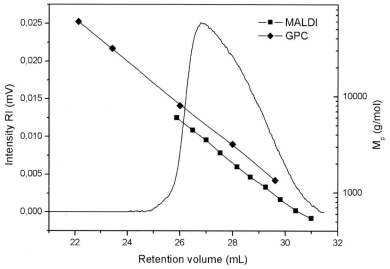

Fig. 8.22 Gel-permeation chromatogram and calibration curves of polybutadiene after epoxidation. Calibration was performed using polyisoprene standards as well as by means of M_p data obtained from MALDI mass spectra after fractionation by GPC. Reprinted by permission of Taylor & Francis Group from Ref. [174]; © Taylor & Francis Group, Abingdon, UK, 2005.

can play the role of an internal standard for the relative quantitation of other oligomers, resulting in the accurate measurement of M_n and M_w. This is not surprising in light of the fact that oligomers have the same backbone structures, with only minor variation in the number of repeat units. Of course, for a broad-polydispersity polymer, there is a large variation in the number of repeat units that cover a broad mass range. As discussed earlier, mass discrimination can take place, resulting in poor representation of actual oligomer distribution in a MALDI spectrum.

For polymer mixtures and copolymers containing polymers with different backbone and/or end-group structures, the relative signal intensities shown in a MALDI spectrum may not accurately reflect the relative contents of the individual components. This is due to the detectability difference among polymers of different structures by MALDI. MALDI is sensitive to small variations of polymer structure. For example, a subtle change at the end-group of a polymer can significantly affect the MALDI signal strength. Thus, great care must be exercised to deduce any quantitative information on the composition of a polymer mixture or a copolymer.

For a polymer mixture consisting of different polymers, one would expect that determination of the relative contents of the individual polymers could be performed using a standard addition method, if standards are available. This approach is similar to that used for biopolymer quantitation by MALDI-MS. However, other techniques such as NMR can be readily used for relative quantitation of polymer mixtures. Thus, MALDI-MS is not necessarily the best approach for this type of application.

One potential area of application that has direct impact in polymer chemistry is the characterization of copolymers. The quantitative analysis of copolymer composition by MALDI has been demonstrated in a limited number of systems, though with varying degrees of success. For example, Abate et al. showed that the relative amounts of hydroxybutyrate and hydroxyvalerate in a random copolymer, poly(β-hydroxybutyrate-co-β-hydroxyvalerate), when determined by MALDI, agree with those expected from theory [175]. Wilczek-Vera et al. have shown that for diblock copolymers of poly(α-methylstyrene)-b-polystyrene the mole fraction of poly(α-methylstyrene) determined by MALDI is similar to that obtained by NMR and GPC, as well as from reagent ratio calculation [176]. The assumption of uniform response from structurally very similar polystyrene and polymethylstyene seems valid in this case. On the other hand, for copolymers with greater monomer structural differences, composition analysis by MALDI can be problematic. For example, for diblock copolymers of poly(α-methylstyrene)-b-poly(4-vinylpyridine), the mole fraction of poly(α-methylstyrene) determined by MALDI can be quite different from that obtained by other methods [177]. A bias in MALDI data was also reported for composition analysis of copolymers of methyl methacrylate and n-butyl methacrylate [178]. Servaty and coworkers showed that quantitative compositional interpretation could be difficult with MALDI data alone for poly(dimethylsiloxane)-co-poly(hydromethysiloxane), and also that the laser power could affect the relative peak intensity in the MALDI spectrum of the copolymer [179].

The difficulty associated with compositional analysis of copolymers can be illustrated in the analysis of an EO/PO copolymer [61]. The structurally similar EO and PO polymers are expected to have very similar detection responses. For example, a mixture of well-blended poly(ethylene glycol) and poly(propylene glycol) homopolymers of similar molecular mass shows similar detection efficiency. However, it was found that experimental conditions in MALDI can significantly influence the mass spectral appearance of the EO/PO copolymer [61]. Specifically, type of matrix and solvent, analyte concentration and laser power can each affect the mass spectral appearance of the EO/PO copolymer. For example, Figure 8.23 shows the expanded MALDI-TOF mass spectra of the copolymer obtained by using different analyte concentrations. In this case, all conditions are the same except that the analyte:matrix ratio is changed. A low analyte concentration favors the detection of peaks labeled as L. The mass dependence of this concentration effect is shown by comparing Figure 8.23(A) with Figure 8.23(B). The exact cause of the concentration effect is difficult to ascertain, but it is plausible that the ion suppression is reduced as the analyte concentration decreases, and this results in increased signal intensities of the minor components (i.e., peaks L). It should be noted that the minor components are completely suppressed throughout the entire mass region studied for the sample prepared at the ratio of 1:10.

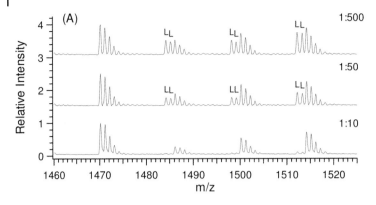

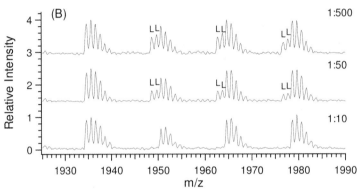

Fig. 8.23 Expanded MALDI mass spectra of an EO/PO copolymer obtained by using different analyte concentrations. (A) Low-mass and (B) high-mass region of the copolymer distribution. *All-trans* retinoic acid was used as the matrix. The number shown refers to the volume ratio of the copolymer sample and 0.1 M matrix/THF solution. The laser power used was at the desorption/ionization threshold. Reprinted by permission of Wiley Interscience from Ref. [61]; © Wiley Interscience, New York, 2000.

Since the quantitative composition analysis of a copolymer relies on the use of relative peak intensities, the mass spectra pattern variation can clearly alter the composition results. The EO/PO copolymer examined in these investigations is composed of structurally similar monomers, but for other types of copolymer, with greater variability in monomer structures, even larger spectral variations are expected. It is clear that a better understanding of how experimental parameters affect the mass spectra pattern will facilitate the search of conditions under which reproducible results can be obtained. Once the reproducible spectra can be obtained, it is then possible to explore the use of other techniques such as NMR to correlate the relative peak intensities in MALDI mass spectra with actual copolymer composition. The validated MALDI method should provide a means of accurate and rapid analysis of copolymer composition.

8.4
Conclusions and Perspectives

Polymer characterization usually requires a combination of several analytical tools such as NMR and GPC. Today, a number of analytical techniques exist that can provide molecular mass, structure, and composition information, and MALDI-MS is now emerging as a powerful method for polymer characterization. Some demonstrated advantages of the technique include the ability to determine average molecular mass and distributions without the need of polymer standards and with high speed, precision, and accuracy, to analyze polymer mixtures with minimum sample work-up, and to provide structural and compositional information via oligomer mass determination and MS/MS. Research into MALDI-MS for polymer analysis is centered on exploring and further developing these unique capabilities in the context of other available analytical methods.

Because of the diversity associated with polymer chemistry, there is no universal approach in MALDI-MS for the analysis of polymers. The major challenge in applying MALDI-MS to characterize a particular polymeric system lies in developing a suitable sample preparation protocol tailored to this polymer. This usually involves screening and selecting a suitable matrix from a list of known matrices used for the analysis of similar polymers, and in some cases this requires new matrices to be identified. Once an appropriate matrix is found, however, attention must still be paid to many details in the MALDI analysis procedure to ensure that the final results reflect the true chemical nature of the polymeric system.

The current MALDI techniques used for polymer analysis have several limitations. For example, they are unable to analyze some narrow-polydispersity polymers, whilst there is a need to develop sample-preparation protocols for the analysis of important polymers such as polyethylene, perfluoropolymer, and polycationic materials. Although the analysis of polystyrenes with molecular masses of up to 1.5 million has been demonstrated, it remains to be seen how the technique can be applied to other high-mass polymers (>500 000 Da).

Sensitive sample preparation methods and improving the overall detection sensitivity of the current MALDI instrument for high-mass analysis are also important requirements. As yet, the MALDI technique has not generated reliable results for the direct analysis of broad-polydispersity polymers, though the instrumental and chemistry problems associated with the analysis of these materials have been widely studied. Today, we have a better understanding of the issues involved, and will surely witness further developments in this area in the near future. Finally, it is difficult, at present, to deduce quantitative information on polymer mixtures from the MALDI spectra, mainly because the overall detection sensitivity for different polymers is not similar. Although there is no direct correlation between the relative peak areas in the spectra and the relative amounts in the mixture, information on relative changes in polymer composition can be obtained if several polymeric systems containing the same polymer mixture, albeit in different proportions, are available for interrogation. In this respect, ESI MS is another important method for polymer characterization [31,57,152,180–191], and can be easily combined with various separation methods for complex mixture analysis. ESI generates multiply charged ions that are more readily fragmented than the singly charged ions produced by MALDI, and this allows the possibility of deducing structural information for higher-mass polymers.

In summary, MALDI-MS is an important tool for polymer characterization, with many attributes that complement those of other analytical techniques. Future advances in analytical method development, as well as of our understanding of the fundamentals relevant to MALDI polymer MS, will undoubtedly enhance the role of this technique in polymer science.

References

1 R.F. Braudy, Jr. (Ed.). *Comprehensive Desk Reference of Polymer Characterization and Analysis*. Oxford University Press, New York, **2003**.

2 R.A. Pethrick, J.V. Dawkins, (Eds.). *Modern Techniques for Polymer Characterization*. John Wiley & Sons, New York, **1999**.

3 G.P. Simon, (Ed.). *Polymer Characterization Techniques and Their Application to Blends*. Oxford University Press, New York, **2003**.

4 T.R. Crompton, (Ed.). *Polymer Reference Book*. Rapra Technology, New York, **2006**.

5 G. Montaudo, R.P. Lattimer, (Eds.). *Mass Spectrometry of Polymers*. CRC, New York, **2001**.

6 H. Pasch, W. Schrepp, (Eds.). *MALDI-TOF Mass Spectrometry of Synthetic Polymers*. Springer, New York, **2003**.

7 D. Briggs, (Ed.). *Surface Analysis of Polymers by XPS and Static SIMS*. Cambridge University Press, New York, **2005**.

8 U. Bahr, A. Deppe, M. Karas, F. Hillenkamp, U. Giessmann. Mass spectrometry of synthetic polymers by UV-matrix-assisted laser desorption/ionization. *Anal. Chem.* **1992**, *64*, 2866–2869.

9 P.O. Danis, D.E. Karr, F. Mayer, A. Holle, C.H. Watson. The analysis of water-soluble polymers by matrix-assisted laser desorption time-of-flight mass spectrometry. *Organic Mass Spectrom.* **1992**, *27*, 843–846.

10 J.A. Castro, C. Koester, C. Wilkins. Matrix-assisted laser

desorption/ionization of high-mass molecules by Fourier-transform mass spectrometry. *Rapid Commun. Mass Spectrom.* **1992**, *6*, 239–241.

11 S.D. Hanton. Mass spectrometry of polymers and polymer surfaces. *Chem. Rev. (Washington, DC)* **2001**, *101*, 527–569.

12 M.W.F. Nielen. MALDI time-of-flight mass spectrometry of synthetic polymers. *Mass Spectrom. Rev.* **1999**, *18*, 309–344.

13 R. Knochenmuss, R. Zenobi. MALDI ionization: The role of in-plume processes. *Chem. Rev. (Washington, DC)* **2003**, *103*, 441–452.

14 J. Gidden, M.T. Bowers, A.T. Jackson, J.H. Scrivens. Gas-phase conformations of cationized poly(styrene) oligomers. *J. Am. Soc. Mass Spectrom.* **2002**, *13*, 499–505.

15 S.D. Hanton, K.G. Owens, C. Chavez-Eng, A.-M. Hoberg, P.J. Derrick. Updating evidence for cationization of polymers in the gas phase during matrix-assisted laser desorption/ionization. *Eur. J. Mass Spectrom.* **2005**, *11*, 23–29.

16 R. Knochenmuss, E. Lehmann, R. Zenobi. Polymer cationization in matrix-assisted laser desorption/ionization. *Eur. Mass Spectrom.* **1998**, *4*, 421–427.

17 P.O. Danis, D.E. Karr. A facile sample preparation for the analysis of synthetic organic polymers by matrix-assisted laser desorption/ionization. *Organic Mass Spectrom.* **1993**, *28*, 923–925.

18 D.C. Schriemer, L. Li. Mass discrimination in the analysis of polydisperse polymers by MALDI time-of-flight mass spectrometry. 1. Sample preparation and desorption/ionization issues. *Anal. Chem.* **1997**, *69*, 4169–4175.

19 A.E. Giannakopulos, S. Bashir, P.J. Derrick. Reproducibility of spectra and threshold fluence in matrix-assisted laser desorption/ionization (MALDI) of polymers. *Eur. Mass Spectrom.* **1998**, *4*, 127–131.

20 R. Arakawa, S. Watanabe, T. Fukuo. Effects of sample preparation on matrix-assisted laser desorption/ionization time-of-flight mass spectra for sodium polystyrene sulfonate. *Rapid Commun. Mass Spectrom.* **1999**, *13*, 1059–1062.

21 S.D. Hanton, P.A. Cornelio Clark, K.G. Owens. Investigations of matrix-assisted laser desorption/ionization sample preparation by time-of-flight secondary ion mass spectrometry. *J. Am. Soc. Mass Spectrom.* **1999**, *10*, 104–111.

22 M.A.R. Meier, B.-J. de Gans, A.M.J. van den Berg, U.S. Schubert. Automated multiple-layer spotting for matrix-assisted laser desorption/ionization time-of-flight mass spectrometry of synthetic polymers utilizing ink-jet printing technology. *Rapid Commun. Mass Spectrom.* **2003**, *17*, 2349–2353.

23 K.L. Walker, M.S. Kahr, C.L. Wilkins, Z. Xu, J.S. Moore. Analysis of hydrocarbon dendrimers by laser desorption time-of-flight and Fourier transform mass spectrometry. *J. Am. Soc. Mass Spectrom.* **1994**, *5*, 731–739.

24 S.J. Wetzel, C.M. Guttman, J.E. Girard. The influence of matrix and laser energy on the molecular mass distribution of synthetic polymers obtained by MALDI-TOF-MS. *Int. J. Mass Spectrom.* **2004**, *238*, 215–225.

25 P. Juhasz, C.E. Costello, K. Biemann. Matrix-assisted laser desorption ionization mass spectrometry with 2-(4-hydroxyphenylazo)benzoic acid matrix. *J. Am. Soc. Mass Spectrom.* **1993**, *4*, 399–409.

26 P.O. Danis, D.E. Karr, W.J. Simonsick, Jr., D.T. Wu. Matrix-assisted laser desorption/ionization time-of-flight characterization of poly(butyl methacrylate) synthesized by group-transfer polymerization. *Macromolecules* **1995**, *28*, 1229–1232.

27 P. Juhasz, C.E. Costello. Generation of large radical ions from oligometallocenes by matrix-assisted laser desorption ionization. *Rapid Commun. Mass Spectrom.* **1993**, *7*, 343–351.

28 A.M. Belu, J.M. DeSimone, R.W. Linton, G.W. Lange, R.M. Friedman. Evaluation of matrix-assisted laser desorption ionization mass spectrometry for polymer characterization. *J. Am. Soc. Mass Spectrom.* **1996**, *7*, 11–24.

29 D.C. Schriemer, L. Li. Detection of high molecular weight narrow polydisperse polymers up to 1.5 million daltons by MALDI mass spectrometry. *Anal. Chem.* **1996**, *68*, 2721–2725.

30 A. Marie, S. Alves, F. Fournier, J.C. Tabet. Fluorinated matrix approach for the characterization of hydrophobic perfluoropolyethers by matrix-assisted laser desorption/ionization time-of-flight MS. *Anal. Chem.* **2003**, *75*, 1294–1299.

31 L. Latourte, J.-C. Blais, J.-C. Tabet, R.B. Cole. Desorption behavior and distribution of fluorinated polymers in MALDI and electrospray ionization mass spectrometry. *Anal. Chem.* **1997**, *69*, 2742–2750.

32 S. Trimpin, A.C. Grimsdale, H.J. Raeder, K. Muellen. Characterization of an insoluble poly(9,9-diphenyl-2,7-fluorene) by solvent-free sample preparation for MALDI-TOF mass spectrometry. *Anal. Chem.* **2002**, *74*, 3777–3782.

33 X. Tang, P.A. Dreifuss, A. Vertes. New matrixes and accelerating voltage effects in matrix-assisted laser desorption/ionization of synthetic polymers. *Rapid Commun. Mass Spectrom.* **1995**, *9*, 1141–1147.

34 M. Mank, B. Stahl, G. Boehm. 2,5-Dihydroxybenzoic acid butylamine and other ionic liquid matrixes for enhanced MALDI-MS analysis of biomolecules. *Anal. Chem.* **2004**, *76*, 2938–2950.

35 P.A. Limbach, S.F. Macha, C. Robins. Analysis of hydrocarbon materials by matrix-assisted laser desorption/ionization mass spectrometry: searching for the perfect matrix. *Preprints – Am.*

Chem. Soc., Division of Petroleum Chemistry **2000**, *45*, 561–563.

36 S.F. Macha, P.A. Limbach, P.J. Savickas. Application of nonpolar matrices for the analysis of low molecular weight nonpolar synthetic polymers by matrix-assisted laser desorption/ionization time-of-flight mass spectrometry. *J. Am. Soc. Mass Spectrom.* **2000**, *11*, 731–737.

37 K. Linnemayr, P. Vana, G. Allmaier. Time-delayed extraction matrix-assisted laser desorption/ionization time-of-flight mass spectrometry of polyacrylonitrile and other synthetic polymers with the matrix 4-hydroxybenzylidene malononitrile. *Rapid Commun. Mass Spectrom.* **1998**, *12*, 1344–1350.

38 R.M. Whittal, D.C. Schriemer, L. Li. Time-lag focusing MALDI time-of-flight mass spectrometry for polymer characterization: oligomer resolution, mass accuracy, and average weight information. *Anal. Chem.* **1997**, *69*, 2734–2741.

39 O. Vorm, P. Roepstorff, M. Mann. Improved resolution and very high sensitivity in MALDI-TOF of matrix surfaces made by fast evaporation. *Anal. Chem.* **1994**, *66*, 3281–3287.

40 Y. Dai, R.M. Whittal, L. Li. Confocal fluorescence microscopic imaging for investigating the analyte distribution in MALDI matrixes. *Anal. Chem.* **1996**, *68*, 2494–2500.

41 S.J. Pastor, C.L. Wilkins. Analysis of hydrocarbon polymers by matrix-assisted laser desorption/ionization-Fourier transform mass spectrometry. *J. Am. Soc. Mass Spectrom.* **1997**, *8*, 225–233.

42 T. Yalcin, D.C. Schriemer, L. Li. Matrix-assisted laser desorption ionization time-of-flight mass spectrometry for the analysis of polydienes. *J. Am. Soc. Mass Spectrom.* **1997**, *8*, 1220–1229.

43 H. Rashidezadeh, B. Guo. Investigation of metal attachment to polystyrenes in matrix-assisted laser desorption ionization. *J. Am. Soc. Mass Spectrom.* **1998**, *9*, 724–730.

44 Y. Wang, H. Rashidzadeh, B. Guo. Structural effects on polyether cationization by alkali metal ions in matrix-assisted laser desorption/

ionization. *J. Am. Soc. Mass Spectrom.* **2000**, *11*, 639–643.

45 B.J. Bauer, W.E. Wallace, B.M. Fanconi, C.M. Guttman. Covalent cationization method for the analysis of polyethylene by mass spectrometry. *Polymer* **2001**, *42*, 09949–09953.

46 J. He, M. He, J. Pei, A. Ding, W. Huang. Analysis of bipyridyl-containing conjugated polymers by matrix-assisted laser desorption/ionization time-of-flight mass spectrometry. *Rapid Commun. Mass Spectrom.* **2001**, *15*, 1239–1243.

47 S.F. Macha, P.A. Limbach, S.D. Hanton, K.G. Owens. Silver cluster interferences in matrix-assisted laser desorption/ionization (MALDI) mass spectrometry of nonpolar polymers. *J. Am. Soc. Mass Spectrom.* **2001**, *12*, 732–743.

48 S. Lin-Gibson, L. Brunner, D.L. Vanderhart, B.J. Bauer, B.M. Fanconi, C.M. Guttman, W.E. Wallace. Optimizing the covalent cationization method for the mass spectrometry of polyolefins. *Macromolecules* **2002**, *35*, 7149–7156.

49 E. Royo, H.-H. Brintzinger. Mass spectrometry of polystyrene and polypropene ruthenium complexes. A new tool for polymer characterization. *J. Organometal. Chem.* **2002**, *663*, 213–220.

50 H. Chen, M. He, X. Wan, L. Yang, H. He. Matrix-assisted laser desorption/ionization study of cationization of PEO-PPP rod-coil diblock polymers. *Rapid Commun. Mass Spectrom.* **2003**, *17*, 177–182.

51 J. Cvacka, A. Svatos. Matrix-assisted laser desorption/ionization analysis of lipids and high molecular weight hydrocarbons with lithium 2,5-dihydroxybenzoate matrix. *Rapid Commun. Mass Spectrom.* **2003**, *17*, 2203–2207.

52 Z. Zhao, S. Xiong, X. Liu, B. Xin, G. Wang. Matrix-assisted laser desorption/ionization time-of-flight mass spectrometry studies of low molecular weight dendrimers. *Eur. J. Mass Spectrom.* **2003**, *9*, 203–211.

53 H.C.M. Byrd, S.A. Bencherif, B.J. Bauer, K.L. Beers, Y. Brun, S. Lin-Gibson, N. Sari. Examination of the covalent cationization method using narrow polydisperse polystyrene. *Macromolecules* **2005**, *38*, 1564–1572.

54 G. Montaudo, F. Samperi, M.S. Montaudo. Characterization of synthetic polymers by MALDI-MS. *Progr. Polymer Sci.* **2006**, *31*, 277–357.

55 A.T. Jackson, H.T. Yayes, W.A. MacDonald, J.H. Scrivens, G. Critchley, J. Brown, M.J. Deery, K.R. Jennings, C. Brookes. Time-lag focusing and cation attachment in the analysis of synthetic polymers by matrix-assisted laser desorption/ionization-time-of-flight-mass spectrometry. *J. Am. Soc. Mass Spectrom.* **1997**, *8*, 132–139.

56 R.J. Goldschmidt, S.J. Wetzel, W.R. Blair, C.M. Guttman. Post-source decay in the analysis of polystyrene by matrix-assisted laser desorption/ionization time-of-flight mass spectrometry. *J. Am. Soc. Mass Spectrom.* **2000**, *11*, 1095–1106.

57 M.J. Deery, K.R. Jennings, C.B. Jasieczek, D.M. Haddleton, A.T. Jackson, H.T. Yates, J.H. Scrivens. A study of cation attachment to polystyrene by means of matrix-assisted laser desorption/ionization and electrospray ionization-mass spectrometry. *Rapid Commun. Mass Spectrom.* **1997**, *11*, 57–62.

58 D. Braun, R. Ghahary, H. Pasch. Triazine-based polymers. 4. MALDI-MS of triazine-based polyamines. *Polymer* **1996**, *37*, 777–783.

59 H. Chen, B. Guo. Use of binary solvent systems in the MALDI-TOF analysis of poly(methyl methacrylate). *Anal. Chem.* **1997**, *69*, 4399–4404.

60 T. Yalcin, Y. Dai, L. Li. Matrix-assisted laser desorption/ionization time-of-flight mass spectrometry for polymer analysis: solvent effect in sample preparation. *J. Am. Soc. Mass Spectrom.* **1998**, *9*, 1303–1310.

61 R. Chen, N. Zhang, A.M. Tseng, L. Li. Effects of matrix-assisted laser desorption/ionization experimental conditions on quantitative compositional analysis of ethylene oxide/propylene oxide copolymers. *Rapid Commun. Mass Spectrom.* **2000**, *14*, 2175–2181.

62 I. Folch, S. Borros, D.B. Amabilino, J. Veciana. Matrix-assisted laser desorption/ionization time-of-flight mass spectrometric analysis of some conducting polymers. *J. Mass Spectrom.* **2000**, *35*, 550–555.

63 S.D. Hanton, K.G. Owens. Using MESIMS to analyze polymer MALDI matrix solubility. *J. Am. Soc. Mass Spectrom.* **2005**, *16*, 1172–1180.

64 A.J. Hoteling, T.H. Mourey, K.G. Owens. Importance of solubility in the sample preparation of poly(ethylene terephthalate) for MALDI-TOF MS. *Anal. Chem.* **2005**, *77*, 750–756.

65 D.V. Zagorevskii, M.J. Nasrullah, V. Raghunadh, B.C. Benicewicz. The effect of tetrahydrofuran as solvent on matrix-assisted laser desorption/ionization and electrospray ionization mass spectra of functional polystyrenes. *Rapid Commun. Mass Spectrom.* **2006**, *20*, 178–180.

66 P.M. Lloyd, K.G. Suddaby, J.E. Varney, E. Scrivener, P.J. Derrick, D.M. Haddleton. A comparison between matrix-assisted laser desorption/ionization time-of-flight mass spectrometry and size exclusion chromatography in the mass characterization of synthetic polymers with narrow molecular-mass distributions: poly(methyl methacrylate) and poly(styrene). *Eur. Mass Spectrom.* **1995**, *1*, 293–300.

67 C.M. Guttman, W.R. Blair, P.O. Danisa. Comparison of molecular weight moments from MALDI-TOF-MS with other absolute methods on a standard reference polymer. *Annual Technical Conference – Society of Plastics Engineers* **1998**, 56th, (Vol. 2), 2109–2113.

68 G. Hagelin, J.M. Arukwe, V. Kasparkova, S. Nordbo, A. Rogstad. Characterization of low molecular weight polymers by matrix-assisted laser desorption/ionization mass spectrometry. A comparison with gel permeation chromatography. *Rapid Commun. Mass Spectrom.* **1998**, *12*, 25–27.

69 C.M. Guttman. The relationship between the signals from size exclusion chromatography and time of flight mass spectrometry to a polymer molecular weight distribution. *Polymer Preprints (Am. Chem. Soc., Division of Polymer Chemistry)* **1996**, *37*, 837–838.

70 N. Sakurada, T. Fukuo, R. Arakawa, K. Ute, K. Hatada. Characterization of poly(methyl methacrylate) by matrix-assisted laser desorption/ionization mass spectrometry. A comparison with supercritical fluid chromatography and gel permeation chromatography. *Rapid Commun. Mass Spectrom.* **1998**, *12*, 1895–1898.

71 C. Jackson, B. Larsen, C. McEwen. Comparison of most probable peak values as measured for polymer distributions by MALDI mass spectrometry and by size exclusion chromatography. *Anal. Chem.* **1996**, *68*, 1303–1308.

72 J. Liu, R.S. Loewe, R.D. McCullough. Employing MALDI-MS on poly(alkylthiophenes): analysis of molecular weights, molecular weight distributions, end-group structures, and end-group modifications. *Macromolecules* **1999**, *32*, 5777–5785.

73 A. Bootz, T. Russ, F. Gores, M. Karas, J. Kreuter. Molecular weights of poly(butyl cyanoacrylate) nanoparticles determined by mass spectrometry and size exclusion chromatography. *Eur. J. Pharmaceut. Biopharmaceut.* **2005**, *60*, 391–399.

74 J.B. Williams, A.I. Gusev, D.M. Hercules. Characterization of polyesters by matrix-assisted laser desorption ionization mass spectrometry. *Macromolecules* **1997**, *30*, 3781–3787.

75 R. Skelton, F. Dubois, R. Zenobi. A MALDI sample preparation method suitable for insoluble polymers. *Anal. Chem.* **2000**, *72*, 1707–1710.

76 M.A.R. Meier, U.S. Schubert. MALDI-TOFMS of synthetic polymers as advanced high-throughput screening tool in combinatorial polymer chemistry. *PMSE Preprints* **2004**, *90*, 344–345.

77 M.A.R. Meier, U.S. Schubert. Evaluation of a new multiple-layer spotting technique for matrix-assisted laser desorption/ionization time-of-flight mass spectrometry of synthetic polymers. *Rapid Commun. Mass Spectrom.* **2003**, *17*, 713–716.

78 M.A.R. Meier, B.G.G. Lohmeijer, U.S. Schubert. Recent advances in the characterization of terpyridine based supramolecular polymers. *Polymer Preprints (Am. Chem. Soc., Division of Polymer Chemistry)* **2004**, *45*, 389–390.

79 M.A.R. Meier, B.G.G. Lohmeijer, U.S. Schubert. Characterization of defined metal-containing supramolecular block copolymers. *Macromol. Rapid Commun.* **2003**, *24*, 852–857.

80 L. Przybilla, J.-D. Brand, K. Yoshimura, H.J. Raeder, K. Muellen. MALDI-TOF mass spectrometry of insoluble giant polycyclic aromatic hydrocarbons by a new method of sample preparation. *Anal. Chem.* **2000**, *72*, 4591–4597.

81 S.D. Hanton, D.M. Parees. Extending the solvent-free MALDI sample preparation method. *J. Am. Soc. Mass Spectrom.* **2005**, *16*, 90–93.

82 S. Trimpin, S. Keune, H.J. Raeder, K. Muellen. Solvent-Free MALDI-MS: Developmental improvements in the reliability and the potential of MALDI in the analysis of synthetic polymers and giant organic molecules. *J. Am. Soc. Mass Spectrom.* **2006**, *17*, 661–671.

83 S. Trimpin, A. Rouhanipour, R. Az, H.J. Rader, K. Mullen. New aspects in matrix-assisted laser desorption/ionization time-of-flight mass spectrometry: a universal solvent-free sample preparation. *Rapid Commun. Mass Spectrom.* **2001**, *15*, 1364–1373.

84 A.P. Gies, W.K. Nonidez, M. Anthamatten, R.C. Cook. Characterization of the imidization of poly(amic acid) to an insoluble polyimide oligomer by matrix assisted laser desorption ionization time-of-flight mass spectrometry. *PMSE Preprints* **2003**, *89*, 392.

85 B.J. Bauer, K.M. Flynn, B.D. Vogt. MALDI of layered polymer films. *PMSE Preprints* **2004**, *90*, 457–458.

86 A.R. Dolan, T.D. Wood. Analysis of polyaniline oligomers by laser desorption ionization and solventless MALDI. *J. Am. Soc. Mass Spectrom.* **2004**, *15*, 893–899.

87 V. Horneffer, M. Glueckmann, R. Krueger, M. Karas, K. Strupat, F. Hillenkamp. Matrix-analyte-interaction in MALDI-MS: Pellet and nano-electrospray preparations. *Int. J. Mass Spectrom.* **2006**, *249/250*, 426–432.

88 D.C. Schriemer, R.M. Whittal, L. Li. Analysis of structurally complex polymers by time-lag-focusing matrix-assisted laser desorption ionization time-of-flight mass spectrometry. *Macromolecules* **1997**, *30*, 1955–1963.

89 R.M. Whittal, L. Li. Characterization of pyrene end-labeled polyethylene glycol by high resolution MALDI time-of-flight mass spectrometry. *Macromol. Rapid Commun.* **1996**, *17*, 59–64.

90 R.P. Lattimer. Tandem mass spectrometry of lithium-attachment ions from polyglycols. *J. Am. Soc. Mass Spectrom.* **1992**, *3*, 225–234.

91 S.J. Pastor, C.L. Wilkins. Laser Fourier transform mass spectrometry (FT-MS). In: G. Montaudo, R.P. Lattimer, (Eds.). *Mass Spectrometry of Polymers.* CRC, New York, **2001**, pp. 389–417.

92 J.A. Castoro, C.L. Wilkins. Fourier transform mass spectrometry for high resolution matrix-assisted laser desorption/ionization MS. *Proceedings, SPIE-The International Society for Optical Engineering* **1993**, 1857, (Lasers and Optics for Surface Analysis), 51–59.

93 J.E. Campana, L.-S. Sheng, S.L. Shew, B.E. Winger. Polymer analysis by photons, sprays, and mass spectrometry. *Trends Anal. Chem.* **1994**, *13*, 239–247.

94 C.G. de Koster, M.C. Duursma, G.J. van Rooij, R.M.A. Heeren, J.J. Boon. Endgroup analysis of polyethylene glycol polymers by matrix-assisted laser desorption/ionization Fourier-transform ion cyclotron resonance mass spectrometry. *Rapid Commun. Mass Spectrom.* **1995**, *9*, 957–962.

95 M. Dey, J.A. Castoro, C.L. Wilkins. Determination of molecular weight distributions of polymers by MALDI-FTMS. *Anal. Chem.* **1995**, *67*, 1575–1579.

96 C.C. Pitsenberger, M.L. Easterling, I.J. Amster. Effects of capacitive coupling on ion remeasurement using quadrupolar excitation in high-resolution FTICR spectrometry. *Anal. Chem.* **1996**, *68*, 4409–4413.

97 G.J. van Rooij, M.C. Duursma, R.M.A. Heeren, J.J. Boon, C.G. de Koster. High resolution end group determination of low molecular weight polymers by matrix-assisted laser desorption ionization on an external ion source Fourier transform ion cyclotron

resonance mass spectrometer. *J. Am. Soc. Mass Spectrom.* **1996**, *7*, 449–457.

98 F.M. White, J.A. Marto, A.G. Marshall. An external source 7 T Fourier transform ion cyclotron resonance mass spectrometer with electrostatic ion guide. *Rapid Commun. Mass Spectrom.* **1996**, *10*, 1845–1849.

99 M.L. Easterling, T.H. Mize, I. Jonathan Amster. MALDI FTMS analysis of polymers: improved performance using an open ended cylindrical analyzer cell. *Int. J. Mass Spectrom. Ion Processes* **1997**, *169/170*, 387–400.

100 P.B. O'Connor, M.C. Duursma, G.J. van Rooij, R.M.A. Heeren, J.J. Boon. Correction of time-of-flight shifted polymeric molecular weight distributions in matrix-assisted laser desorption/ionization Fourier transform mass spectrometry. *Anal. Chem.* **1997**, *69*, 2751–2755.

101 P.O. Staneke, N.M.M. Nibbering. Matrix-assisted laser desorption ionization and electrospray ionization combined with Fourier transform ion cyclotron resonance mass spectrometry. *Spectroscopy (Amsterdam)* **1997**, *13*, 145–150.

102 E.R.E. van der Hage, M.C. Duursma, R.M.A. Heeren, J.J. Boon, M.W.F. Nielen, A.J.M. Weber, C.G. de Koster, N.K. de Vries. Structural analysis of polyoxyalkyleneamines by matrix-assisted laser desorption/ionization on an external ion source FT-ICR-MS and NMR. *Macromolecules* **1997**, *30*, 4302–4309.

103 S.J. Pastor, C.L. Wilkins. Sustained off-resonance irradiation and collision-induced dissociation for structural analysis of polymers by MALDI-FTMS. *Int. J. Mass Spectrom. Ion Processes* **1998**, *175*, 81–92.

104 G.J. van Rooij, M.C. Duursma, C.G. de Koster, R.M.A. Heeren, J.J. Boon, P.J.W. Schuyl, E.R.E. van der Hage. Determination of block length distributions of poly(oxypropylene) and poly(oxyethylene) block copolymers by MALDI-FTICR mass spectrometry. *Anal. Chem.* **1998**, *70*, 843–850.

105 M.L. Easterling, I.J. Amster, G.J. van Rooij, R.M.A. Heeren. Isotope beating effects in the analysis of polymer distributions by Fourier transform mass spectrometry. *J. Am. Soc. Mass Spectrom.* **1999**, *10*, 1074–1082.

106 S.D.H. Shi, J.J. Drader, C.L. Hendrickson, A.G. Marshall. Fourier transform ion cyclotron resonance mass spectrometry in a high homogeneity 25 tesla resistive magnet. *J. Am. Soc. Mass Spectrom.* **1999**, *10*, 265–268.

107 A.J. Jaber, C.L. Wilkins. Hydrocarbon polymer analysis by external MALDI Fourier transform and reflectron time of flight mass spectrometry. *J. Am. Soc. Mass Spectrom.* **2005**, *16*, 2009–2016.

108 J. Jaber Arwah, J. Kaufman, R. Liyanage, E. Akhmetova, S. Marney, C.L. Wilkins. Trapping of wide range mass-to-charge ions and dependence on matrix amount in internal source MALDI-FTMS. *J. Am. Soc. Mass Spectrom.* **2005**, *16*, 1772–1780.

109 F.J. Cox, K. Qian, A.O. Patil, M.V. Johnston. Microstructure and composition of ethylene-carbon monoxide copolymers by matrix-assisted laser desorption/ionization mass spectrometry. *Macromolecules* **2003**, *36*, 8544–8550.

110 T.H. Mize, W.J. Simonsick, Jr., I.J. Amster. Characterization of polyesters by matrix-assisted laser desorption/ionization and Fourier transform mass spectrometry. *Eur. J. Mass Spectrom.* **2003**, *9*, 473–486.

111 R. Chen, T. Yalcin, W.E. Wallace, C.M. Guttman, L. Li. Laser desorption ionization and MALDI time-of-flight mass spectrometry for low molecular mass polyethylene analysis. *J. Am. Soc. Mass Spectrom.* **2001**, *12*, 1186–1192.

112 T. Yalcin, W.E. Wallace, C.M. Guttman, L. Li. Metal powder substrate-assisted laser desorption/ionization mass spectrometry for polyethylene analysis. *Anal. Chem.* **2002**, *74*, 4750–4756.

113 C. McEwen, C. Jackson, B. Larsen. The fundamentals of characterizing polymers using MALDI mass spectrometry. *Polymer Preprints (Am. Chem. Soc., Division of Polymer Chemistry)* **1996**, *37*, 314–315.

114 B. Guo, H. Rashidzadeh. Major origins of mass discrimination encountered in the MALDI-TOF analysis of polydisperse

polymers. *Annual Technical Conference – Society of Plastics Engineers* **1998**, 56th, (Vol. 2), 2096–2100.

115 H. Rashidzadeh, B. Guo. Use of MALDI-TOF to measure molecular weight distributions of polydisperse poly(methyl methacrylate). *Anal. Chem.* **1998**, *70*, 131–135.

116 J.B. Williams, T.M. Chapman, D.M. Hercules. Matrix-assisted laser desorption/ionization mass spectrometry of discrete mass poly(butylene glutarate) oligomers. *Anal. Chem.* **2003**, *75*, 3092–3100.

117 G. Montaudo, M.S. Montaudo, C. Puglisi, F. Samperi. Characterization of polymers by matrix-assisted laser desorption ionization-time of flight mass spectrometry. End group determination and molecular weight estimates in poly(ethylene glycols). *Macromolecules* **1995**, *28*, 4562–4569.

118 A.T. Jackson, H.T. Yates, C.I. Lindsay, Y. Didier, J.A. Segal, J.H. Scrivens, G. Critchley, J. Brown. Utilizing time-lag focusing matrix-assisted laser desorption/ionization mass spectrometry for the end group analysis of synthetic polymers. *Rapid Commun. Mass Spectrom.* **1997**, *11*, 520–526.

119 A.T. Jackson, H.T. Yates, C.I. Lindsay, Y. Didier, J.A. Segal, J.H. Scrivens, G. Critchley, J. Brown. Utilizing time-lag focusing ultraviolet-matrix-assisted laser desorption/ionization-mass spectrometry for the end group analysis of synthetic polymers. *Analusis* **1998**, *26*, M31–M35.

120 H. Zhu, T. Yalcin, L. Li. Analysis of the accuracy of determining average molecular weights of narrow polydispersity polymers by matrix-assisted laser desorption ionization time-of-flight mass spectrometry. *J. Am. Soc. Mass Spectrom.* **1998**, *9*, 275–281.

121 S.J. Wetzel, C.M. Guttman, K.M. Flynn, J.J. Filliben. Significant parameters in the optimization of MALDI-TOF-MS for synthetic polymers. *J. Am. Soc. Mass Spectrom.* **2006**, *17*, 246–252.

122 P.O. Danis, D.E. Karr. Analysis of poly(styrenesulfonic acid) by matrix-assisted laser desorption/ionization time-of-flight mass spectrometry. *Macromolecules* **1995**, *28*, 8548–8551.

123 P.J. Savickas. *Poster presentation at the Sanibel Conference on Mass Spectrometry, Chicago, IL,* **1994**.

124 R.J. Wenzel, U. Matter, L. Schultheis, R. Zenobi. Analysis of megadalton ions using cryodetection MALDI time-of-flight mass spectrometry. *Anal. Chem.* **2005**, *77*, 4329–4337.

125 A.T. Jackson, K.R. Jennings, J.H. Scrivens. Generation of average mass values and end group information of polymers by means of a combination of matrix-assisted laser desorption/ionization-mass spectrometry and liquid secondary ion-tandem mass spectrometry. *J. Am. Soc. Mass Spectrom.* **1997**, *8*, 76–85.

126 A. Adhiya, C. Wesdemiotis. Poly(propylene imine) dendrimer conformations in the gas phase: a tandem mass spectrometry study. *Int. J. Mass Spectrom.* **2002**, *214*, 75–88.

127 C.S. Creaser, J.C. Reynolds, A.J. Hoteling, W.F. Nichols, K.G. Owens. Atmospheric pressure matrix-assisted laser desorption/ionisation ion trap mass spectrometry of synthetic polymers: A comparison with vacuum matrix-assisted laser desorption/ionization time-of-flight mass spectrometry. *Eur. J. Mass Spectrom.* **2003**, *9*, 33–44.

128 P. Rizzarelli, C. Puglisi, G. Montaudo. Sequence determination in aliphatic poly(ester amide)s by matrix-assisted laser desorption/ionization time-of-flight and time-of-flight/time-of-flight tandem mass spectrometry. *Rapid Commun. Mass Spectrom.* **2005**, *19*, 2407–2418.

129 K.E. Warburton, M.R. Clench, M.J. Ford, J. White, D.A. Rimmer, V.A. Carolan. Characterisation of derivatised monomeric and prepolymeric isocyanates by matrix-assisted laser desorption/ionisation time-of-flight mass spectrometry and structural elucidation by tandem mass spectrometry. *Eur. J. Mass Spectrom.* **2005**, *11*, 565–574.

130 V. Havlicek, C. Kieburg, P. Novak, K. Bezouska, T.K. Lindhorst. Structure analysis of trivalent glyco-clusters by post-source decay matrix-assisted laser desorption/ionization mass spectrometry. *J. Mass Spectrom.* **1998**, *33*, 591–598.

131 L. Przybilla, H.J. Rader, K. Mullen. Post-source decay fragment ion analysis of polycarbonates by matrix-assisted laser desorption/ionization time-of-flight mass spectrometry. *Eur. Mass Spectrom.* **1999**, *5*, 133–143.

132 H. Nonami, F. Wu, R.P. Thummel, Y. Fukuyama, H. Yamaoka, R. Erra-Balsells. Evaluation of pyridoindoles, pyridylindoles and pyridylpyridoindoles as matrices for ultraviolet matrix-assisted laser desorption/ionization time-of-flight mass spectrometry. *Rapid Commun. Mass Spectrom.* **2001**, *15*, 2354–2373.

133 L. Przybilla, V. Francke, H.J. Raeder, K. Muellen. Block length determination of a poly(ethylene oxide)-b-poly(p-phenylene ethynylene) diblock copolymer by means of MALDI-TOF mass spectrometry combined with fragment-ion analysis. *Macromolecules* **2001**, *34*, 4401–4405.

134 I. Fournier, A. Marie, D. Lesage, G. Bolbach, F. Fournier, J.C. Tabet. Post-source decay time-of-flight study of fragmentation mechanisms of protonated synthetic polymers under matrix-assisted laser desorption/ionization conditions. *Rapid Commun. Mass Spectrom.* **2002**, *16*, 696–704.

135 S. Keki, M. Nagy, G. Deak, Z. Miklos, P. Herczegh. Matrix-assisted laser desorption/ionization mass spectrometric study of bis(imidazole-1-carboxylate) endfunctionalized polymers. *J. Am. Soc. Mass Spectrom.* **2003**, *14*, 117–123.

136 O. Laine, S. Trimpin, H.J. Raeder, K. Muellen. Changes in post-source decay fragmentation behavior of poly(methyl methacrylate) polymers with increasing molecular weight studied by matrix-assisted laser desorption/ionization time-of-flight mass spectrometry. *Eur. J. Mass Spectrom.* **2003**, *9*, 195–201.

137 S.D. Hanton, D.M. Parees, K.G. Owens. MALDI PSD of low molecular weight ethoxylated polymers. *Int. J. Mass Spectrom.* **2004**, *238*, 257–264.

138 A.J. Hoteling, K.G. Owens. Improved PSD and CID on a MALDI-TOFMS. *J. Am. Soc. Mass Spectrom.* **2004**, *15*, 523–535.

139 U.E.C. Berndt, T. Zhou, R.C. Hider, Z.D. Liu, H. Neubert. Structural characterization of chelator-terminated dendrimers and their synthetic intermediates by mass spectrometry. *J. Mass Spectrom.* **2005**, *40*, 1203–1214.

140 R. Chen, L. Li. Reactions of atomic transition-metal ions with long-chain alkanes. *J. Am. Soc. Mass Spectrom.* **2001**, *12*, 367–375.

141 R. Chen, T. Yalcin, W.E. Wallace, C.M. Guttman, L. Li. Laser desorption ionization and MALDI time-of-flight mass spectrometry for low molecular mass polyethylene analysis. *J. Am. Soc. Mass Spectrom.* **2001**, *12*, 1186–1192.

142 S.C. Moyer, R.J. Cotter. Atmospheric pressure MALDI. *Anal. Chem.* **2002**, *74*, 468A–476A.

143 V.M. Doroshenko, V.V. Laiko, N.I. Taranenko, V.D. Berkout, H.S. Lee. Recent developments in atmospheric pressure MALDI mass spectrometry. *Int. J. Mass Spectrom.* **2002**, *221*, 39–58.

144 S.D. Hanton, D.M. Parees, J. Zweigenbaum. The fragmentation of ethoxylated surfactants by AP-MALDI-QIT. *J. Am. Soc. Mass Spectrom.* **2006**, *17*, 453–458.

145 R.P. Lattimer, M.J. Polce, C. Wesdemiotis. MALDI-MS analysis of pyrolysis products from a segmented polyurethane. *J. Analyt. Appl. Pyrolysis* **1998**, *48*, 1–15.

146 R. Murgasova, E.L. Brantley, D.M. Hercules, H. Nefzger. Characterization of polyester-polyurethane soft and hard blocks by a combination of MALDI, SEC, and chemical degradation. *Macromolecules* **2002**, *35*, 8338–8345.

147 R. Murgasova, D.M. Hercules, J.R. Edman. Characterization of polyimides by combining mass spectrometry and selective chemical reaction. *Macromolecules* **2004**, *37*, 5732–5740.

148 D.C. Schriemer, L. Li. Mass discrimination in the analysis of polydisperse polymers by MALDI time-of-flight mass spectrometry. 2. Instrumental issues. *Anal. Chem.* **1997**, *69*, 4176–4183.

149 W.E. Wallace, C.M. Guttman. Data analysis methods for synthetic polymer mass spectrometry: autocorrelation.

150 W.E. Wallace, A.J. Kearsley, C.M. Guttman. An operator-independent approach to mass spectral peak identification and integration. *Anal. Chem.* **2004**, *76*, 2446–2452.

151 C.M. Guttman, S.J. Wetzel, W.R. Blair, B.M. Fanconi, J.E. Girard, R.J. Goldschmidt, W.E. Wallace, D.L. Van der Hart. NIST-sponsored interlaboratory comparison of polystyrene molecular mass distribution obtained by matrix-assisted laser desorption/ionization time-of-flight mass spectrometry: statistical analysis. *Anal. Chem.* **2001**, *73*, 1252–1262.

152 R. Chen, A.M. Tseng, M. Uhing, L. Li. Application of an integrated matrix-assisted laser desorption/ionization time-of-flight, electrospray ionization mass spectrometry and tandem mass spectrometry approach to characterizing complex polyol mixtures. *J. Am. Soc. Mass Spectrom.* **2001**, *12*, 55–60.

153 H.C.M. Byrd, C.N. McEwen. The limitations of MALDI-TOF mass spectrometry in the analysis of wide polydisperse polymers. *Anal. Chem.* **2000**, *72*, 4568–4576.

154 G. Montaudo, M.S. Montaudo, C. Puglisi, F. Samperi. Characterization of polymers by matrix-assisted laser desorption/ionization time-of-flight mass spectrometry: molecular weight estimates in samples of varying polydispersity. *Rapid Commun. Mass Spectrom.* **1995**, *9*, 453–460.

155 J. Axelsson, E. Scrivener, D.M. Haddleton, P.J. Derrick. Mass discrimination effects in an ion detector and other causes for shifts in polymer mass distributions measured by matrix-assisted laser desorption/ionization time-of-flight mass spectrometry. *Macromolecules* **1996**, *29*, 8875–8882.

156 K. Martin, J. Spickermann, H.J. Raeder, K. Muellen. Why does matrix-assisted laser desorption/ionization time-of-flight mass spectrometry give incorrect results for broad polymer distributions? *Rapid Commun. Mass Spectrom.* **1996**, *10*, 1471–1474.

J. Res. Natl. Inst. Standards Technol. **2002**, *107*, 1–17.

157 M. Domin, R. Moreea, M.J. Lazaro, A.A. Herod, R. Kandiyoti. Effect of polydispersity on the characterization of coal-derived liquids by matrix-assisted laser desorption/ionization time-of-flight mass spectrometry: inferences from results for mixtures of polystyrene molecular mass standards. *Rapid Commun. Mass Spectrom.* 1997, *11*, 1845–1852.

158 C.N. McEwen, C. Jackson, B.S. Larsen. Instrumental effects in the analysis of polymers of wide polydispersity by MALDI mass spectrometry. *Int. J. Mass Spectrom. Ion Processes* 1997, *160*, 387–394.

159 K. Shimada, M.A. Lusenkova, K. Sato, T. Saito, S. Matsuyama, H. Nakahara, S. Kinugasa. Evaluation of mass discrimination effects in the quantitative analysis of polydisperse polymers by matrix-assisted laser desorption/ionization time-of-flight mass spectrometry using uniform oligostyrenes. *Rapid Commun. Mass Spectrom.* 2001, *15*, 277–282.

160 P. Malvagna, G. Impallomeni, R. Cozzolino, E. Spina, D. Garozzo. New results on matrix-assisted laser desorption/ionization mass spectrometry of widely polydisperse hydrosoluble polymers. *Rapid Commun. Mass Spectrom.* 2002, *16*, 1599–1603.

161 P. Mineo, D. Vitalini, E. Scamporrino, S. Bazzano, R. Alicata. Effect of delay time and grid voltage changes on the average molecular mass of polydisperse polymers and polymeric blends determined by delayed extraction matrix-assisted laser desorption/ionization time-of-flight mass spectrometry. *Rapid Commun. Mass Spectrom.* 2005, *19*, 2773–2779.

162 G. Montaudo, D. Garozzo, M.S. Montaudo, C. Puglisi, F. Samperi. Molecular and structural characterization of polydisperse polymers and copolymers by combining MALDI-TOF mass spectrometry with GPC fractionation. *Macromolecules* 1995, *28*, 7983–7989.

163 G. Montaudo, M.S. Montaudo, C. Puglisi, F. Samperi. Molecular weight determination and structural analysis in polydisperse polymers by hyphenated gel permeation chromatography/matrix-assisted laser desorption ionization-time-of-flight mass spectrometry. *Int. J. Polymer Analysis Characterization* 1997, *3*, 177–192.

164 M.W.F. Nielen, S. Malucha. Characterization of polydisperse synthetic polymers by size-exclusion chromatography/matrix-assisted laser desorption/ionization time-of-flight mass spectrometry. *Rapid Commun. Mass Spectrom.* 1997, *11*, 1194–1204.

165 D. Vitalini, P. Mineo, E. Scamporrino. Further application of a procedure for molecular weight and molecular weight distribution measurement of polydisperse polymers from their matrix-assisted laser desorption/ionization time-of-flight mass spectra. *Macromolecules* 1997, *30*, 5285–5289.

166 M.S. Montaudo, C. Puglisi, F. Samperil, G. Montaudo. Application of size exclusion chromatography matrix-assisted laser desorption/ionization time-of-flight to the determination of molecular masses in polydisperse polymers. *Rapid Commun. Mass Spectrom.* 1998, *12*, 519–528.

167 M.W.F. Nielen. Polymer analysis by size-exclusion microchromatography/MALDI time-of-flight mass spectrometry with a robotic interface. *Anal. Chem.* 1998, *70*, 1563–1568.

168 C. Puglisi, F. Samperi, S. Carroccio, G. Montaudo. Analysis of poly(bisphenol a carbonate) by size exclusion chromatography/matrix-assisted laser desorption/ionization. 1. End group and molar mass determination. *Rapid Commun. Mass Spectrom.* 1999, *13*, 2260–2267.

169 E. Esser, C. Keil, D. Braun, P. Montag, H. Pasch. Matrix-assisted laser desorption/ionization mass spectrometry of synthetic polymers. 4. Coupling of size exclusion chromatography and MALDI-TOF using a spray-deposition interface. *Polymer* 2000, *41*, 4039–4046.

170 X. Lou, J.L.J. van Dongen, E.W. Meijer. Off-line size-exclusion chromatographic fractionation-matrix-assisted laser desorption ionization time-of-flight mass spectrometry for polymer characterization. Theoretical and experimental study. *J. Chromatogr. A* **2000**, *896*, 19–30.

171 G.E. Kassalainen, S.K.R. Williams. Coupling thermal field-flow fractionation with matrix-assisted laser desorption/ionization time-of-flight mass spectrometry for the analysis of synthetic polymers. *Anal. Chem.* **2003**, *75*, 1887–1894.

172 X.M. Liu, E.P. Maziarz, E. Quinn, Y.-C. Lai. A perspective on relative quantitation of a polydisperse polymer using chromatography and mass spectrometry. *Int. J. Mass Spectrom.* **2004**, *238*, 227–233.

173 T.H. Mourey, A.J. Hoteling, S.T. Balke, K.G. Owens. Molar mass distributions of polymers from size exclusion chromatography and matrix-assisted laser desorption/ionization time-of-flight mass spectrometry: Methods for comparison. *J. Appl. Polymer Sci.* **2005**, *97*, 627–639.

174 B. Kona, S.M. Weidner, J.F. Friedrich. Epoxidation of polydienes investigated by MALDI-TOF mass spectrometry and GPC-MALDI coupling. *Int. J. Polymer Analysis Characterization* **2005**, *10*, 85–108.

175 R. Abate, A. Ballistreri, G. Montaudo, D. Garozzo, G. Impallomeni, G. Critchley, K. Tanaka. Quantitative applications of matrix-assisted laser desorption/ionization with time-of-flight mass spectrometry: determination of copolymer composition in bacterial copolyesters. *Rapid Commun. Mass Spectrom.* **1993**, *7*, 1033–1036.

176 G. Wilczek-Vera, P.O. Danis, A. Eisenberg. Individual block length distributions of block copolymers of polystyrene-b-poly-a-methylstyrene. *Polymer Preprints (Am. Chem. Soc., Division of Polymer Chemistry)* **1996**, *37*, 294–295.

177 G. Wilczek-Vera, Y. Yu, K. Waddell, P.O. Danis. Eisenberg, A., Detailed structural analysis of diblock copolymers by matrix-assisted laser desorption/ionization time-of-flight mass spectrometry. *Rapid Commun. Mass Spectrom.* **1999**, *13*, 764–777.

178 K.G. Suddaby, D.M. Haddleton, J.J. Hastings, S.N. Richards, J.P. O'Donnell. Catalytic chain transfer for molecular weight control in the emulsion polymerization of methyl methacrylate and methyl methacrylate-styrene. *Macromolecules* **1996**, *29*, 8083–8091.

179 S. Servaty, W. Koehler, W.H. Meyer, C. Rosenauer, J. Spickermann, H.J. Raeder, G. Wegner, A. Weier. MALDI-TOF-MS copolymer analysis: Characterization of a poly(dimethylsiloxane)-*co*-poly(hydromethylsiloxane) as a precursor of a functionalized silicone graft copolymer. *Macromolecules* **1998**, *31*, 2468–2474.

180 S.M. Hunt, M.M. Sheil, P.J. Derrick. Comparison of electrospray ionization mass spectrometry with matrix-assisted laser desorption ionization mass spectrometry and size exclusion chromatography for the characterization of polyester resins. *Eur. Mass Spectrom.* **1998**, *4*, 475–486.

181 W. Yan, D.M. Ammon, Jr., J.A. Gardella, Jr., E.P. Mariarz, III, A.M. Hawkridge, G.L. Grobe, III, T.D. Wood. Quantitative mass spectrometry of technical polymers: a comparison of several ionization methods. *Eur. Mass Spectrom.* **1998**, *4*, 467–474.

182 S. Koster, M.C. Duursma, J.J. Boon, R.M.A. Heeren. End group determination of synthetic polymers by electrospray ionization Fourier transform ion cyclotron resonance mass spectrometry. *J. Am. Soc. Mass Spectrom.* **2000**, *11*, 536–543.

183 L. Shan, R. Murgasova, D.M. Hercules, M. Houalla. Electrospray and matrix-assisted laser desorption/ionization mass spectral characterization of linear single nylon-6 oligomers. *J. Mass Spectrom.* **2001**, *36*, 140–144.

184 R. Murgasova, D.M. Hercules. Polymer characterization by combining liquid chromatography with MALDI and ESI mass spectrometry. *Anal. BioAnal. Chem.* **2002**, *373*, 481–489.

185 X. Jiang, P.J. Schoenmakers, J.L.J. van Dongen, X. Lou, V. Lima, J. Brokken-Zijp. Mass spectrometric characterization of functional poly(methyl methacrylate) in combination with critical liquid chromatography. *Anal. Chem.* **2003**, *75*, 5517–5524.

186 A.T. Jackson, J.H. Scrivens, J.P. Williams, E.S. Baker, J. Gidden, M.T. Bowers. Microstructural and conformational studies of polyether copolymers. *Int. J. Mass Spectrom.* **2004**, *238*, 287–297.

187 A.T. Jackson, S.E. Slade, J.H. Scrivens. Characterization of poly(alkyl methacrylate)s by means of electrospray ionisation-tandem mass spectrometry (ESI-MS/MS). *Int. J. Mass Spectrom.* **2004**, *238*, 265–277.

188 C.A. Jackson, W.J. Simonsick, Jr. Application of mass spectrometry to the characterization of polymers. *Curr. Opinion Solid State Mater. Sci.* **1997**, *2*, 661–667.

189 R. Chen, X. Yu, L. Li. Characterization of poly(ethylene glycol) esters using low energy collision-induced dissociation in electrospray ionization mass spectrometry. *J. Am. Soc. Mass Spectrom.* **2002**, *13*, 888–897.

190 R. Chen, L. Li. Lithium and transition metal ions enable low energy collision-induced dissociation of polyglycols in electrospray ionization mass spectrometry. *J. Am. Soc. Mass Spectrom.* **2001**, *12*, 832–839.

191 T. Yalcin, W. Gabryelski, L. Li. Structural analysis of polymer end groups by electrospray ionization high-energy collision-induced dissociation tandem mass spectrometry. *Anal. Chem.* **2000**, *72*, 3847–3852.

9
Small-Molecule Desorption/Ionization Mass Analysis

Lucinda Cohen, Eden P. Go, and Gary Siuzdak

9.1
Introduction

At its advent, the application of MALDI mass spectrometry (MS) to low-molecular mass (LMM) compounds (so-called "small molecules") seemed quite unlikely, due to saturation by matrix ions signals below 500 Da. Although MALDI-MS has literally transformed the analysis of high-molecular weight biomolecules, its application to LMM compounds has lagged behind. The advantages of MALDI for soft and efficient ionization of various fragile and non-volatile samples should, in theory, circumvent the problems previously exhibited by laser mass spectrometry, including severe fragmentation for even low-molecular mass organic molecules, the need for high laser energy, and reduced ionization efficiency [1–3]. Other advantages of MALDI, including tolerance for contamination and buffers, uncomplicated spectra as most ions are singly charged ions, very high absolute sensitivity [4], rapid analysis compared to electrospray ionization mass spectrometry (ESI-MS), and relatively simple instrumentation, serve to reinforce a logical rationale to utilize this technique in the lower mass arena.

Until recently, the use of MALDI to characterize small molecules has been "suppressed" by multiple factors, including the low resolution of first-generation linear TOF-MALDI instruments, matrix ion interference and detector saturation in the low mass range, complex coupling of MALDI with on-line techniques such as liquid and planar chromatography, and – most importantly – strong competition from ESI-MS technique in general. As a result, most mass spectrometric practitioners have for some time considered "conventional" MALDI to be inappropriate or even useless for the determination of LMM compounds. However, interest in the application of MALDI to small molecules has continued to grow over the past decade [5]. Improvements in TOF systems – and specifically the utilization of delayed extraction (time lag focusing) – has led in turn to remarkable improvements in MALDI-TOF resolution [6,7]. With recent technological advances in MS instrumentation, a range of mass spectrometers has been interfaced with MALDI sources, the most popular and impacting to date being tandem mass spectrometers. MALDI-MS/MS approaches have been driven by advances in two

MALDI MS. A Practical Guide to Instrumentation, Methods and Applications.
Edited by Franz Hillenkamp and Jasna Peter-Katalinić
Copyright © 2007 Wiley-VCH Verlag GmbH & Co. KGaA, Weinheim
ISBN: 978-3-527-31440-9

key instrument components: (i) high-repetition lasers which can be satisfactorily used with existing triple quadrupole instruments; or (ii) atmospheric pressure (AP) MALDI sources (see Chapter 1), which can be coupled to a range of mass analyzers including ion traps, TOFs, and quadrupole instruments. As will be described in greater detail in this chapter, the progress achieved in the use of structured solid support materials, such as those used in Desorption Ionization on Silicon (DIOS), is also increasing the inherent utility of MALDI for small-molecule analysis. In addition, a continued demand for high-throughput methods in drug discovery and biotechnology, as well as the analysis of complex mixtures in high-salt matrices and buffers, has regenerated efforts to utilize the full power of MALDI-MS over the entire mass range of interest.

This chapter will explore the qualitative and quantitative investigations of MALDI-MS for molecules with a mass less than 1500 Da. As an exhaustive exploration of studies involving any molecular ions detected at 1500 Da or less could yield thousands of articles, we have concentrated on highlighting recent relevant studies that focus on LMM compounds. More detailed information on LMM biomolecules and polymers can be found in Chapters 3–8. In addition to exploring alternative sample preparation approaches, coupling MALDI to various separation techniques will be examined as mechanisms to strengthen the utility of MALDI in the low mass range.

9.2
Matrix Choices for Small-Molecule MALDI

The successful application of MALDI in the analysis of a wide variety of molecules lies in its ability to generate intact ions of thermally labile molecules using a UV-absorbing matrix. A variety of compounds has been used as MALDI matrices, depending on the application. In general, a good MALDI matrix should have the following properties: (i) the ability to absorb at the wavelength of the laser used; (ii) the solubility of the matrix with the analyte; and (iii) the ability to transfer protons during the ionization process. As most of these matrices have molecular weights below 500 Da, the presence of matrix-related ions interferes with LMM analyte detection. Thus, careful choice of the matrix is crucial to the successful application of MALDI to small-molecule analysis. Essentially, if a MALDI matrix provides efficient ionization, minimal or controllable fragmentation and exhibits lack of mass interferences, it can be used for the analysis of LMM compounds. Those matrices found to be useful for small- molecule analysis are discussed in the following sections.

9.2.1
Organic Matrices

The most widely used conventional matrices in the analysis of LMM compounds are crystalline organic molecules that have strong UV absorbance. Based on their chemical structures, two main classes of these conventional matrices are used in MALDI: cinnamic acid derivatives; and aromatic carbonyl derivatives [8]. The use of cinnamic acid derivatives – in particular, α-cyano-4-hydroxycinnamic acid (α-CHCA) and aromatic carbonyl derivatives such as 2,5-dihydroxy benzoic acid (DHB) – has proven to be effective in the analysis of a variety of small molecules [9–11]. Both α-CHCA and DHB yields spectra with high signal-to-noise (S/N) ratio and resolution. The use of these matrices in the analysis of LMM relies heavily on the matrix-to-analyte molar ratio. Unlike MALDI analysis of high-molecular weight compounds, which require an optimal ratio in the order of 10^3 to $10^5 : 1$ [12], the matrix-to-analyte molar ratio is much lower for LMM compounds (10^{-1} to $10^3 : 1$) [10,13].

As most MALDI applications rely on solid (crystalline) organic matrices, which may create considerable background in the low mass range and complicate the characterization of LMM compounds, a few organic crystalline matrices (e.g., 2-hydroxy-1-naphthoic acid [14]) have been introduced specifically to address the characterization of LMM samples such as peptides and porphyrins. One clever approach to circumvent matrix interference was to use a higher molecular weight matrix, which does not interfere in the low-mass region. To this end, some porphyrins have been used as MALDI matrices [15–17].

The use of multicomponent matrices with accelerated sample drying along with a modified instrumental set-up was also considered when preparing samples for quantitative MALDI analysis. This method has been shown to improve the measured relative standard deviation (RSD) to better than 10% of analyte to internal standard (IS) intensity ratio [18]. Both, signal reproducibility and precision of the standard curve slope were improved by a factor of two when a DHB/fucose/5-methoxysalicylic acid multicomponent matrix was used compared to DHB alone. Similar results were observed when a ferulic acid/fucose multicomponent matrix was used compared to the use of ferulic acid alone.

9.2.2
Inorganic Matrices

An alternative approach to organic crystalline matrices involves the use of inorganic matrices. This was originally introduced by Tanaka et al. in 1988, and utilizes ultrafine 30-nm diameter cobalt powder suspended in glycerol [19]. This approach was applied to the analysis of polyethylene glycol (PEG) 200 and methyl stearate using metal or metal oxide powders suspended in liquid paraffin or glycerol [20]. Paraffin exhibited a lower background than glycerol and, interestingly, the analytes were cationized with Na^+ or K^+ ions, not the metal species suspended in the liquid matrix. Unfortunately, higher laser fluences were needed for ionization, which caused increased analyte fragmentation. Despite this, the investigators still observed the "sweet spot" effect with these suspension-type matrices and problems with the vertically mounted sample holder, which caused the sample physically to shift to the bottom of the holder.

Sunner et al. have introduced the surface-assisted desorption/ionization (SALDI) technique, which utilizes graphite particles suspended in a mixture of glycerol, sucrose, and methanol [21]. An optimized sample preparation involved evaporation of the graphite suspension onto a solid substrate, followed by spotting the analyte solution onto the suspension. It is important to note that the size of graphite particles is between 2 and 150 μm, or a few orders of magnitude larger than those used by Tanaka et al. [19]. The applicability of the SALDI method was demonstrated for small organic molecules and peptides, and the results obtained for LMM analytes were similar to those obtained by fast atom bombardment (FAB) MS. These authors noted that SALDI mass spectra exhibit a relatively low chemical background, and are well suited to the characterization of LMM compounds. Graphite powder was originally used for SALDI, but at a later stage a micrometer-sized activated carbon powder was found to provide even better results.

The use of surfactant additive to a SALDI matrix system was shown to enhance SALDI sensitivity [22]. A marked improvement in signal intensities of small organic molecules was observed using appropriate amounts of charged surfactants with acidic characteristics. A detection limit of 100 pg was achieved in the analysis of methylephedrine with a 0.5 M *p*-toluenensulfonic acid (PTSA) added to the SALDI sample preparation. It was suggested that PTSA not only acts as a proton source for the analytes; rather, its interaction with the protonated analyte species also facilitates their migration to the surface, thereby enhancing the signal intensity. Thus, the ionic interaction between surfactant and the analyte was proposed as a possible mechanism for the demonstrated surfactant-enhancing effect. In a subsequent study, use of the surfactant additive was extended in the TLC-SALDI analysis of porphines [23]. In order to obtain homogeneous carbon particle deposition onto the TLC plate, carbon particles ablated from a 1-mm pencil line drawn along the track of the developing sample before the separation were used in the SALDI matrix system. This combination of uniform analyte distribution and the addition of PTSA to the SALDI matrix system increased the S/N ratio of the analyte peaks and increased the sensitivity 50-fold compared to the conventional TLC-SALDI method. In another study, the use of surfactant-modified carbon powder as solid-phase extraction (SPE) sorbents and carbon particles in the SALDI matrix system demonstrated an increase in sensitivity of the SPE SALDI analysis of trace organic compounds in water [24]. In SPE SALDI, the samples are passed through the SPE cartridge, which is subsequently dried and the trace organic compounds adsorbed onto the surfactant-modified SPE sorbent are analyzed directly. In this way, detection limits of trace organic compounds as low as 25 ppt from 100-mL water samples were obtained.

In addition to the use of graphite in SALDI, alternative carbon particles such as carbon nanotubes (CNTs) have been shown to be effective inorganic matrices for MALDI analysis. CNTs obtained from coal by arc discharge were used as MALDI matrix in the analysis of small peptides, organic molecules, β-cyclodextrin, and small proteins [25]. The matrix is prepared by suspending the CNTs in ethanol, followed by sonication for a few minutes to fully disperse them. Sample preparation involves the deposition and drying of the CNT matrix on the MALDI probe, followed by deposition of the analyte of interest. Mass spectra obtained with CNT matrices were characterized by alkali metal ion adducts of the analyte peaks, with few or no background ions. Interestingly, CNTs require a lower laser fluence for the desorption/ionization of peptides, and this results in a higher detection sensitivity (low femtomole range) than with conventional organic matrices. In order to reduce alkali metal ion adducts and to observe more protonated species, citrate buffer – which serves as a proton source as well as an alkali metal ion-chelating reagent – is added to the CNT matrix system [26]. An inherent problem with CNTs, however, is their tendency to fly off from the target plate when subjected to a laser pulse; this can result in contamination of the ion source, time-limited analyte signals, and a time-consuming search for "sweet spots". Ren et al. described a method to overcome this limitation by immobilizing the CNTs with polyurethane adhesive for more prolonged analysis [27]; such immobilization of the CNTs did not affect their properties as matrix. As shown in Figure 9.1, immobilized CNTs as matrix were successfully applied to monitor glucose in urine samples from healthy and diabetic patients. In a recent follow-up study, the same group used chemically modified CNTs in the analysis of neutral carbohydrates, small peptides, and proteins [28]. Chemically modified CNTs terminated with hydroxyl and carboxyl groups were prepared by treating oxidized CNTs with dilute nitric acid. This modification resulted in an increased surface polarity, which provided modified CNTs with a better shot-to-shot reproducibility than for unmodified CNTs. An analysis of simple neutral carbohydrates with mass <500 Da showed analyte peaks with high S/N ratios and detection sensitivity in the low femtomole range.

9.2.3
Liquid Matrices

Some unfortunate drawbacks of using crystalline matrices in MALDI analysis are poor shot-to-shot and sample-to-sample signal reproducibility and "sweet-spot" phenomena. An apparent alternative to crystalline UV MALDI matrices is the utilization of liquid matrices [29–31]; these provide a long-lasting signal, do not have "sweet-spots", exhibit higher signal reproducibility, and are miscible with both polar and nonpolar analytes. A number of organic UV-absorbing additives were tested to widen the applicability range of liquid matrices [32]. However, the use of liquid matrices is still associated with low mass resolution, high chemical background, potential instrument contamination and poor ionization efficiency.

The SALDI approach with two-phase matrices was investigated by Zenobi and coworkers [33]. In addition to glycerol, a few liquid matrices [e.g., 3-nitrobenzyl alcohol (NBA), nitrophenyl octyl ether (NPOE) and thioglycerol] were successfully used. Detection sensitivity (5–50 fmol) was comparable to that obtained by conventional crystalline matrices. In a follow-up report [34], it was also demonstrated that the selection of liquid phases can be guided by criteria developed previously for FAB – that is, the more-acidic glycerol promotes protonization whereas the more-basic diethanolamine favors deprotonization. Minimal background (chemical noise) was observed in the low mass range.

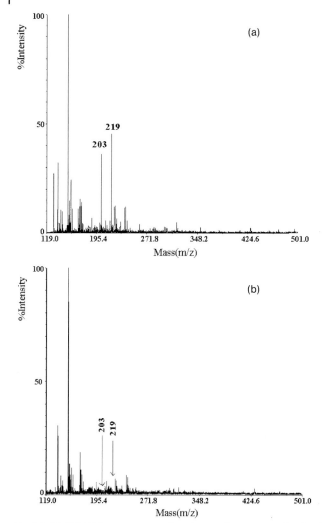

Fig. 9.1 MALDI-TOF mass spectra for urine sample of: (a) a diabetic patient; and (b) a healthy person. Unlabeled peaks arose from a normal urine sample. Ion peaks at *m/z* 203 and 219 correspond to urinary glucose. Ion peaks for urinary glucose were observed in the spectrum for the diabetic patient, while there was no visible glucose signal on the spectrum for the healthy person. Reprinted by permission of Elsevier from Ref. [27]; © American Society for Mass Spectrometry, 2005.

9.2.4
Matrix-Free Approaches

The introduction of a matrix-free desorption/ionization on electrochemically etched silicon surfaces has allowed the analysis of analytes with mass as low as 100 Da [35,36]. In desorption/ionization on silicon mass spectrometry (DIOS-MS), analytes are deposited onto the porous silicon (pSi) surface and desorbed/ionized by the irradiation of a UV-emitting laser. The morphology of pSi provides a scaffold for retaining solvent and analyte molecules, and its UV absorptivity affords a mechanism for transfer of the laser energy to the analyte. This unique combination of characteristics allows DIOS to be useful for a large variety of biomolecules including peptides, carbohydrates, and small organic compounds of various types. Unlike other direct, matrix-free desorption techniques, DIOS enables desorption/ionization of LMM compounds with little or no analyte fragmentation. Recently, Trauger et al. showed that silylation of the oxidized pSi using a variety of commercially available silylating agents results in a much-improved DIOS-MS performance [37]. A dramatic improvement in sensitivity, significant advantages in shelf-life, ease of modification, and analyte specificity was demonstrated from the DIOS analysis from the silylated pSi surface. Figure 9.2 illustrates some examples of mass spectra obtained using the silylated pSi surfaces. Note that minimal background is observed in the low mass range. DIOS and MALDI post-source decay (PSD) have also been compared to ESI data (see Fig. 9.3) [36]. An analysis of identical samples confirmed that PSD with MALDI and DIOS produced frag-

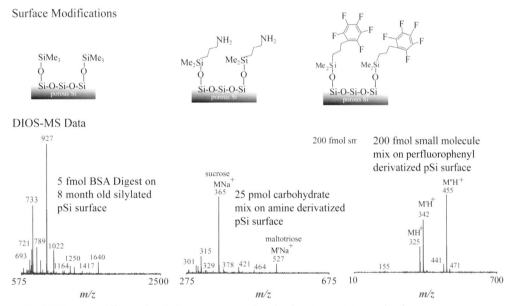

Fig. 9.2 By using different chemical functionalities, pSi surfaces can be tailored to selectively adsorb and efficiently ionize analytes. Left to right: DIOS-MS spectra of 500 fmol bovine serum albumin (BSA) digest analyzed on an 8-month-old TMS-derivatized DIOS chip, carbohydrate mix containing sucrose (MNa$^+$ 365) and maltotriose (M'Na$^+$ 527) on an amine silylated pSi surface, and small molecule mix containing midazolam (MH$^+$ 325), propafenone (M'H$^+$ 342), and verapamil (M"H$^+$ 455) on a perfluorophenyl silylated pSi surface. The hydrophobic TMS- and perfluorophenyl-derivatized surface is amenable to hydrophobic molecules, while the amine derivatized surface is more amenable to hydrophilic molecules.

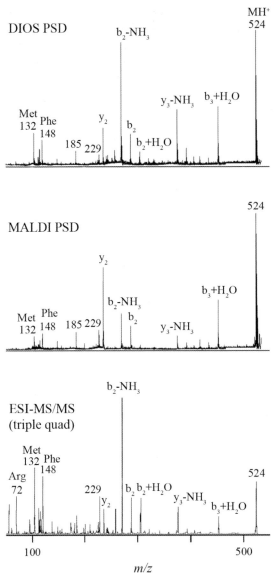

Fig. 9.3 Comparison of DIOS and MALDI post source decay (PSD) data, in addition to electrospray triple quadrupole tandem mass characterization of the peptide MRFA. In the characterization of smaller molecules (*m/z* < 500) by MALDI PSD (not shown), interference from background ions was observed in the fragmentation spectrum.

ments similar to those observed with ESI low-energy CID MS/MS, and can be used for structural elucidation.

In a very recent development, silicon nanowires (SiNWs) have been used as an alternative platform for DIOS-MS [38]. Unlike pSi surfaces, SiNWs are grown on the surface of a solid Si-substrate. In this study, SiNWs are prepared by vapor-liquid-solid (VLS) growth mechanism using Au nanoclusters as growth catalyst [39]. Size-selected Au colloid particles were deposited onto silicon wafers to produce high-quality SiNWs with a narrow diameter distribution. By tailoring the nanowire density, size, and growth orientation, the DIOS performance of SiNWs as a platform for matrix-free MS was examined using peptides and small drug molecules as model compounds. A recently developed chemical modification on ozone-oxidized pSi, which has proven essential to achieve high sensitivities, was employed on the SiNWs [37]. Remarkably, DIOS from SiNWs required much less laser fluence to desorb/ionize analytes compared to pSi and MALDI. As a result, very few surface-related background ions were observed from SiNWs. Although the detection limit obtained for small molecules was comparable with that for silylated pSi surfaces, the detection sensitivity of peptides was six orders of magnitude lower than for pSi. The detection can be further improved by optimizing the fabrication parameters and silylation chemistry. In addition to its ability to support laser desorption and ionization, SiNW has strong fluid wicking properties, these being driven by capillary action generated in the interstitial spaces between SiNWs. This property was exploited in the chromatographic separation and subsequent MS analysis of endogenous analytes from human serum and small drug molecules in biofluids (see Section 9.6).

The applicability of ordered nanocavity arrays as SALDI substrate in the analysis of peptides and plant metabolites has been reported recently [40]. These Si nanocavities are about 200 nm in diameter and 200 nm deep, and are fabricated by nanosphere lithography coupled with reactive ion etching. Desorption/ionization characteristics of the arrays were found to depend heavily on the surface morphology. An analysis of small peptides ($m/z < 500$) from this arrayed surface shows analyte peaks with S/N ratios comparable to those for pSi surfaces. However, for large peptides more background ions were observed due to the higher laser fluence needed for desorption. Plant metabolites from *Arabidopsis thaliana* root extract were successfully detected from this arrayed surface with minimal background ions.

9.3
Sample Preparation

Sample preparation plays a critical role in obtaining good-quality mass spectral data. Optimal MALDI sample preparation should provide good sensitivity as well as reproducibility. A number of sample preparation methods have been developed, depending on the type of analyte. Among the most commonly used MALDI sample preparation method (mainly due to its simplicity) is the dried droplet method; this entails mixing the analyte solution with the matrix material dissolved in an organic solvent/water mixture with an appropriate matrix:analyte ratio (≥10:1), deposition on the MALDI probe, and then air-drying. The evaporation of the solvent allows for co-crystallization of the analyte and matrix. This crystallization process usually results in the heterogeneous distribution of analyte within the sample spot, which in turn leads to poor shot-to-shot and sample-to-sample re-producibility. Several sample preparation methods are discussed in the following section to address this issue.

9.3.1
Electrospray Sample Deposition

Numerous developments in MALDI sample preparation methods have focused on improving sample homogeneity. Among these methods, electrospray sample deposition (ESD) is the most effective technique to achieve sample homogeneity. ESD, when coupled with MALDI-MS, has been reported to markedly improve sample homogeneity and result in enhanced sensitivity and signal reproducibility [41,42]. In ESD, the analyte solution is drawn into a syringe and fed to a needle mounted on an XY translation stage. To generate the electrospray, a voltage difference is applied between the needle and the MALDI target plate. Electrospraying results in the formation of a very fine mist of positively charged droplets that rapidly dry and are dispersed into a homogeneous circular pattern on the target plate. Spray stability is dependent on the flow rate, the shape and size of the capillary tip, the potential difference, the solvent system, and the distance between the needle and the target plate. Typical electrospray conditions include low flow rates (1–20 μL min^{-1}), an electrospray needle with inner diameter <10 μm, a potential difference of about 2–10 kV, analyte and matrix solutions in volatile solvent, and a separation distance of ~1 mm between the needle and the target plate. By optimizing these conditions, ESD yields a uniformly distributed analyte within the sample spot.

MALDI sample preparation for ES deposition involves mixing a saturated solution of the matrix with the analyte solution at a ratio of 1:1 (v/v) in a volatile solvent. Samples are introduced into the electrospray capillary and deposited onto a MALDI target. The spray conditions, including voltage, spray distance and flow rate, are optimized such that analyte particles arrive at the MALDI target plate with some solvent. It has been reported that dry deposition produces small or no crystal on the target plate, and this results in a weak MALDI signal [43]. In the same study, wet deposition was demonstrated to yield detection limits in the low attomole range for peptides.

9.3.2
Analyte Derivatization

Despite the successful application of MALDI and DIOS in the analysis of LMM compounds, the analysis remains a challenge. This stems primarily from low ionization efficiencies due to a lack of functional groups with high proton affinity, matrix suppression effects, isobaric overlay of the matrix with the analyte, and the high volatility of many analytes. One approach to improve the analysis of LMM compounds is by derivatization. Derivatization of samples prior to MS analysis has been widely employed in gas chromatography-mass spectrometry (GC-MS) [44–53] and ESI-MS [47,54–57], but very few MALDI [58–60] and DIOS [61] studies have been reported. The derivatization of LMM compounds prior to mass analysis offers the advantage of increased ionization efficiencies due to the incorporation of high proton affinity functional groups, differentiating matrix from analyte signals due to the increase in observable mass, stabilization of analytes, and facile introduction of isotopic labels for analyte quantitation. Several straightforward one-pot derivatization schemes have been reported in the analysis of biochemically relevant small organic molecules such as alcohols, carboxylic acid, keto-carboxylic acids, aldehydes, ketones, and amines [62]. This simple derivatization procedure allowed for the rapid and efficient analysis of a wide variety of small organic molecules, and provided clear distinction among isobaric compounds due to a specific increase in the mass of the analytes. The limit of detection for each class of organic compound was found to be similar for both derivatized and underivatized samples. Another derivatization procedure reported recently for small-molecule MALDI is the addition of a large, positively charged tag to the analyte using *N*-hydroxysuccinimide ester [63]. This derivatization reagent reacts specifically with LMM compounds with mass <500 Da containing either primary or secondary amine functional group, and results in a mass shift of 573 Da. The addition of a large positive tag to the analytes significantly improved sensitivity and the detection limit in the low femtomole range compared to underivatized samples, and also allowed for the analysis of samples without further purification. In addition, due to the ease of incorporating isotopically labeled tag, antibiotic mixtures were successfully quantified.

9.3.3
Analyte Pre-Concentration

9.3.3.1 **Prestructured Sample Supports**
One of the key steps in the dried droplet method is crystallization of the analyte-matrix mixture. During the crystallization process, a heterogeneous distribution of analyte containing crystals is formed over a spot size of ~5–15 mm^2. With only small portions of the dried sample irradiated by the laser, "sweet-spots" are observed which impart a limitation in the quantitative analysis of LMM and sample throughput. One way of solving this problem is to deposit the sample onto nitrocellulose substrates [64]. Indeed, Russell et al. have shown that the use of nitrocellulose as MALDI substrate resulted in an increased yield of protonated peptide signal and significant improvements in reproducibility and the precision of quantification. These authors suggested that nitrocellulose substrate modified the crystallization of the matrix-analyte solution and allowed more uniform coverage over the sample surface. Use of the nitrocellulose substrate produced a linear standard curve with a correlation coefficient of 0.991 in the range of 4 to 200 pmol during the analysis of bradykinin.

An alternative approach is to reduce the sample spot size close to the diameter of the laser spot. One way of doing this is to apply a small volume of the sample solution (typically a few nanoliters) onto a MALDI target plate. This sample preparation technique requires state-of-the-art liquid handling systems, with the ability to pipette volumes in the low nanoliter range to enhance sample utilization by pre-concentration. Sample pre-concentration provides the practical benefits of improved sensitivity, better shot-to-shot reproducibility, and less signal suppression.

Another sample pre-concentration technique that can be utilized in the analysis of LMM compounds is the use of MALDI Anchor-Chip™ (Bruker Biosciences) [65,66]. This sample support system consists of an array of small gold hydrophilic center (diameters from 200 to 800 µM) surrounded by a Teflon-coated hydrophobic exterior region. Aqueous samples deposited are exclusively held in place in the hydrophilic center due to the surrounding Teflon material, which repels the sample droplet. Samples are enriched before matrix crystallization, and this results in an improved detection sensitivity, provided that the also increased concentration of impurities does not interfere with the MALDI process. For example, MALDI analysis of peptides using Anchor-Chip™ showed detection limits in the low attomole range.

An alternative, less-expensive method to produce disposable sample supports has been described by Owen et al. [67], in which blank stainless-steel plates were coated with Teflon and Scotch Guard™ hydrophobic coatings. Enhancement in detection sensitivity correlated with reduced spot sizes when using the hydrophobic coatings, which could be removed by cleaning and reapplied between uses [67].

9.3.3.2 DIOS with Solid Liquid Extraction

The ease of incorporating functional groups to modify the pSi surface for mass spectrometric application has allowed for the development of a convenient and simple method for enriching the concentration of analytes. This technique, termed DIOS solid-liquid extraction (SLE), uses the property of differential adsorption selectively to capture analytes from a solution containing contaminants that impede MS analysis. Hydrophobically modified DIOS surfaces could be readily used selectively to remove interferences prior to analysis. DIOS SLE simply involves the deposition of a droplet containing analytes onto the chemically modified DIOS surface; then, after approximately 3 s, the sample is aspirated with the same pipette (Fig. 9.4). This short-term deposition allows for any molecule with a propensity for adsorption onto the surface to attach itself, yet any potential hydrophilic contaminants such as salts and buffers remain in solution, leaving the surface free of such contaminants. The van der Waals interaction between the analyte and surface selectively extract the small molecules, while the hydrophilic contaminants such as salts are removed with the droplet. DIOS analysis of peptides and small molecules from complex matrices are greatly enhanced by differential adsorption [37].

The applicability of DIOS SLE is demonstrated in the DIOS analysis of endogenous analytes in blood plasma. Using a simple cold methanol extraction to remove the proteins followed by a butyl ester derivatization, amino acids such as phenylalanine, alanine, isoleucine/leucine, glutamic acid, and the doped deuterated internal standard of phenylalanine, used to quantify the amino acid,

Solid Liquid Extraction

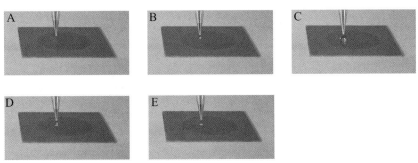

Fig. 9.4 Analyte clean up with DIOS solid-liquid extraction (SLE). This simply involves the deposition of a droplet containing analytes such as a protein digest and its subsequent removal with a pipettor. The Van der Waals forces between the analyte and surface selectively extract the peptides or small molecules, while the hydrophilic contaminants such as salts are removed with the droplet.

were detected with good sensitivity (Fig. 9.5). Using DIOS PSD analysis, the detection of the amino acid in blood plasma was confirmed by monitoring the MS/MS fragmentation pattern of the butyl ester derivative of phenylalanine and its deuterated analog. DIOS PSD analysis shows a characteristic neutral loss of 102 Da associated with butyl formate resulting in the product ions at 120 and 125 Da from the collision-induced dissociation (CID) of butylated phenylalanine and its deuterated internal standard, respectively (Fig. 9.5).

The value of DIOS SLE on silylated pSi surfaces was further demonstrated in the analysis of drug molecules from drug–protein complexes. The presence of high quantities of salt and protein often results in the formation of a crust on the target, which inhibits both MALDI and DIOS-MS analyses. In addition, the dried droplet method containing the denatured protein always fails to release the drug molecule in both DIOS and MALDI. By employing DIOS SLE, most contaminants such as polar salts can be removed by differentially adsorbing the analytes onto the silylated pSi surface. The selective extraction of the potent kinase inhibitor, staurosporine, in the presence of two other unbound molecules by using a 10 kDa MW cut-off filter to isolate the Rho-kinase II–staurosporine complex, spotting the sample onto the DIOS target, and then removing polar interferences prior to MS analysis is shown in Figure 9.6. The hydrophobic interaction of the drug with the perfluoroalkyl-modified DIOS surface allowed the drug to be extracted, even in the presence of protein, salts and stabilizing agents for the active enzyme (25% glycerol, 75 mM NaCl, 0.5 mM benzamidine, 0.1 mM PMSF, 0.05 mM EGTA, 0.015% Brij 35, 0.05% 1-mercaptoethanol in the original mixture). Thus, SLE serves as a useful tool for monitoring drugs in biological fluids in the presence of large amounts of proteins and soluble non-volatile contaminants.

(a)

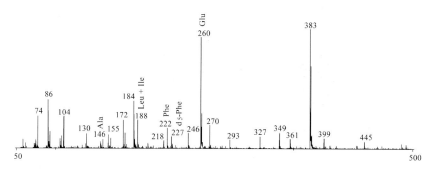

(b)

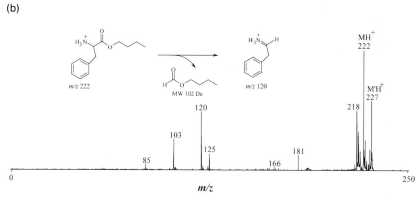

m/z

Fig. 9.5 (A) Solid-liquid extraction (SLE) followed by DIOS mass analysis of amino acids in human plasma after esterification. (B) DIOS PSD profile of phenylalanine (M′H⁺ 222) and d₅-phenylalanine (MH⁺ 227) showing a characteristic neutral loss of 102 Da corresponding to butyl formate. This results in the product ions of the butyl ester derivatives of phenylalanine (MH⁺ 120) and its corresponding deuterated internal standard (M′H⁺ 125). Reprinted with permission from Ref. [37]; © American Chemical Society, 2004.

9.3.4
Matrix Suppression

Sample preparation can be tailored to suppress excess matrix background peaks by controlling the matrix:analyte ratio. Although numerous groups observed this phenomenon soon after the introduction of MALDI, it has only recently been exploited to achieve its full potential for small-molecule analysis. The matrix:analyte molar ratio within the matrix crystals has been found to be critical for MALDI quantification [10]. Critical parameters for successful execution of the matrix suppression effect (MSE) include sufficient analyte and optimized laser intensity. As in the case of FAB-MS, the use of surfactants such as cetrimonium bromide (CTAB) can be used to substantially or even completely suppress the matrix-

(a) DIOS-MS on a Mixture of Rho-kinase II and Potential Inhibitors

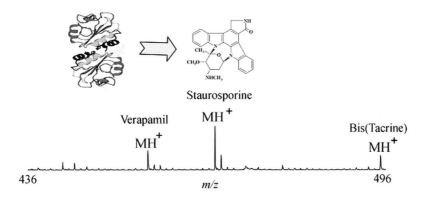

(b) DIOS-MS on the Mixture After 10 kDa MW Cutoff Filter

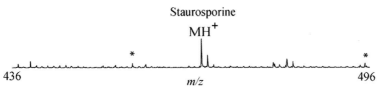

Fig. 9.6 Monitoring of enzyme inhibitors using SLE with DIOS-MS. Solid-liquid extraction (SLE) of a mixture of Rho-kinase II, staurosporine (a potent kinase inhibitor), verapamil and bis(tacrine) before (A) and after (B) passage through a 10 kDa MW cut-off filter in the presence of a stabilizing buffer allowed for the identification of a known inhibitor. The asterisks mark the position of verapamil (m/z 455) and bis(tacrine) (m/z 496) after depletion. Reprinted with permission from Ref. [37]; © American Chemical Society, 2004.

related ion background [68]. The use of CTAB surfactant also resulted in improved mass resolution for low-molecular weight molecules including amino acids, peptides, drugs, and cyclodextrins.

Knochenmuss and coworkers have extensively studied the MSE [69,70], and have introduced a system to calculate the extent of matrix suppression and its effect on qualitative spectral interpretation for LMM analytes. Figure 9.7 shows three MALDI-TOF mass spectra of caffeine, each exhibiting different MSE scores. These scores are calculated using the equation:

$$\text{MSE score} = \frac{\sum A}{\sum A + \sum M},$$

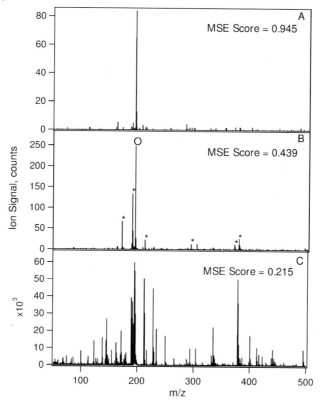

Fig. 9.7 Three MALDI-TOF mass spectra of caffeine, including the matrix suppression effect (MSE) scores. The matrix used was α-cyano-4-hydroxycinnamic acid (α-CHCA). Peaks marked with asterisks in (B) indicate matrix signals included in the calculation of the MSE score, while the signal from the protonated analyte is marked with "O". MSE is nearly complete in (A). In (C), many more matrix signals appear than in (B). These were not included in calculating the score, which would otherwise be even lower. The matrix: analyte mole ratio was 3 in spectra (A) and (C), and 27 in spectrum (B). Low laser intensity was applied for measurements in (A) and (B), and a high laser intensity was used for spectrum (C). Reprinted with permission from Ref. [71]; © American Chemical Society, 2004.

where *M* is the sum intensity of predefined matrix peaks and *A* is the intensity of the predefined analyte peak. MALDI images can be generated and filtered using MSE scores to monitor analyte homogeneity across sample spots and identify areas where optimal signals can be generated. Filtered MALDI images of a mixture of yohimbine and caffeine are shown in Figure 9.8. The green pixels represent areas where the correct ratio of yohimbine to caffeine was determined. Isolated areas of elevated caffeine signal can be identified indicating local fractionation during preparation, particularly for samples prepared with DHB [71].

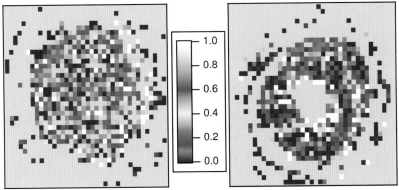

Fig. 9.8 Filtered MALDI images of samples containing a mixture of yohimbine and caffeine analytes with matrixes α-CHCA (A) and DHB (B). The ratio of yohimbine to caffeine to matrix was 4:1:36. In the absence of analyte suppression effects, the ratio of caffeine to yohimbine signals is expected to be 0.25. The yohimbine MSE score was used as a filter. The ratio of caffeine to yohimbine was calculated only if the MSE score was greater than 0.9; otherwise, the pixel was left blank (pink areas). The expected ratio was 0.25 (green pixels). Local regions of unexpectedly low (blue pixels) and high caffeine signal (white pixels) are observed. This may be an indication of local fractionation during sample drying. Reprinted with permission from Ref. [71]; © American Chemical Society, 2004.

9.4
Qualitative Characterization of LMM Molecules

Arguably, small-peptide analysis is one of the largest areas of MALDI application to LMM compounds. Chen et al. [72] initially demonstrated the applicability of MALDI to study small peptides (800–1500 Da). A variety of excellent examples of MALDI for the analysis of small peptides can be found in almost every MALDI publication [73], and it is probably safe to say that MALDI has become the method of choice for peptide analysis. Furthermore, some small peptides such as angiotensin and substance P have become "gold standards" for the verification of matrix and instrument performance [4]. The analysis of peptide maps for the characterization of proteins is discussed in detail in Chapter 3.

One other well-established area of MALDI is that of carbohydrate analysis. It has been shown that MALDI is applicable for the analysis of underivatized and derivatized carbohydrates, and can also provide complementary data to other techniques [74–77]. MALDI matrices and sample preparation techniques used in carbohydrates analysis have been reviewed in detail by Harvey [76]. Interestingly, carbohydrates themselves have been used to improve MALDI signal reproducibility and mass resolution of other analytes via improved matrix crystal homogeneity and potentially gas-phase cooling of analyte molecules [78–80]. Further details relating to the MALDI-MS of carbohydrates and glycoconjugates can be found in Chapter 6.

In one of the earliest reports utilizing MALDI for LMM analytes, Duncan and coworkers [81] examined a variety of LMM compounds using α-CHCA and DHB matrices. A range of compounds, including quaternary ammonium salts, sterols, nucleosides, purine and pyrimidine bases, amino acids, catecholamines, opiods, antibiotics, prostaglandins, macrocyclic metal complexes of porphyrins, and phthalocyanines, was successfully analyzed. Although these authors utilized a linear TOF system with limited mass resolution and conventional sample preparation with crystalline matrices, the report indicated that the majority of compounds studied gave interpretable mass spectra. In another application of MALDI, a series of tetrathiofulvalene compounds was successfully studied and characterized [82], again using conventional α-CHCA and DHB matrices. The results obtained for 26 compounds showed MALDI to be more viable than other mass spectrometric methods for a broad spectrum of chemical materials.

MALDI methodology has been successfully applied to both polar and non-polar retinoids [83]. Although MALDI spectra exhibited background ions from the DHB matrix, the analyte ions were easily identified. Furthermore, the specific structural fragments observed provided additional data to assist in retinoid characterization. Interestingly, only 4-oxoretinoic acid was detected as a protonated species (versus radical molecular ions), even under LDI conditions. The authors suggested that this might be the result of a higher proton affinity of the oxo-retinoic acid compared to the other retinoic acids.

UV MALDI has also been used to characterize complex mixtures of hazardous waste [13]. Three classes of compounds were studied, namely organic acids, salts of oxyanions, and amine-based chelators. The organic acids exhibited strong signals in negative-ion mode, whereas the chelating compounds showed strong signals in positive-ion mode. It was also found that a high sodium concentration could quench MALDI signals. The implementation of an off-line clean up with ion-exchange column prior to MALDI analysis yielded a 100-fold sensitivity increase.

The early applications of MALDI for pharmaceutical compounds have focused primarily on the characterization of compounds not amenable to ESI-MS determinations. MALDI has been utilized for simple identification and accurate mass determination of the anti-hypertensive prazosine and its synthetic analogues [84], the anti-malarial double deuterium-labeled ferrochloroquine [85], and a glucuronide metabolite of an H_1 antihistamine, dimethindene [86].

Metabolism studies, though not numerous, have highlighted the problem-solving ability of MALDI. In one study, Morvan et al. monitored the fate of an oligonucleotide dodecathymidine prodrug in cell extracts, whereby the kinetics and metabolic half-life of the prodrug were determined, and the prodrug was fully metabolized to T_{12} phosphorothioate [87]. An unknown metabolite of the antiviral acyclic nucleotide analogue cidofovir was isolated from rat kidney and confirmed by MALDI and NMR to be cidofovir-phosphocholine [88]. Olson and Fabris have shown that MALDI can also be used as a tool to determine metabolites *in vitro*, via CYP_{450} enzymes in liver microsomes, of tamoxifen, promethazine, and diphenhydramine [89].

Siuzdak and coworkers have examined the intracellular transport of cationic drugs (notably the tetraphenylphosphonium cation) by using MALDI-MS. The cation was quantified at subpicomole levels in cell lysates, and the biodistribution-based drug resistance characterized. The results obtained by MALDI were comparable to those produced by scintillation counting [90]. The use of MALDI PSD and CID analyses and MALDI PSD/CID has been compared to ESI trap characterization for 4-quinolone antibiotics and oleandomycin. The choice of matrix was found to control the amount of fragmentation, with the TiO_2 matrix exhibiting less background interference but also fewer fragments for identification [91]. Nicotine- and cotinine- (a nicotine metabolite) adducts of melanin have been studied as a model to investigate the incorporation of drugs into hair. Adducts of monomeric melanin dopaquinone (DOPAQ) with nicotine and cotinine were determined and compared to DOPAQ adducts with deuterated analogues of nicotine and cotinine [92]. Not surprisingly, one logical pharmaceutical application of MALDI has been in the realm of phototherapy, though few systems have been investigated; photofrin, a porphyrin derivative used to treat tumors [93], temoporfin, a tumor-localizing photosensitizer [94], and a dye-dendrimer complex, pheophorbide-a-substituted diaminobutane polypropyleneimine [95].

MALDI may also be a method of choice for the rapid and automated analysis of combinatorial libraries in the presence of buffers and contamination [96–98]. One appealing aspect of the MALDI method is that the UV laser light can be used simultaneously to promote photolytic cleavage of the analyte from the solid support and its gas phase ionization for subsequent mass spectral analysis. The general utility of termination synthesis with MALDI analytical methods has been illustrated with a non-linear, non-peptide cyclic oligocarbamate library [99]. Siuzdak and Lewis have shown that peptides and carbohydrates, covalently linked to a polymeric support through a photolabile linker, can be directly identified by MALDI in a single step which requires no pretreatment of the sample to induce cleavage from the support [100]. In another application [101], MALDI was used for the photolytic release and determination of an active glycopeptide from the resin support. The resulting mass spectral data contained the ladder of glycopeptide fragments and yielded the active glycopeptide sequence. A small library of chymostatin derivatives has been synthesized, with 22 compounds identified by MALDI [102]. The "one-bead–one-compound" (OBOC) combinatorial method has also been successfully supported by MALDI analysis using an isotopically tagged encoding strategy to drive unambiguous identification based on bromine and chlorine patterns within the data generated [103,104]. A robust, high-throughput method was also validated for a model 12 288-member library that was screened against streptavidin [104].

In general, surfactants are confounding mixture components which interfere with biomolecule characterization by MS, and sample clean-up steps are necessary prior to analysis. MALDI analysis of LMM ionic and nonionic surfactants has been conducted by a number of groups. Several different detergents including Triton X-100 and 114, Tween 20 and Brij 35 were analyzed by TLC, reverse-phase HPLC and MALDI with comparable results being obtained among the techniques

[105]. In the same studies, Mega 8 *n*-octylglycoside, Chaps, Chapso, sulfobetaine, and the zwitterionic surfactant SB14 were examined, with additional methylene group-containing impurities being detected in some samples. Attempts to remove matrix interference by addition of silver salt were unsuccessful.

Ayorinde and coworkers have analyzed several types of surfactants, including nonionic polysorbates Tween 20, 40, 60, 80, all of which are fatty acid esters of polyethoxysorbitan. Complex mixtures of PEGs, PEG esters, and isosorbide polyetholuxlate were detected. During the analysis, seven different series of oligomers ions were present due to mixture of fatty acid esters [106]. Nonylphenylethoxlate surfactants (Surfonic®) which can serve as a carbon source/substrate to produce polyhydroxybutyrate for bacterial degradation have been investigated by this group [107]. Either a higher molecular weight (non-interfering) porphyrin matrix or α-CHCA were utilized as matrices for the Surfonic samples, although higher sample concentrations were necessary for the α-CHCA matrix. Monomer and dimer peaks were not observed in the MALDI spectra of these samples. Other classes of surfactants analyzed by MALDI include Surfynol [108] and quaternary ammonium compounds [109].

9.5
Analyte Quantitation by MALDI

The use of MALDI technique in quantitative analysis of LMM compounds has been well documented and reviewed [110]. However, despite numerous reports [11,73,111–115], analyte quantitation by MALDI is still considered to be a highly unreliable procedure. Some of the critical issues associated with quantitative MALDI-MS include: (i) instrumentation, related mass resolution and accuracy; (ii) linearity and precision of the data acquisition system; and (iii) sample preparation and signal reproducibility. Mass resolution and linearity of acquisition systems had been one of the major obstacles in the early applications of MALDI [116]. However, technological advances in MS instrumentation – including developments in delayed extraction [6,7,117], powerful ion detectors, high-repetition lasers, AP-MALDI interfaces, and fast speed acquisition systems – have adequately addressed most of the instrumental issues. The main limiting factor in quantitative MALDI is sample preparation, and associated with this are *intra* (sample-to-sample) and *inter* (point-to-point and shot-to-shot signal) experimental reproducibility. The best way to compensate for signal deviation in quantitative MALDI is to use an internal standard (IS); some examples summarizing the major approaches for sample preparation and selection and utilization of IS for quantitative characterization in the LMM range are provided in the following sections.

9.5.1
Selection of IS

The correct selection of an IS is critical in achieving accurate analyte quantitation and acceptable standard curve linearity of biologically interesting LMM compounds. The use of an IS can minimize variability in analyte signal intensities and improve experimental reproducibility [113]. One requirement needed for an IS is that its physico-chemical characteristics should be identical to those of the analyte of interest during the measurement, in order to ensure similar ionization behavior. In addition, the mass separation between the analyte and IS must not overlap for a straightforward measurement of peak intensity. Traditionally, isotopically labeled compounds are used as internal standards, but in some cases when these are unavailable then structural homologues or analogues have been used as an alternative.

The quantitative characterization of LMM biomolecules (<500 Da) has been demonstrated by Duncan and coworkers [81]. Different types of IS for the stable isotope-labeled analyte (^{13}C and deuterated) and structural analogues were evaluated. Three different systems were examined: (i) 3,4-dihydroxyphenylalanine (DOPA) for the ^{13}C-labeled system; (ii) acetocholine for deuterated analogues as IS; and (iii) two synthetic peptides, differing only slightly in amino acid composition. The acetocholine system was shown to provide the best linearity, with a correlation coefficient of 0.98 for the standard curve. The analyte and IS peak were well-resolved, and no interference occurred with the DHB matrix peaks. These authors also noted that use of the ^{13}C label on six carbons compromised the results for DOPA system due to poor mass resolution. Thus, although a linear standard curve was obtained (correlation coefficient 0.967) for the ^{13}C-DOPA system, resolution of the IS from the analyte was crucial for quantitative characterization.

9.5.2
Methods for Improving Quantitative Performance

An instrumental protocol for MALDI data collection can also be a powerful tool for minimizing signal deviations [18,69,71,118]. Data acquisition at constant laser energy was compared to data acquisition with constant ion abundance [118]. The latter studies showed improved reproducibility, as well as allowing more consecutive laser shots at the same position. A linear standard curve ($r = 0.9997$ from 0.05 to 30 ng mL^{-1}) was demonstrated for chlormequat using an isotopically labeled IS and data collection using constant ion abundance.

The dependence of the RSD of the analyte:IS ratio on the number of laser shots has been studied by Gusev et al. [18,119]. These authors found that the maximum signal intensity was obtained for the first 40 laser shots, and the lowest RSD was achieved for the 40 laser shots after the first 10. A higher percentage RSD for the first 10 shots was explained by the presence of contamination at the surface, which causes deviations in the analyte:IS ratio. In order to avoid these deviations, the authors suggested discarding the first 10 shots, which halved the RSD of analyte:IS ratio. A similar protocol was shown to be effective for fast evaporation [120] and dried droplet sample preparations [121]. An instrumental data collection protocol can also be automated using on-line correlation analysis in order simultaneously to improve mass resolution and to minimize signal deviations [120].

9.5.3
Quantitation of Pharmaceutical Compounds

In recent years, MALDI quantitation has shown the most progress in the pharmaceutical arena, due mainly to the ever-present demand for higher throughput analytical methods. On the basis of the availability of high-repetition lasers and AP MALDI sources, reports continue to emerge in this area. For example, LeRiche et al. showed that, for 4-quinolone antibiotics and oleandomycin, MALDI with PSD and PSD/CID generated a higher number of different product ions than ESI trap MS^n (with $n \leq$) [91]. In a recent report, AP-DIOS was utilized in the forensic analysis of tablets seized during drug raids [122].

The use of high-repetition lasers has facilitated its coupling to analyzers other than TOFs, and has also driven real-time analysis of three or more samples per minute, which is the current benchmark for reasonable throughput analysis by ESI-LC/MS/MS. Hatsis et al. have demonstrated successful quantitation of benzodiazepines using either quadrupole-TOF or triple quadrupole mass spectrometers with a frequency-tripled (355 nm) Nd:YAG laser operating at 1 kHz [9]. Sleno and Volmer have optimized various instrumentation and sample preparation parameters for LMM quantitation; here, α-CHCA was found to be a better matrix compared to DHB or sinapinic acid for the compounds examined, the data correlating with reports from other groups [123,124]. Sample consumption was in the 1 femtomole range, offering significant advantage over ESI LC/MS/MS.

Gobey et al. have reported an exhaustive comparison of ESI and MALDI quantitation as applied to high-throughput *in-vitro* screens [124]. In this study, 53 diverse compounds were assayed through a metabolic stability screen using human liver microsomes and split for MALDI and ESI-LC/MS/MS analyses. A typical data collection profile is shown in Figure 9.9. In this experiment, test compounds were subjected to incubation with human liver microsomes, an *in-vitro* model that simulated how analytes are metabolized in humans. If the test compound is liable to undergo biotransformation via the enzymes found in human liver microsomes, then the amount of test compound remaining over the course of the experiment represents how susceptible the compound is to metabolism. Thus, plotting the amount of analyte remaining as a function of time as monitored by either analyte peak area or analyte:IS peak area ratio allows an assessment to be made of analyte metabolic stability and half-life.

Samples were prepared by first terminating the incubation reaction by the addition of acetonitrile, followed by SPE using an Oasis HLB 96-well extraction procedure. Both the α-CHCA matrix and IS were added as part of the SPE elution solvent mixture. After extraction, samples were spotted onto the target plate, with contiguous standards and aliquots from each incubation time point. The entire sample set was then rastered past the laser beam and collected into a single chromatogram (see Fig. 9.9). Half-lives were calculated using the more accurate and reproducible analyte:IS peak area ratios. A comparison of half-lives determined by MALDI and ESI is shown in Figure 9.10. In general, good agreement was seen between the two techniques. MALDI quantitation was also applied to other

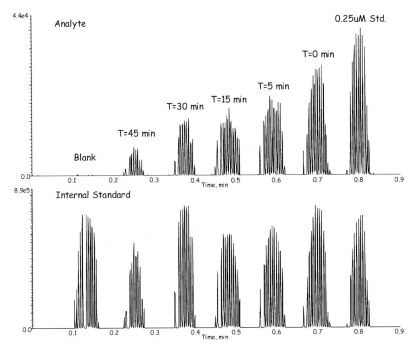

Fig. 9.9 Example of microsome incubation time course.
Reprinted with permission from Ref. [124]; © American
Chemical Society, 2005.

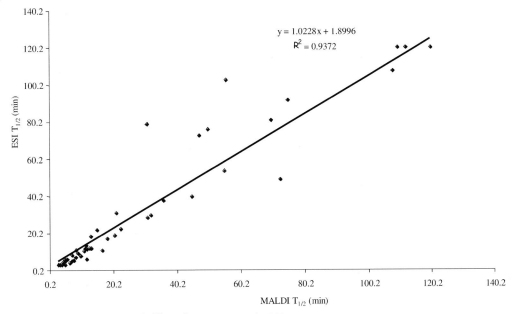

Fig. 9.10 Human microsome half-lives for 53 compounds: ESI
versus MALDI. Reprinted with permission from Ref. [124]; ©
American Chemical Society, 2005.

in-vitro assays such as human hepatocytes, Caco-2, and MDR Pgp transport. However, ESI demonstrated better universality than MALDI in ionization success (15.9% failure rate for MALDI versus 6.7% for ESI) when analyzing a set of 208 compounds [124].

9.5.4
Enzyme Activity and Inhibition Studies

The development of efficient methods to assay molecules for catalytic enzyme activity is of considerable importance in drug discovery. In this respect, MS has emerged as a versatile tool in monitoring enzymatic reactions because it can directly detect most natural substrates and/or products of enzymatic reactions in a time-resolved fashion. To date, MALDI-MS has been employed in a number of quantitative analyses of enzyme-catalyzed reactions by monitoring LMM substrates and products [10,71,125]. In most of these studies, MALDI analysis is performed without additional purification and desalting steps. For example, quantitative MALDI has been employed in the enzymatic conversion of glucose to gluconolactone using DHB as matrix [125]. Using an optimized sample prepara-tion method and ^{13}C-glucose as IS, sodiated peaks of the substrate and product were used for the analysis. A linear standard curve was obtained with an %RSD value of 6% out of 40 measurements. Moreover, enzyme activity was determined by monitoring the time course of disappearance of the substrate and appearance of the products, and compared with HPLC for method validation. Results from the time-course experiments obtained from MALDI and HPLC were in good agree-ment. However, unlike HPLC, MALDI analysis required small sample amount and allowed the direct analysis of enzymatic reaction mixture, without further purification.

In taking advantage of the simple and straightforward analysis of enzymatic reactions, these authors extended the utility of quantitative MALDI in screening enzyme activity using different variants of pyranose oxidase, an enzyme that con-verts glucose to glucosone, as a model system [125]. Using ^{13}C-glucose as IS and a liquid ionic matrix, DHB-pyridine (DHB-Py), which was synthesized by mixing DHB in acetonitrile/water/TFA (50:50:0.1, v/v) with an equimolar amount of pyridine for MALDI sample preparation, the glucose concentration was deter-mined from samples obtained via the enzymatic reaction. The activity of 10 en-zyme variants was monitored over a 1-h period, with six of the 10 enzyme variants used showing enzyme activity. In addition, other monosaccharides and sugar alcohols were detected from the reaction mixture. This study demonstrates the direct analysis of substrates and products of enzyme reactions in liquid ionic matrix, without a prepurification step.

The same MALDI-based assay approach with DHB butylamine (DHBB) was implemented to monitor the desialylation reaction of 3′-sialyllactose [126]. DHBB is a liquid ionic matrix prepared by mixing equimolar amounts of DHB and butylamine, and can be used in lieu of an organic solvent in the enzyme reaction, thereby allowing rapid monitoring of the enzyme-catalyzed reaction.

In another study, MALDI was employed in the quantitative analysis of LMM species generated from enzyme-catalyzed reactions using substrates as IS [10]. Because the substrate is consumed during the reaction, the measured signal ratios of the product and substrate are amplified, thereby reducing the experimental errors on the estimated kinetics. By optimizing the pH, matrix:analyte ratio, solvent composition, and the target plate, a good linear response ($r^2 = 0.985$, 0.999) was obtained for the analysis. This MALDI approach is potentially useful in enzyme screening, as IS is not required.

Recently, the applicability of DIOS-MS to monitor enzyme activity and inhibition studies has been reported [127,128]. The viability of DIOS-MS to monitor enzyme activity was demonstrated in the reaction time course of α-(2,6)-sialyltransferase in the sialylation of *N*-acetyllactosamine (LacNAc). Sialyltransferase is an important class of glycotransferase that exhibits high acceptor-substrate specificity, and also plays an important role in cell adhesion and molecular recognition events. By optimizing the reaction conditions, the time-course production of α-(2,6)-sialylated trisaccharide (*m/z* 475) was monitored from a single-step reaction catalyzed by α-(2,6)-sialyltransferase from a corresponding lactoside (*m/z* 788) in a period of 15 min. DIOS was also effective in monitoring the inhibition of acetylcholine esterase (AChE), a biologically relevant target enzyme used for enzyme screening inhibition studies. The pseudo first-order reaction of the hydrolytic catalysis of the neurotransmitter acetylcholine (*m/z* 146) to choline (*m/z* 104) was measured quantitatively using d_9-choline as internal standard and electrospray sample deposition. Once the kinetics was obtained, screening for potential AChE inhibitors was examined by monitoring the AChE activity in the presence of potential inhibitors from a small-molecule library. Known inhibitors were identified from the library through the observation of high acetylcholine:choline ratios; no false positives or false negatives were observed. By employing a standard commercial instrument with a 200 Hz laser, a total of 4000 samples was analyzed during the course of a 4-h period. This screening method is inherently sensitive, simple and quick, and should greatly facilitate inhibitor discovery.

9.5.5
Quantitative Analysis of Samples from Complex Biological Matrices

The characterization of complex biological samples presents a major challenge for any analytical technique. Low analyte concentrations, co-extracted endogenous materials, and nonvolatile buffers result in signal interference, signal suppression, and complicate qualitative and quantitative characterization. Despite significant strides in instrumentation, reliable MALDI quantitation without extensive sample pretreatment remains elusive for analytes in complex biological samples such as blood. Compounds, which perform well in neat solution, may exhibit significant losses in sensitivity when analyzed from blood or plasma. Even so, a handful of reports has emerged for quantitative MALDI analysis of amperozide [112], tacrolimus [113] and its metabolites [129], and cyclosporine A [130] and its metabolites [116] in plasma or blood. In general, MALDI results showed good correlation with results from other techniques such as ESI LC-MS/MS and HPLC/UV. Angiotensin II has been successfully quantitated in renal microdialysate fluid, a somewhat cleaner matrix than blood [80]. Desiderio et al. [111] have demonstrated the quantitative analysis of opioid peptide with pentadeuterated IS in ovine plasma samples. Based on the data obtained, the time profile in sheep plasma and the corresponding pharmacokinetic data were calculated.

Sample clean-up requirements still remain strict for MALDI analysis, with even low protein/lipid/carbohydrate matrices such as human urine demanding a two-step liquid/liquid extraction protocols [36]. However, recent reports of MALDI analysis using CNTs as matrix have shown the potential to circumvent multi-step sample clean-up, due to the sample extraction properties of the CNT matrix [131,132]. Pan et al. have demonstrated the analysis of a mixture of propranolol, quinine, and cinchonine in urine by mixing the analyte-containing urine sample with the CNT matrix, glycerol, and sucrose for maximum dispersion and adsorption of the analytes onto the CNT matrix [132]. Oxidized CNTs have been utilized in the analysis of three bioactive alkaloids, berberine, jatrorrhizine and palmatine, in extracts of traditional Chinese medicines using the method of standard additions [131].

9.5.6
Environmental Applications of Quantitative MALDI

The initial application of quantitative MALDI to environmental samples has focused on surfactants. Willetts et al. have studied the quantitation of Synperonic NP9, a potential contaminant in surface water [133]. The quantification of non-phenylethoxylates (NPEs) carries significant environmental importance due to their biodegradation to nonylphenol, which is known to cause an estrogenic response in breast cancer cells. LiCl was added to the sample preparation to simplify the observed oligomer distribution, and a quantification range of 10–50 mg L^{-1} was obtained. Due to problems quantifying across the molecular weight distribution of the oligomers, separate calibration curves were obtained for each individual peak. The limit of detection was 80 μg L^{-1} (compared to 10 μg L^{-1} by HPLC). No Synperonic NP9 was detected in an actual water sample, but spiked samples were appropriately quantified by MALDI [133]. Ayorinde and Elhila have found that instrumental detection limits for NPEs are directly dependent on molecular weight. Low-molecular weight NPEs were detected at 10 μg L^{-1}, while higher-molecular weight NPEs showed detection limits as high as 4.5 mg L^{-1} [134].

As the agricultural use of antibacterial drugs continues to rise, concerns have arisen regarding the accumulation of antibiotics in the soil and ground water, leading in time to increased drug resistance in human and veterinary medicine [135,136]. MALDI quantitation combined with Solid Phase ImmunoExtraction (SPIE) has been used for the analysis of sulfamethazine and its major metabolite, N^4-acetylsulfamethazine. The SPIE-MALDI procedure allowed rapid purification and detection of low ppb levels in water, soil, and manure samples [137].

Sleno and Volmer have utilized the enhanced capabilities of a MALDI-triple quadrupole mass spectrometer with a high-repetition laser to monitor spirolide toxins in phytoplankton [138]. A comprehensive analytical study was successfully

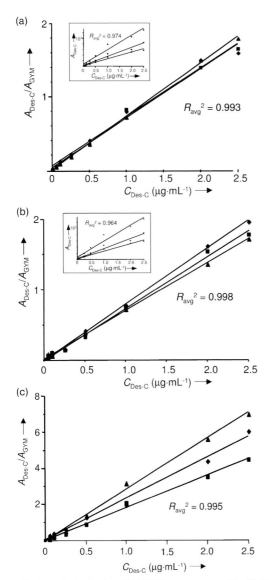

Fig. 9.11 Optimization of MALDI quantitation. Each experiment was performed in triplicate (interday variations: ▲, day 1; ◆, day 2; ■, day 3). (a) SIM detection with constant matrix concentration (inset without IS); (b) SIM detection with constant analyte : matrix ratio (inset without IS); and (c) MRM detection with constant matrix concentration. Reprinted with permission from Ref. [138]; © American Chemical Society, 2005.

conducted, encompassing screening for known and unknown spirolide analogues, quantitation of toxins in the 0.01 to 1.75 μg mL^{-1} concentration range, and preliminary structural characterization based on low-energy CID spectra. The optimization of MALDI quantitation for 13-desmethylspirolide C using gymnodimine as IS is shown in Figure 9.11. Results using selected ion monitoring (SIM, measurement of the mass spectral intensity of a specific m/z value) with either a set matrix concentration (Fig. 9.11(a)) or a constant analyte:matrix ratio (Fig. 9.11(b)) are contrasted with data from multiple reaction monitoring (MRM, measurement of the mass spectral intensity of a specific precursor to product ion m/z transition) with a set matrix concentration (Fig. 9.11(c)). Not surprisingly, greater precision was observed when the analyte:IS peak area ratio was used rather than simply the analyte peak area. Using either SIM or MRM, good linearity and r^2 values were observed on three separate days.

9.6
Separation Methods Coupled with MALDI and DIOS

Although MALDI is used primarily as a stand-alone analytical method, substantial efforts and research have been directed toward its coupling with various column and planar separation methods. The use of a separation technique prior to MS analysis greatly reduces signal suppression of low-abundance ions in the analysis of complex mixtures, thus enhancing its sensitivity. Most MALDI coupling schemes have been accomplished off-line – that is, detection is performed after the entire separation procedure has been completed. The off-line approach involves subsequent (but separate) MALDI or DIOS analysis of fractions collected from HPLC, gel-permeation chromatography (GPC) and capillary electrophoresis (CE), or spots scraped and extracted from thin-layer chromatograms and polyacrylamide gels [139–143]. A variety of examples of off-line fraction collection-MALDI analysis can be found in reviews by Murray [144] and Gusev [145].

9.6.1
TLC-MALDI

Meldal and coworkers [142] have demonstrated rapid identification of organic reaction products using off-line TLC-MALDI. Crude mixtures of peptides, glycopeptides, carbohydrate reactions, and classical organic reactions were separated on the TLC plate, scraped off, extracted and analyzed by MALDI. The authors proposed using a relatively low matrix:analyte ratio (15:1) in order to suppress matrix peaks and minimize interference. Low matrix:analyte ratios can significantly reduce matrix peaks to the point where the analyte signal become the most abundant in the spectrum. These authors also found that doping both matrix and analyte solutions with Cs^+ ions resulted in the suppression of both Na^+ and K^+ cationized peaks, which simplified the identification of LMM samples.

One novel approach in the off-line planar separation method coupled with DIOS is the use of SiNWs as a platform for the separation of molecules [38]. The capability of a SiNW to separate a sample mixture lies in its high surface-to-volume ratio and in the differences in analyte–surface interactions. When combined with an ability to support laser desorption/ionization MS, chromatographic separation followed by MS analysis with SiNWs provides a simple, inexpensive, rapid and qualitative means of separating and analyzing sample mixtures. The ability of SiNWs to separate analytes from complex mixture is demonstrated in the analysis of a mixture of small-drug molecules and endogenous analytes from human serum and mouse spinal cord tissue. With further optimization of SiNW dimensions, this planar separation technique, followed by subsequent MS analysis, is attractive compared to existing planar chromatography coupled with MALDI-MS because it offers a one-step procedure without a matrix deposition step.

The direct coupling of MALDI with planar separation yields one- and two-dimensional scans of TLC where each pixel is a mass spectrum obtained directly from the plate. Several approaches have been developed to directly couple planar separations to MALDI: (i) the introduction of a solid MALDI matrix in solution [146,147]; (ii) spraying the MALDI matrix solution onto the surface [148]; and (iii) the transfer of analyte to another membrane or support [149,150]. Although both TLC-MALDI and the optimization of the coupling protocol have been extensively reviewed [144,145], more recent reports demonstrate a continued interest in this area and advancements in methodology. Based on the availability of AP-MALDI sources, the developed TLC plates can be easily and quickly analyzed after mounting with tape. The structural identification of even incompletely separated analytes has been demonstrated using either MALDI PSD [151] or MALDI/CID with ion trap mass analyzers [152]. In one report, MALDI mass spectra allowed the detection of a mixture of chlortetracycline and tetracycline not resolved during the TLC plate development [153]. TLC coupling with AP-MALDI also offers the advantage of better mass accuracy compared to TLC-MALDI, as the changes in ion flight path from the TLC plate based on surface morphology will generate a longer detection time interval and thus lower mass resolution. In addition, a broader variety of matrixes including particle suspensions [152] or ionic liquids [153] can be directly applied to the TLC plate and analyzed on-line. Using this approach, small molecules such as antibiotics, alkaloids, anesthetics, and bacterial siderophobes (LMM Fe(III)-specific binding molecules) have been successfully separated by TLC and analyzed on-spot [152,154]. On-line TLC-MALDI quantitation has also been evaluated, with various approaches being explored for the application of IS. For example, Crecelius et al. have investigated: (i) mixing the analyte and IS prior to spotting and TLC plate development; (ii) adding IS to the mobile phase and pre-developing the TLC plate; (iii) electrospraying the IS onto the plate prior to matrix application; and (iv) mixing the IS with matrix and electrospraying the mixture. Of these four approaches, pre-developing the IS in the mobile phase was deemed to be the most successful. Results for the quantitation of piroxicam were somewhat disappointing, however, with a limited dynamic range (400–800 ng on plate) and extensive ion suppression of the analyte by IS, and IS by analyte [155].

On-line high-performance TLC (HPTLC) MALDI-FTMS at elevated pressures was recently described in the direct analysis of gangliosides from TLC plates using solid matrices [156]. Gangliosides are glycosphingolipids that contain one or more sialic acids which are easily cleaved during MS analysis. The degree of metastable fragmentation can be reduced significantly by employing a high-pressure gas pulse, with typical gas pressures ranging from 1 to 10 mbar being employed. Under this pressure regime vibrational cooling is achieved, thereby reducing ganglioside fragmentation and allowing its analysis. In addition, intact gangliosides were observed even when "hot UV MALDI matrices" were employed. HPTLC UV MALDI-MS analysis of gangliosides results in a detection limit of ~100 fmol before separation, a mass resolution of >1:50 000, and a mass accuracy better than 1.5 ppm. However, when IR laser and IR MALDI matrices were used, a significant fragmentation of gangliosides was observed. Moreover, the sensitivity was low due to poor incorporation of the matrices on the silica gel. A key improvement in HPTLC IR MALDI in the analysis of ganglioside mixtures from cultured Chinese hamster ovary cells was reported by Dreisewerd and coworkers [157]. In this study, a liquid matrix (glycerol) was employed for the homogeneous wetting of the silica gel, an IR laser for softer ionization and efficient desorption of gangliosides from the TLC plates, and an orthogonal TOF mass spectrometer for high mass accuracy. Results obtained from MS analysis of a mixture of gangliosides directly from HPTLC plates obtained in both positive- and negative-ion modes were characterized by high mass accuracy and low detection limit. The method also allows the simultaneous detection of various GM3 (II^3-α-Neu5Ac-LacCer) ganglioside species from analyte bands, with high relative sensitivity and lateral resolution.

A recent report from Kostiainen and co-workers utilizing ultra TLC (UTLC) separations coupled with MALDI appears to offer significant improvement over high-performance TLC approaches. The fabrication of monolithic silica UTLC plates provides 10- to 100-fold greater sensitivity than HPTLC, lower sample consumption, and faster separation. Some caveats include less resolving power due to shorter elution distances (2 cm versus 5 cm) and a smaller overall adsorption surface area available for analytes. This method has been successfully applied to the analysis, purity estimates and qualitative identification components of crude mixtures of compounds synthesized via combinatorial chemistry [158]. Although at present TLC-MALDI analysis remains only semi-quantitative in nature, its primary utility as a qualitative tool for mixture separation and structural identification is currently on the rise.

9.6.2
Capillary and Frontal Affinity Liquid Chromatography

To date, the vast majority of experiments that have coupled liquid separation techniques with MALDI have utilized off-line fraction collection. The availability of MALDI sample preparation robots has facilitated the preparation of hundreds of samples as discrete spots on the target plate in a short time frame. If the fraction collection is triggered by detection of a UV signal (or other analytical detector), then MS analysis can be focused only on the chromatographic regions which contain the most abundant analytes. In addition, fraction collection allows separation and detection of small sample volumes, with the potential to re-analyze these precious samples if necessary. Furthermore, MALDI target anchor plates can provide additional concentrating effect to further improve sensitivity.

Capillary HPLC has been hyphenated with MALDI-MS for nanoscale screening of single-bead combinatorial libraries [159]. A small test library of nine compounds showed optimum results using DHB as matrix with on-line mixing of HPLC effluent and matrix solution followed by ESD. Seven corticosteroids have been identified using 2,4-dinitrophenylhydrazine as both derivatizing agent and matrix. Derivatization of the analytes was accomplished on-line during ESD and drying time [160].

Off-line collection of LC effluent can also be accomplished into continuous channels or "tracks" on the MALDI target plate. In this manner, the ion signals from rastering the laser along this channel can be used to generate a reconstructed chromatogram from a more continuous process than spotting discrete fractions. This approach has been successfully applied to the screening of compound libraries using frontal affinity chromatography (FAC) [161]. FAC relies upon continuous infusion of a known concentration of compound (rather than a single injection as in HPLC) to drive equilibration of the ligand between free and bound states with an immobilized protein stationary phase. The breakthrough time of the compound should correspond to the ligand affinity for the immobilized protein. Potent inhibitors should exhibit greater affinity and thus be retained longer on the protein stationary phase. Coupling MALDI with FAC circumvents some inherent problems with FAC ESI-MS as the need for high ionic strength eluents in FAC causes ion suppression in ESI. FAC MALDI-MS was applied to the screening of small molecules against entrapped dihydrofolate reductase (DHFR). Figure 9.12 shows FAC MALDI-MS/MS traces from columns with and without DHFR. Only results from the first two tracks on the target plate are shown; once all the analyte signals reach a plateau further analysis is unnecessary. Potent inhibitors trimethoprim and pyrimethamine exhibited significantly greater affinity (and thus retention) than fluorescein, glucosamine and folic acid. Results using MALDI were comparable to those obtained by ESI, and offer greater throughput and good sensitivity.

Although an off-line approach simplifies the coupling and allows for the separation and MALDI-MS steps to occur independently, this provides only the mass spectral data for several discrete times (HPLC, GPC, and CE) or spots (TLC and gels). A few experimental approaches for continuous monitoring of the entire separation include: (i) aerosol MALDI liquid introduction [162–164], analogous to the particle beam interface; (ii) a continuous-flow MALDI probe with liquid matrix or continuous solid matrix crystallization [165–170], analogous to continuous-flow FAB; and (iii) rotational ball or wheel MALDI interfaces [171,172], analogous to a moving-belt interface. Although the on-line coupling of MALDI and liquid column separation is currently in the experimental stages, it may yet revolutionize the MALDI-MS field and have a major impact on the characterization of LMM samples in complex mixtures or biological and environmental matrices.

(a)

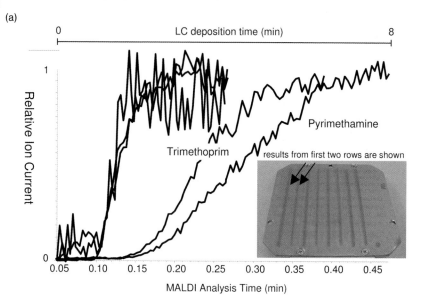

(b)

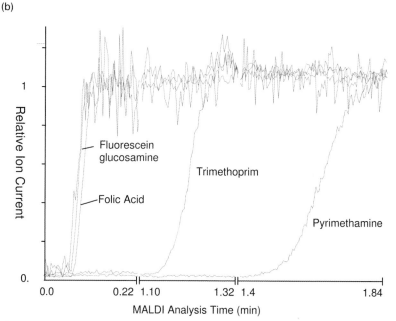

Fig. 9.12 FAC MALDI-MS/MS traces obtained using protein-loaded and blank polymer-coated macroporous silica monolithic columns. (A) Blank column containing no protein. (B) Column containing 25 pmol DHFR (initial loading) showing breakthrough of *N*-acetylglucosamine, fluorescein, and folic acid at early times, then trimethoprim, and finally pyrimethamine. All compounds were infused at 50 nM. All traces are normalized to the maximum signal obtained after compound breakthrough. Note that the MALDI analysis time is 19-fold faster than the LC deposition time. All FAC traces were obtained using a fast laser translation speed, 3.8 mm s^{-1}) and represent the average of five re-runs over a given sample region. Reprinted with permission from Ref. [161]; © American Chemical Society, 2005.

9.7
Conclusions

The wide variety of both numerous and diverse applications described in this chapter clearly demonstrates the utility and future potential of MALDI for both qualitative and quantitative characterization of LMM compounds. Many new instrumentation and sample preparation techniques are now available for automated and painless MALDI characterization. Arguably, sample preparation from complex milieu such as blood still requires considerable attention and effort. The sample clean-up procedures prior to MALDI analysis need not be more complicated or lengthy than those conducted for ESI-MS analysis. Realistically, problems such as shot-to-shot and sample-to-sample reproducibility still exist, but can be overcome by the techniques discussed earlier. Other criticisms of quantitative MALDI such as poor sensitivity, limited linear range and imprecision (higher RSD) are primarily compound-dependent and have been significantly improved by instrumental advancements.

The greatest potential for utilizing the full analytical power of MALDI in the low mass range exists for mixtures of both high- and low-molecular weight components, as well as both polar and nonpolar compounds. At present, mixtures require tedious optimization of the type of matrix and either sequential analysis by MALDI or complementary analysis by electrospray MS. Novel sample preparation approaches such as SALDI or DIOS, which simplify the process, serve as an extraction step during sample preparation, and remove the matrix altogether, are being increasingly applied to more complex problems. We can envision exciting opportunities for the application of MALDI to simultaneous analysis of biomarkers in biological media, including both large molecules (proteins) and small molecules such as pharmaceutical compounds or glucose.

The somewhat surprising number and scope of publications on MALDI characterization of LMM compounds not only reinforces the utility and practical applicability of this technique, but also indicates the substantial efforts being made by several research groups. These efforts are transforming what might be pigeonholed as a "high-molecular mass biomolecule only" analytical technique into a flexible analytical tool.

Acknowledgments

The authors thank Steve Michael and Arkady Gusev for helpful review and discussion, and Mark Cole, Jason Gobey, Richard Knochenmuss, Dietrich Volmer, and Yinlong Guo who provided figures for this chapter. The authors appreciate funding from DOE grant DE-AC02-05CH11231.

Abbreviations/Acronyms

α-CHCA	α-cyano-4-hydroxycinnamic acid
AChE	acetylcholine esterase

AP/MALDI	atmospheric pressure/matrix-assisted laser desorption/ionization
BOC	t-butyloxycarbonyl
CE	capillary electrophoresis
CID	collision-induced dissociation
CNT	carbon nanotube
CsA	cyclosporine A
CTAB	cetrimonium bromide
DHB	2,5-dihydroxy benzoic acid
DHBB	2,5-dihydroxy benzoic acid butylamine
DHB-Py	2,5-dihydroxy benzoic acid pyridine
DIOS	desorption/ionization on silicon
DOPA	3,4-dihydroxyphenylalanine
DOPAQ	dopaquinone
ESD	electrospray sample deposition
ESI	electrospray ionization
FAC	frontal affinity chromatography
GM3	II^3-α-Neu5Ac-LacCer
HABA	2-(4-hydroxyphenylazo)benzoic acid
HPLC	high-performance liquid chromatography
IR	infrared
IS	internal standard
LacNac	N-acetyllactosamine
LDI	laser desorption ionization
LMM	low molecular mass
MALDI	matrix-assisted laser desorption/ionization
MET	2-methoxy-N-[(1R)-1-phenylethyl]acetamide
MRM	multiple reaction monitoring
MS	mass spectrometry
MSE	matrix suppression effect
NBA	3-nitrobenzyl alcohol
NMR	nuclear magnetic resonance
NPE	nonphenylethoxylates
NPOE	nitrophenyl octyl ether
OBOC	one-bead–one-compound
PEA	rac-a-phenylethylamine
PEG	polyethylene glycol
PMMA	polymethylmethacrylate
PSD	post source decay
pSi	porous silicon
PTSA	p-toluenensulfonic acid
RSD	relative standard deviation
SALDI	surface-assisted laser desorption/ionization
SIM	single ion monitoring
SiNW	silicon nanowire
SLE	solid-liquid extraction

S/N	signal-to-noise
SPE	solid-phase extraction
SPIE	solid-phase immunoextraction
TAC	tacrolimus
TCNQ	7,7,8,8-tetraccyanoquinodemethane
THF	tetrahydrofuran
TLC	thin-layer chromatography
TOF	time-of-flight
UTLC	ultra thin-layer chromatography
UV	ultraviolet
VLS	vapor liquid solid

References

1 M.L. Alexander, P.H. Hemberger, M.E. Cisper, N.S. Nogar. *Anal. Chem.* **1993**, *65*, 1609–1614.

2 G.L. Glish, D.E. Goeringer, K.G. Asano, S.A. Mcluckey. *Int. J. Mass Spectrom. Ion Processes* **1989**, *94*, 15–24.

3 E.W. Schlag, J. Grotemeyer, R.D. Levine. *Chem. Phys. Lett.* **1992**, *190*, 521–527.

4 B.O. Keller, L. Li. *J. Am. Soc. Mass Spectrom.* **2001**, *12*, 1055–1063.

5 L.H. Cohen, A.I. Gusev. *Anal. Bioanal. Chem.* **2002**, *373*, 571–586.

6 R.S. Brown, J.J. Lennon. *Anal. Chem.* **1995**, *67*, 3990–3999.

7 M.L. Vestal, P. Juhasz, S.A. Martin. *Rapid Commun. Mass Spectrom.* **1995**, *9*, 1044–1050.

8 J. Krause, M. Stoeckli, U.P. Schlunegger. *Rapid Commun. Mass Spectrom.* **1996**, *10*, 1927–1933.

9 P. Hatsis, S. Brombacher, J. Corr, P. Kovarik, D.A. Volmer. *Rapid Commun. Mass Spectrom.* **2003**, *17*, 2303–2309.

10 M.J. Kang, A. Tholey, E. Heinzle. *Rapid Commun. Mass Spectrom.* **2000**, *14*, 1972–1978.

11 R. Lidgard, M.W. Duncan. *Rapid Commun. Mass Spectrom.* **1995**, *9*, 128–132.

12 M. Karas, F. Hillenkamp. *Anal. Chem.* **1988**, *60*, 2299–2301.

13 S.C. Goheen, K.L. Wahl, J.A. Campbell, W.P. Hess. *J. Mass Spectrom.* **1997**, *32*, 820–828.

14 M.G. Bartlett, K.L. Busch, C.A. Wells, K.L. Schey. *J. Mass Spectrom.* **1996**, *31*, 275–279.

15 F.O. Ayorinde, P. Hambright, T.N. Porter, Q.L. Keith. *Rapid Commun. Mass Spectrom.* **1999**, *13*, 2474–2479.

16 N. Srinivasan, C.A. Haney, J.S. Lindsey, W.Z. Zhang, B.T. Chait. *J. Porphyrins Phthalocyanines* **1999**, *3*, 283–291.

17 R.M. Jones, J.H. Lamb, C.K. Lim. *Rapid Commun. Mass Spectrom.* **1995**, *9*, 968–969.

18 A.I. Gusev, W.R. Wilkinson, A. Proctor, D.M. Hercules. *Anal. Chem.* **1995**, *67*, 1034–1041.

19 K. Tanaka, H. Waki, Y. Ido, S. Akita, Y. Yoshida, T. Yoshida. *Rapid Commun. Mass Spectrom.* **1988**, *2*, 151–153.

20 T. Kinumi, T. Saisu, M. Takayama, H. Niwa. *J. Mass Spectrom.* **2000**, *35*, 417–422.

21 J. Sunner, E. Dratz, Y.C. Chen. *Anal. Chem.* **1995**, *67*, 4335–4342.

22 Y.C. Chen, M.F. Tsai. *J. Mass Spectrom.* **2000**, *35*, 1278–1284.

23 J.Y. Wu, Y.C. Chen. *J. Mass Spectrom.* **2002**, *37*, 85–90.

24 Y.C. Chen, M.F. Tsai. *Rapid Commun. Mass Spectrom.* **2000**, *14*, 2300–2304.

25 S. Xu, Y. Li, H. Zou, J. Qiu, Z. Guo, B. Guo. *Anal. Chem.* **2003**, *75*, 6191–6195.

26 W.Y. Chen, L.S. Wang, H.T. Chiu, Y.C. Chen, C.Y. Lee. *J. Am. Soc. Mass Spectrom.* **2004**, *15*, 1629–1635.

27 S.F. Ren, L. Zhang, Z.H. Cheng, Y.L. Guo. *J. Am. Soc. Mass Spectrom.* **2005**, *16*, 333–339.

28 S.F. Ren, Y.L. Guo. *Rapid Commun. Mass Spectrom.* **2005**, *19*, 255–260.

29 T.W.D. Chan, I. Thomas, A.W. Colburn, P.J. Derrick. *Chem. Phys. Lett.* **1994**, *222*, 579–585.

30 D.S. Cornett, M.A. Duncan, I.J. Amster. *Org. Mass Spectrom.* **1992**, *27*, 831–832.

31 D.S. Cornett, M.A. Duncan, I.J. Amster. *Anal. Chem.* **1993**, *65*, 2608–2613.

32 J.B. Williams, A.I. Gusev, D.M. Hercules. *Macromolecules* **1996**, *29*, 8144–8150.

33 M.J. Dale, R. Knochenmuss, R. Zenobi. *Anal. Chem.* **1996**, *68*, 3321–3329.

34 R. Dale, R. Knochenmuss, R. Zenobi. *Rapid Commun. Mass Spectrom.* **1997**, *11*, 136–142.

35 J. Wei, J.M. Buriak, G. Siuzdak. *Nature* **1999**, *399*, 243–246.

36 Z.X. Shen, J.J. Thomas, C. Averbuj, K.M. Broo, M. Engelhard, J.E. Crowell, M.G. Finn, G. Siuzdak. *Anal. Chem.* **2001**, *73*, 612–619.

37 S.A. Trauger, E.P. Go, Z.X. Shen, J.V. Apon, B.J. Compton, E.S.P. Bouvier, M.G. Finn, G. Siuzdak. *Anal. Chem.* **2004**, *76*, 4484–4489.

38 E.P. Go, J.V. Apon, G. Luo, A. Saghatelian, R.H. Daniels, V. Sahi, R. Dubrow, A. Vertes, G. Siuzdak. *Anal. Chem.* **2005**, *77*, 1641–1646.

39 Y. Cui, L.J. Lauhon, M.S. Gudiksen, J.F. Wang, C.M. Lieber. *Appl. Phys. Lett.* **2001**, *78*, 2214–2216.

40 N.H. Finkel, B.G. Prevo, O.D. Velev, L. He. *Anal. Chem.* **2005**, *77*, 1088–1095.

41 J. Axelsson, A.M. Hoberg, C. Waterson, P. Myatt, G.L. Shield, J. Varney, D.M. Haddleton, P.J. Derrick. *Rapid Commun. Mass Spectrom.* **1997**, *11*, 209–213.

42 R.R. Hensel, R.C. King, K.G. Owens. *Rapid Commun. Mass Spectrom.* **1997**, *11*, 1785–1793.

43 H. Wei, K. Nolkrantz, D.H. Powell, J.H. Woods, M.C. Ko, R.T. Kennedy. *Rapid Commun. Mass Spectrom.* **2004**, *18*, 1193–1200.

44 W.W. Christie. *Lipids* **1998**, *33*, 343–353.

45 A. Fox, G.E. Black. *Mass Spectrometry for the Characterization of Microorganisms* **1994**, *541*, 107–131.

46 H. Frauendorf, R. Herzschuh. *Eur. Mass Spectrom.* **1998**, *4*, 269–278.

47 W.J. Griffiths, S. Liu, G. Alvelius, J. Sjovall. *Rapid Commun. Mass Spectrom.* **2003**, *17*, 924–935.

48 J.M. Halket, V.G. Zaikin. *Eur. J. Mass Spectrom.* **2003**, *9*, 1–21.

49 J.M. Halket, V.G. Zaikin. *Eur. J. Mass Spectrom.* **2004**, *10*, 1–19.

50 J.M. Halket, D. Waterman, A.M. Przyborowska, R.K.P. Patel, P.D. Fraser, P.M. Bramley. *J. Exptl. Bot.* **2005**, *56*, 219–243.

51 K. Shimada, K. Mitamura. *J. Chromatogr. B – Biomedical Applications* **1994**, *659*, 227–241.

52 D. Tsikas. *J. Biochem. Biophys. Methods* **2001**, *49*, 705–731.

53 V.G. Zaikin, J.M. Halket. *Eur. J. Mass Spectrom.* **2004**, *10*, 555–568.

54 A.P. Bruins. *J. Chromatogr. A* **1998**, *794*, 345–357.

55 M. Larsson, R. Sundberg, S. Folestad. *J. Chromatogr. A* **2001**, *934*, 75–85.

56 Y. Suzuki, N. Tanji, C. Ikeda, A. Honda, K. Ookubo, K. Citterio, K. Suzuki. *Anal. Sci.* **2004**, *20*, 475–482.

57 M.A. Watkins, B.E. Winger, R.C. Shea, H.I. Kenttamaa. *Anal. Chem.* **2005**, *77*, 1385–1392.

58 S. Broberg, A. Broberg, J.O. Duus. *Rapid Commun. Mass Spectrom.* **2000**, *14*, 1801–1805.

59 J.W. Gouw, P.C. Burgers, M.A. Trikoupis, J.K. Terlouw. *Rapid Commun. Mass Spectrom.* **2002**, *16*, 905–912.

60 D.J. Harvey. *J. Am. Soc. Mass Spectrom.* **2000**, *11*, 900–915.

61 E.P. Go, Z.X. Shen, K. Harris, G. Siuzdak. *Anal. Chem.* **2003**, *75*, 5475–5479.

62 A. Tholey, C. Wittmann, M.J. Kang, D. Bungert, K. Hollemeyer, E. Heinzle. *J. Mass Spectrom.* **2002**, *37*, 963–973.

63 P.J. Lee, W.B. Chen, J.C. Gebler. *Anal. Chem.* **2004**, *76*, 4888–4893.

64 L.M. Preston, K.K. Murray, D.H. Russell. *Biol. Mass Spectrom.* **1993**, *22*, 544–550.

65 E. Nordhoff, M. Schurenberg, G. Thiele, C. Lubbert, K.D. Kloeppel, D. Theiss, H. Lehrach, J. Gobom. *Intl. J. Mass Spectrom.* **2003**, *226*, 163–180.

66 M. Schuerenbeg, C. Luebbert, H. Eickhoff, M. Kalkum, H. Lehrach,

E. Nordhoff. *Anal. Chem.* **2000**, *72*, 3436–3442.

67 S.J. Owen, F.S. Meier, S. Brombacher, D.A. Volmer. *Rapid Commun. Mass Spectrom.* **2003**, *17*, 2439–2449.

68 Z. Guo, Q. Zhang, H. Zou, B. Guo, J. Ni. *Anal. Chem.* **2002**, *74*, 1637–1641.

69 R. Knochenmuss, F. Dubios, M.J. Dale, R. Zenobi. *Rapid Commun. Mass Spectrom.* **1996**, *10*, 871–877.

70 R. Knochenmuss, V. Karbach, U. Weisle, K. Breuker, R. Zenobi. *Rapid Commun. Mass Spectrom.* **1998**, *12*, 529–534.

71 G. McCombie, R. Knochenmuss. *Anal. Chem.* **2004**, *76*, 4990–4997.

72 M.J. Kang, A. Tholey, E. Heinzle. *Rapid Commun. Mass Spectrom.* **2001**, *15*, 1327–1333.

73 Y.F. Zhu, K.L. Lee, K. Tang, S.L. Allman, N.I. Taranenko, C.H. Chen. *Rapid Commun. Mass Spectrom.* **1995**, *9*, 1315–1320.

74 D.J. Harvey. *Rapid Commun. Mass Spectrom.* **1993**, *7*, 614–619.

75 D.J. Harvey, P.M. Rudd, R.H. Bateman, R.S. Bordoli, K. Howes, J.B. Hoyes, R.G. Vickers. *Org. Mass Spectrom.* **1994**, *29*, 753–766.

76 D.J. Harvey. *Mass Spectrom. Rev.* **1999**, *18*, 349–450.

77 G. Talbo, M. Mann. *Rapid Commun. Mass Spectrom.* **1996**, *10*, 100–103.

78 T.M. Billeci, J.T. Stults. *Anal. Chem.* **1993**, *65*, 1709–1716.

79 C. Koster, J.A. Castoro, C.L. Wilkins. *J. Am. Chem. Soc.* **1992**, *114*, 7572–7574.

80 A.J. Nicola, A.I. Gusev, A. Proctor, E.K. Jackson, D.M. Hercules. *Rapid Commun. Mass Spectrom.* **1995**, *9*, 1164–1171.

81 M.W. Duncan, G. Matanovic, A. Cerpa Poljak. *Rapid Commun. Mass Spectrom.* **1993**, *7*, 1090–1094.

82 S.X. Xiong, D. Pu, B. Xin, G.H. Wang. *Rapid Commun. Mass Spectrom.* **2001**, *15*, 1885–1889.

83 T. Wingerath, D. Kirsch, B. Spengler, W. Stahl. *Anal. Biochem.* **1999**, *272*, 232–242.

84 C. Andalo, P. Bocchini, R. Pozzi, G.C. Galletti. *Rapid Commun. Mass Spectrom.* **2001**, *15*, 665–669.

85 C. Biot, S. Caron, L.A. Maciejewski, J.S. Brocard. *J. Labelled Compd. Radiopharm.* **1998**, *41*, 911–918.

86 J. Rudolf, G. Blaschke. *Enantiomer* **1999**, *4*, 317–323.

87 F. Morvan, J.C. Bres, I. Lefebvre, J.J. Vasseur, A. Pompon, J.L. Imbach. *Nucleosides Nucleotides Nucleic Acids* **2001**, *20*, 1159–1163.

88 E.J. Eisenberg, G.R. Lynch, A.M. Bidgood, K. Krishnamurty, K.C. Cundy. *J. Pharm. Biomed. Anal.* **1998**, *16*, 1349–1356.

89 M. Olson, D. Fabris. Proceedings of the American Chemical Society 22nd National Meeting, Washington D.C. USA Abstract ANYL-054. **2000**.

90 D. Rideout, A. Bustamante, G. Siuzdak. *Proc. Natl. Acad. Sci. USA* **1993**, *90*, 10226–10229.

91 T. LeRiche, J. Osterodt, D.A. Volmer. *Rapid Commun. Mass Spectrom.* **2001**, *15*, 608–614.

92 D.L. Dehn, D.J. Claffey, M.W. Duncan, J.A. Ruth. *Chem. Res. Toxicol.* **2001**, *14*, 275–279.

93 M.M. Siegel, K. Tabei, R. Tsao, M.J. Pastel, R.K. Pandey, S. Berkenkamp, F. Hillenkamp, M.S. de Vries. *J. Mass Spectrom.* **1999**, *34*, 661–669.

94 M. Angotti, B. Maunit, J.F. Muller, L. Bezdetnaya, F. Guillemin. *Rapid Commun. Mass Spectrom.* **1999**, *13*, 597–603.

95 S. Hackbarth, V. Horneffer, A. Wiehe, F. Hillenkamp, B. Roder. *Chem. Phys.* **2001**, *269*, 339–346.

96 B.J. Egner, G.J. Langley, M. Bradley. *J. Org. Chem.* **1995**, *60*, 2652–2653.

97 N.J. Haskins, D.J. Hunter, A.J. Organ, S.S. Rahman, C. Thom. *Rapid Commun. Mass Spectrom.* **1995**, *9*, 1437–1440.

98 D.A. Lake, M.V. Johnson, C.N. McEwen, B.S. Larsen. *Rapid Commun. Mass Spectrom.* **2000**, *14*, 1008–1013.

99 R.S. Youngquist, G.R. Fuentes, C.M. Miller, G.M. Ridder, M.P. Lacey, T. Keough. *Adv. Mass Spectrom.* **1998**, *14*, 423.

100 G. Siuzdak, J.K. Lewis. *Biotechnol. Bioeng.* **1998**, *61*, 127–134.

101 P.M. St. Hilaire, T.L. Lowary, M. Meldal, K. Bock. *J. Am. Chem. Soc.* **1998**, *120*, 13312–13320.

102 S. Mathur, M. Hassel, F. Steiner, K. Hollemeyer, R.W. Hartmann. *J. Biomol. Screen.* **2003**, *8*, 136–148.

103 S.H. Hwang, A. Lehman, X. Cong, M.M. Olmstead, K.S. Lam, C.B. Lebrilla, M.J. Kurth. *Org. Lett.* **2004**, *6*, 3829–3832.

104 X.B. Wang, J.H. Zhang, A.M. Song, C.B. Lebrilla, K.S. Lam. *J. Am. Chem. Soc.* **2004**, *126*, 5740–5749.

105 G.A. Cumme, E. Blume, R. Bublitz, H. Hoppe, A. Horn. *J. Chromatogr. A* **1997**, *791*, 245–253.

106 F.O. Ayorinde, S.V. Gelain, J.H. Johnson, L.W. Wan. *Rapid Commun. Mass Spectrom.* **2000**, *14*, 2116–2124.

107 F.O. Ayorinde, B.E. Eribo, J.H. Johnson, E. Elhilo. *Rapid Commun. Mass Spectrom.* **1999**, *13*, 1124–1128.

108 D.M. Parees, S.D. Hanton, P.A.C. Clark, D.A. Willcox. *J. Am. Soc. Mass Spectrom.* **1998**, *9*, 282–291.

109 A.P. Morrow, O.O. Kassim, F.O. Ayorinde. *Rapid Commun. Mass Spectrom.* **2001**, *15*, 767–770.

110 D.C. Muddiman, A.I. Gusev, D.M. Hercules. *Mass Spectrom. Rev.* **1995**, *14*, 383–429.

111 D.M. Desiderio, U. Wirth, J.L. Lovelace, G. Fridland, E.S. Umstot, T.M.D. Nguyen, P.W. Schiller, H.S. Szeto, J.F. Clapp. *J. Mass Spectrom.* **2000**, *35*, 725–733.

112 S. Jespersen, W.M.A. Niessen, U.R. Tjaden, J. vanderGreef. *J. Mass Spectrom.* **1995**, *30*, 357–364.

113 D.C. Muddiman, A.I. Gusev, A. Proctor, D.M. Hercules, R. Venkataramanan, W. Diven. *Anal. Chem.* **1994**, *66*, 2362–2368.

114 M. Petkovic, J. Schiller, J. Muller, M. Muller, K. Arnold, J. Arnhold. *Analyst* **2001**, *126*, 1042–1050.

115 K. Tang, S.L. Allman, R.B. Jones, C.H. Chen. *Anal. Chem.* **1993**, *65*, 2164–2166.

116 D.C. Muddiman, A.I. Gusev, K. Stoppeklangner, A. Proctor, D.M. Hercules, P. Tata, R. Venkataramanan, W. Diven. *J. Mass Spectrom.* **1995**, *30*, 1469–1479.

117 S.V. Kovtoun. *Rapid Commun. Mass Spectrom.* **1997**, *11*, 810–815.

118 J. Horak, W. Werther, E.R. Schmid. *Rapid Commun. Mass Spectrom.* **2001**, *15*, 241–248.

119 J.T. Mehl, A.I. Gusev, D.M. Hercules. *Chromatographia* **1997**, *46*, 358–364.

120 A.J. Nicola, A.I. Gusev, A. Proctor, D.M. Hercules. *Anal. Chem.* **1998**, *70*, 3213–3219.

121 A.I. Gusev, W.R. Wilkinson, A. Proctor, D.M. Hercules. *Appl. Spectrosc.* **1993**, *47*, 1091–1092.

122 K. Pihlainen, K. Grigoras, S. Franssila, R. Ketola, T. Kotiaho, R. Kostiainen. *J. Mass Spectrom.* **2005**, *40*, 539–545.

123 L. Sleno, D.A. Volmer. *Rapid Commun. Mass Spectrom.* **2005**, *19*, 1928–1936.

124 J.S. Gobey, M. Cole, J. Janiszewski, T. Covey, T. Chau, P. Kovarik, J. Corr. *Anal. Chem.*, **2005**, *77*, 5643–5654.

125 D. Bungert, E. Heinzle, A. Tholey. *Anal. Biochem.* **2004**, *326*, 167–175.

126 M. Mank, B. Stahl, G. Boehm. *Anal. Chem.* **2004**, *76*, 2938–2950.

127 Z.X. Shen, E.P. Go, A. Gamez, J.V. Apon, V. Fokin, M. Greig, M. Ventura, J.E. Crowell, O. Blixt, J.C. Paulson, R.C. Stevens, M.G. Finn, G. Siuzdak. *Chembiochem* **2004**, *5*, 921–927.

128 D.B. Wall, J.W. Finch, S.A. Cohen. *Rapid Commun. Mass Spectrom.* **2004**, *18*, 1403–1406.

129 E.J. Zaluzec, D.A. Gage, J. Allison, J.T. Watson. *J. Am. Soc. Mass Spectrom.* **1994**, *5*, 230–237.

130 J.Y. Wu, K. Chatman, K. Harris, G. Siuzdak. *Anal. Chem.* **1997**, *69*, 3767–3771.

131 C. Pan, S. Xu, L. Hu, X. Su, J. Ou, H. Zou, Z. Guo, Y. Zhang, B. Guo. *J. Am. Soc. Mass Spectrom.* **2005**, *16*, 883–892.

132 C. Pan, S. Xu, H. Zou, Z. Guo, Y. Zhang, B. Guo. *J. Am. Soc. Mass Spectrom.* **2005**, *16*, 263–270.

133 M. Willetts, M.R. Clench, R. Greenwood, G. Mills, V. Carolan. *Rapid Commun. Mass Spectrom.* **1999**, *13*, 251–255.

134 F.O. Ayorinde, E. Elhilo. *Rapid Commun. Mass Spectrom.* **1999**, *13*, 2166–2173.

135 W. Witte. *Science* **1998**, *279*, 996–997.

136 M.C. Roberts. *Mol. Biotechnol.* **2002**, *20*, 261–283.

137 G.A. Grant, S.L. Frison, P. Sporns. *J. Agric. Food Chem.* **2003**, *51*, 5367–5375.

138 L. Sleno, D.A. Volmer. *Anal. Chem.* **2005**, *77*, 1509–1517.

139 J.C. Dunphy, K.L. Busch, R.L. Hettich, M.V. Buchanan. *Anal. Chem.* **1993**, *65*, 1329–1335.

140 C. Eckerskorn, K. Strupat, M. Karas, F. Hillenkamp, F. Lottspeich. *Electrophoresis* **1992**, *13*, 664–665.

141 F. Li, M.Q. Dong, L.J. Miller, S. Naylor. *Rapid Commun. Mass Spectrom.* **1999**, *13*, 464–466.

142 P.M. St. Hilaire, L. Cipolla, U. Tedebark, M. Meldal. *Rapid Commun. Mass Spectrom.* **1998**, *12*, 1475–1484.

143 H. Therisod, V. Labas, M. Caroff. *Anal. Chem.* **2001**, *73*, 3804–3807.

144 K.K. Murray. *Mass Spectrom. Rev.* **1997**, *16*, 283–299.

145 A.I. Gusev. *Fresenius J. Anal. Chem.* **2000**, *366*, 691–700.

146 Y.C. Chen, J. Shiea, J. Sunner. *J. Chromatogr. A* **1998**, *826*, 77–86.

147 Y.C. Chen. *Rapid Commun. Mass Spectrom.* **1999**, *13*, 821–825.

148 J. Guittard, X.L. Hronowski, C.E. Costello. *Rapid Commun. Mass Spectrom.* **1999**, *13*, 1838–1849.

149 A.I. Gusev, O.J. Vasseur, A. Proctor, A.G. Sharkey, D.M. Hercules. *Anal. Chem.* **1995**, *67*, 4565–4570.

150 A.I. Gusev, A. Proctor, Y.I. Rabinovich, D.M. Hercules. *Anal. Chem.* **1995**, *67*, 1805–1814.

151 A. Crecelius, M.R. Clench, D.S. Richards, D. Evason, V. Parr. *J. Chromatogr. Sci.* **2002**, *40*, 614–620.

152 A. Crecelius, M.R. Clench, D.S. Richards, V. Parr. *J. Pharm. Biomed. Anal.* **2004**, *35*, 31–39.

153 L.S. Santos, R. Haddad, N.F. Hoehr, R.A. Pilli, M.N. Eberlin. *Anal. Chem.* **2004**, *76*, 2144–2147.

154 H. Hayen, D.A. Volmer. *Rapid Commun. Mass Spectrom.* **2005**, *19*, 711–720.

155 A. Crecelius, M.R. Clench, D.S. Richards, V. Parr. *J. Chromatogr. A* **2002**, *958*, 249–260.

156 V.B. Ivleva, Y.N. Elkin, B.A. Budnik, S.C. Moyer, P.B. O'Connor, C.E. Costello. *Anal. Chem.* **2004**, *76*, 6484–6491.

157 K. Dreisewerd, J. Muthing, A. Rohlfing, I. Meisen, Z. Vukelic, J. Peter-Katalinic, F. Hillenkamp, S. Berkenkamp. *Anal. Chem.* **2005**, *77*, 4098–4107.

158 P.K. Salo, H. Salomies, K. Harju, R.A. Ketola, T. Kotiaho, J. Yli-Kauhaluoma, R. Kostiainen. *J. Am. Soc. Mass Spectrom.* **2005**, *16*, 906–915.

159 O. Keil, T. LeRiche, H. Deppe, D.A. Volmer. *Rapid Commun. Mass Spectrom.* **2002**, *16*, 814–820.

160 S. Brombacher, S.J. Owen, D.A. Volmer. *Anal. Bioanal. Chem.* **2003**, *376*, 773–779.

161 P. Kovarik, R.J. Hodgson, T. Covey, M.A. Brook, J.D. Brennan. *Anal. Chem.* **2005**, *77*, 3340–3350.

162 K.K. Murray, D.H. Russell. *Anal. Chem.* **1993**, *65*, 2534–2537.

163 K.K. Murray, D.H. Russell. *J. Am. Soc. Mass Spectrom.* **1994**, *5*, 1–9.

164 X. Fei, K.K. Murray. *Anal. Chem.* **1996**, *68*, 3555–3560.

165 L. Li, A.P.L. Wang, L.D. Coulson. *Anal. Chem.* **1993**, *65*, 493–495.

166 D.S. Nagra, L. Li. *J. Chromatogr. A* **1995**, *711*, 235–245.

167 A.I. Gusev, D.M. Hercules. *US Patent 6,140,639*, **1998**.

168 R.M. Whittal, L.M. Russon, L. Li. *J. Chromatogr. A* **1998**, *794*, 367–375.

169 Q. Zhan, A. Gusev, D.M. Hercules. *Rapid Commun. Mass Spectrom.* **1999**, *13*, 2278–2283.

170 S.J. Lawson, K.K. Murray. *Rapid Commun. Mass Spectrom.* **2000**, *14*, 129–134.

171 H. Orsnes, T. Graf, H. Degn, K.K. Murray. *Anal. Chem.* **2000**, *72*, 251–254.

172 J. Preisler, F. Foret, B.L. Karger. *Anal. Chem.* **1998**, *70*, 5278–5287.

Index

MALDI MS. A Practical Guide to Instrumentation, Methods and Applications.
Edited by Franz Hillenkamp and Jasna Peter-Katalinić
Copyright © 2007 Wiley-VCH Verlag GmbH & Co. KGaA, Weinheim
ISBN: 978-3-527-31440-9

Related Titles

Lifshitz, C., Laskin, J.

Principles of Mass Spectrometry Applied to Biomolecules

2006
ISBN-13: 978-0-471-72184-0
ISBN-10: 0-471-72184-0

Lee, M. S.

Integrated Strategies for Drug Discovery Using Mass Spectrometry

2005
ISBN-13: 978-0-471-46127-2
ISBN-10: 0-471-46127-X

Kaltashov, I. A.

Mass Spectrometry in Biophysics – Conformation and Dynamics of Biomolecules

2005
ISBN-13: 978-0-471-45602-5
ISBN-10: 0-471-45602-0

Ardrey, R. E.

Liquid Chromatography – Mass Spectrometry
An Introduction

2003
ISBN-13: 978-0-471-49801-8
ISBN-10: 0-471-49801-7